Alphonso Wood

Leaves and flowers

Object lessons in botany with a flora

Alphonso Wood

Leaves and flowers
Object lessons in botany with a flora

ISBN/EAN: 9783337271671

Printed in Europe, USA, Canada, Australia, Japan

Cover: Foto ©berggeist007 / pixelio.de

More available books at **www.hansebooks.com**

Leaves and Flowers;

OR,

OBJECT LESSONS

IN

BOTANY.

The Mexican Sage,—the Pink,—Phlox,—Bell-flower,—Sweet Pea,—Lily,—Morning-glory, —Heather,—Rose,—Ear-drop,—Monk's-hood, &c.,—illustrating the several classes of corolla forms. See Lesson XV.

Leaves and Flowers;

OR,

OBJECT LESSONS IN BOTANY

WITH

A FLORA.

PREPARED FOR BEGINNERS IN ACADEMIES AND PUBLIC SCHOOLS.

BY ALPHONSO WOOD, A.M.,

AUTHOR OF THE CLASS-BOOK OF BOTANY, ETC.

WITH 665 ILLUSTRATIONS.

NEW YORK:

A. S. BARNES & CO., 111 & 113 WILLIAM STREET.

TROY: MOORE & NIMS.

1869.

THE NATIONAL SERIES

OF

STANDARD TEXTS IN THE SCIENCES.

THESE STANDARD WORKS

ARE FOR SALE BY ALL BOOKSELLERS,

Or may be procured from the Publishers by Mail, postpaid, on receipt of price.

A. S. BARNES & CO.,

NEW YORK.

Entered according to Act of Congress, in the year 1860,
By A. S. BARNES & BURR,
In the Clerk's Office of the United States District Court for the Southern District of New York.

PREFACE.

AMONG the happiest days of our childhood were those devoted to the study of Botany. Pure sunshine rests upon the memory of those rambles in the fields and woods, amid the opening flowers of Spring, and then in the gay profusion of advancing Summer, in which we made acquaintance with many a floral gem before unknown. We love to think of that wild woodland lake where first we saw the sparkling Sundew, the quaint Sarracenia, and the fair Nymphæa, resting on the bosom of the waters; or of that lowly dell by the brookside, where the Yellow Violet, the Hepatica, and the Bloodroot bloomed; or of that craggy mountain, where, among the rocks, the Columbine hung out its scarlet flowers. Then returning home with our gathered treasures, how we entered with a will upon the work of Analysis, toiling for hours as no schoolmaster could have compelled us to do, being attracted to the task by the very love of it alone. Here, then, we have at least one department in learning whose earnest pursuit is so congenial to the affections and tastes of the mind as to be no irksome task, but a pastime,—a perpetual feast; and this not only to maturer minds, but to the season of early youth even in a higher degree, since then the objects of nature are especially invested with the charms of novelty.

Let it not be said, however, that Botany attracts such willing votaries because it requires no labor, no persevering effort. No science is more intricate or profound. It cannot be understood except by vigorous and persevering effort. Consequently, in its successful pursuit there is *discipline* for the mind as well as for the body; and since the subject itself is replete with refinement and beauty, and fresh from the hand of God, its pursuit must also conduce to the invigoration of the moral nature.

If, then, it be desirable to preoccupy the minds of our children with controlling ideas of purity, refinement, and moral beauty,—with exalted thoughts of God, habits of mental activity, strength of judgment, and decision of character; and, moreover, to do all this by means of a study whose path, in a double sense, is strown with flowers, then is the study of Botany desirable; and that labor is not in vain which is bestowed upon the preparation of a work designed, like the present, for *primary* classes, from the ages of ten to fourteen.

As the title implies, we have aimed to represent to the eye nearly every subject or form treated of, or described in these Lessons. But, notwithstanding the copiousness of these illustrations, neither the teacher nor the pupil will be satisfied to rely upon their aid alone. Nature alone can afford the proper illustrations in the study of Nature's works, and it is only by comparison with the living specimen that either the picture or the description becomes intelligible. Therefore let specimens in unlimited number accompany every botanical recitation.

Most of the figures are original. Others have been derived from Maout, Payer, Richard, Balfour, Lindley, and a few, by permission, from Darlington's "Weeds and Useful Plants."

Finally, to the children and youth of our country, gathered in schools of every name, this humble volume is dedicated, with confident belief that it will prove to many of them a source of intellectual and moral culture as well as of pure and rational delight.

BROOKLYN, N. Y., February 23, 1863.

BOTANICAL APPARATUS.

The Publishers have recently provided and have on sale a set of *apparatus* of the most approved form for the use of the student in botanical pursuits, and as described in the Class-Book, page 15. It consists of a *knife-trowel* for digging and cutting specimens, a *microscope* and *tweezers* for analysis, a *tin-box* for preserving them fresh, and a *press* for drying them. The Set, securely packed, will be sent by Express to order, at a moderate price.

IN PRESS.

"*The Botanical Index*," a work for Schools and Seminaries, altogether *new* and *peculiar*, in which the study of the entire flora of the country and city is reduced to the last degree of simplicity and precision. Its use will lighten the labors of the teacher and add still new pleasures to botanical pursuits for every one. It will be issued in June or July.

Also, *in the domain of Science*, the Publishers offer
    Steele's 14 Weeks' Course in Astronomy.
      "   "   "   Chemistry (with Apparatus).
      "   "   "   Philosophy (in Press).
Most interesting and valuable Text-Books.

CONTENTS.

OBJECT LESSONS IN BOTANY.

LESSON I.

THE LEAF, AND ITS PARTS.

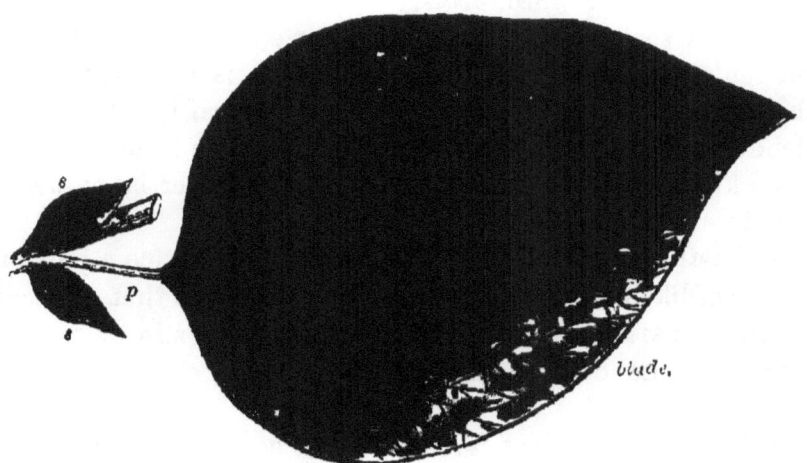

Fig. 1. Leaf of the Quince.

1. WE have before us the picture of a Quince leaf, care-fully drawn and colored. It is of a rich green color, very pleasant to the eye. Its outlines are full, even, and grace-fully curved, and its upper surface is smooth and naked. Although it is indeed but one leaf, yet it seems to be made up of three parts—*b*, *p*, *ss*.

2. The upper part, *b*, is broad and thin, and is called the

BLADE. The upper end of the blade is the *apex*, and the lower end is the *base*. You see at once that the outline of this blade represents a certain form or figure, with an even margin, rounded, and broader at the base than at the apex, like the figure of an egg. So it is called an egg-shaped leaf: or, to use a softer word, *ovate*.

3. Now see how this blade is supported. At the base it is suddenly narrowed to a foot-stalk, which is properly called the PETIOLE. You see that this part of the leaf is narrow and slender, and in this leaf *terete*, or *cylindrical*, in form. But in some kinds of leaves it is flattened. Remember its name,— petiole.

4. Lastly, at the base of the petiole you notice a pair of little leaf-like bodies, one on this side and one on that. These we call the STIPULES. Stipules, then, are always in pairs, and placed at the base of the petiole. Their shape is quite various.

5. Thus, when a leaf is complete, it consists of a blade, a petiole, and a pair of stipules. But you will not find every kind of leaf complete. Many sorts have no stipules at all. Can you find stipules on the leaves of the Lilac? Some leaves, moreover, have not even a petiole. See the leaves of Phlox. Such leaves are said to be *sessile*, that is, sitting.

---

1. What is the color of the leaf of the Quince bush? What is the color of leaves generally? *Ans.* Green, of lighter or darker shade. What of the outline of this leaf?—its upper surface?

2. What is the *blade?*—the apex?—the base? What is the figure of the blade?

3. How is the blade supported? Describe the foot-stalk. Tell its real name.

4. Describe the stipules.

5. Now state the three parts of a complete leaf. Do all kinds of leaves have stipules? Do the leaves of the Lilac?—of St. Johnswort, &c.?—of the

LESSON II.

VEINS AND VENATION OF THE LEAF.

6. THE blade of the Quince leaf (Fig. 2) shows many veins running through it, and branching all over it. Examine

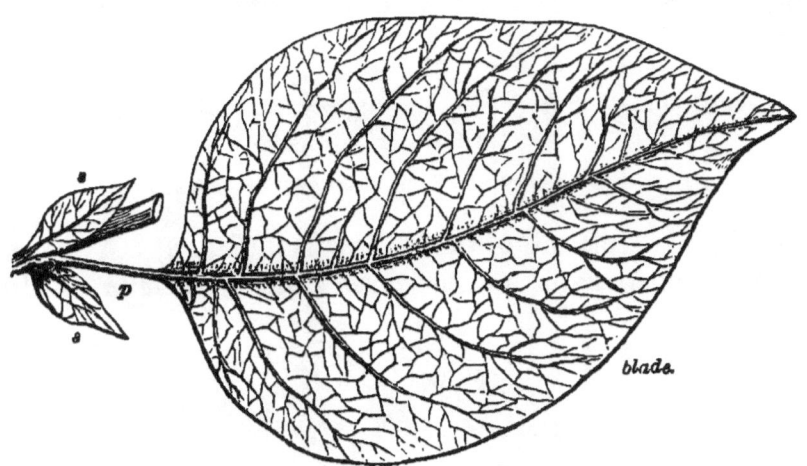

Fig. 2. Leaf of the Quince, showing the veins.

them. First, the petiole seems to be extended and continued right through, from the base to the apex, forming the largest vein in the leaf. This is the *midvein.*

7. Next observe several large branches sent off from this midvein on both sides, right and left. These are the *veinlets.* Now, looking at these veins, their arrangement reminds us of a feather, and we call such leaves *feather-veined.* Therefore,

---

Violet? Do all kinds of leaves have petioles?—of Phlox, for example? What do you understand by *sessile* leaves?

6. Describe the midvein of the Quince leaf.

7. Describe the veinlets. What is the feather-veined venation?

we may say that the feather-veined venation consists of *one midvein branching into veinlets*. This is very common.

8. Thirdly, the veinlets themselves send off little branches (branchlets) on their right and left, and we call these the *veinulets*. These again and again may divide, and finally, all the little divisions unite again, forming a complete net-work all over the leaf. Thus we learn what a *net-veined* leaf is.

Fig. 8. The Willow leaf. Some of the veinulets are shown.

9. Here is a picture of the Willow leaf (Fig. 3). You can point out all its parts, and the three kinds of veins in it. In

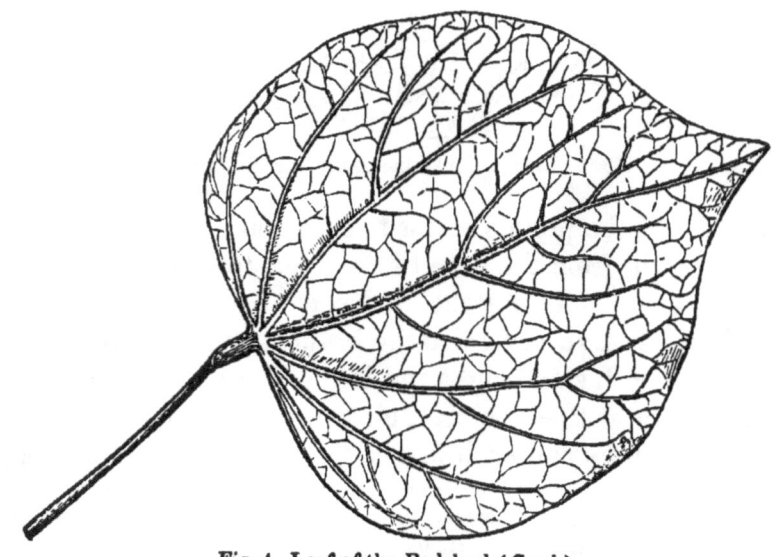

Fig. 4. Leaf of the Red-bud (*Cercis*).

8. What are the veinulets? When is a leaf said to be net-veined?

the next cut (Fig. 4), representing the Judas-tree or Red-bud leaf, you see a different venation.

10. At the base of the blade the petiole seems to divide all at once into five large veins, each running through, one to the apex, and four to the margin. In this case the vein-ing (that is, the *venation*) is compared to the division of the hand (or *palm* of the hand) into fingers, and so named *pal-*

Fig. 4 a. Leaf of Sweet-gum (*Liquidambar*).

mate venation. Therefore, you may say that the *palmate venation* consists of about five veins starting together at the base of the blade, each one branching into veinlets and veinulets. Fig. 4 *a* is a *lobed* leaf of the Liquidambar tree,

9. Note the parts of this Willow leaf. Point out its midvein. Its veinlets —veinulets.

10. Define the palmate venation. What are the veins?

with palmate venation.　Maple leaves are very familiar examples of the same.

11. Our next figure (5) represents the leaf of the Tulip. How very different is its venation! How smooth, even, and polished its surface! The veins all run side by side, from base to apex, in graceful and regular curves. They are so nearly parallel, that this kind of venation is called the *parallel venation*. Look at the grass leaves, the Corn leaves, and the Palm leaves, and see the same kind of venation.

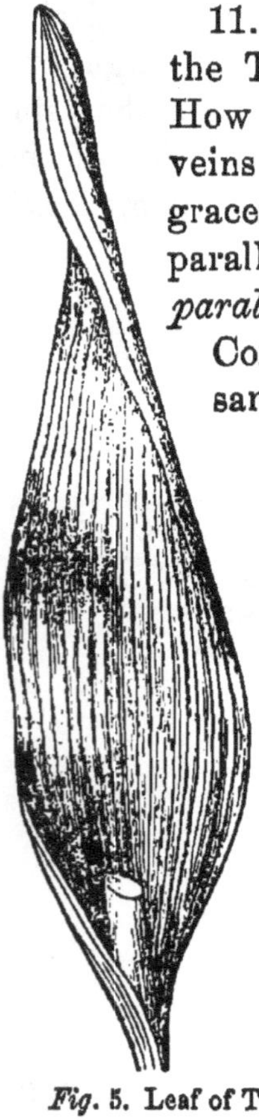

Fig. 5. Leaf of Tulip. 　　　　*Fig.* 6. Leaf of the Climbing Fern.

12. Let us examine one other kind of venation, and the list will be complete. Here is a cut showing the leaf of

11. Define the parallel venation. Mention examples.
12. The forked venation. Examples.
Now repeat the names of the five or six kinds of venation.

Climbing Fern (Fig. 6). To say nothing of the five veins (which are palmate, as in the leaves of Sweet-gum, Fig. 4 *a*), you may notice the veinlets, how they fork and run to the margin, without uniting again to form a net-work. This is the *forked venation*. You will find this sort in nearly all the Ferns.

LESSON III.

FORMS AND FIGURES OF LEAVES.

13. You have already noticed that the form of the Quince leaf, and of most others, is a thin, flat expansion, presenting a large surface to the air. A few plants have thick, solid leaves, as the leaves of the Live-forevers and Ice-plants.

Fig. 7. Represents a branch of Juniper, with awl-shaped leaves (subulate).
Fig. 8. Leaves of the Fleur-de-lis (*Iris*); they are sword-shaped (ensiform).
Fig. 9. Leaves of the Scotch Pine; they are needle-shaped (acerose).

Other plants have slender leaves, as the Pines. See Figs. 7, 8, and 9.

18. What is the general form of leaves? What plants have thick and solid leaves? What form of leaves has the Pine? the Iris? the Juniper?

14. We also spoke of the figure of the outline of the Quince leaf, which is *ovate*. But you must have observed that there is a very great variety in the figure of leaves, affording a very interesting study. First, we will examine, one by one, the figures of the feather-veined leaves (Figs. 10–21.)

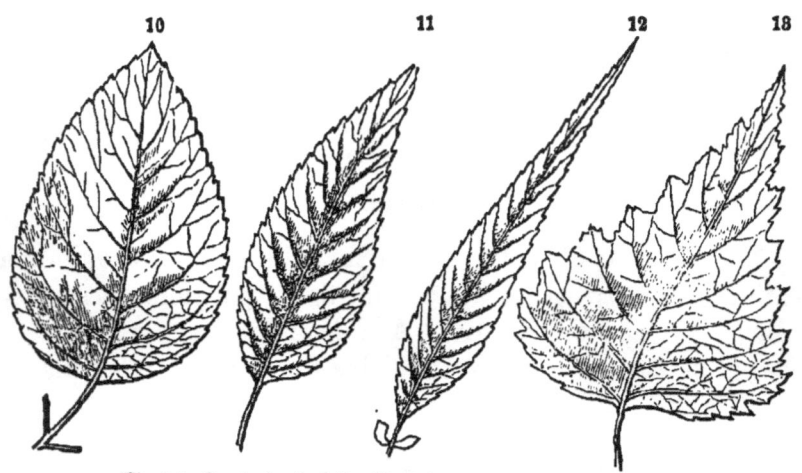

Fig. 10. Ovate leaf of the Pear-tree.
Fig. 11. Lanceolate leaf of the Flowering Almond.
Fig. 12. Narrow lanceolate leaf of the Weeping Willow.
Fig. 13. Deltoid leaf of the White Birch.

15. The leaf of the Flowering Almond (Fig. 11) is *lanceolate*. It is narrower than ovate, shaped like a lance, having the lower half wider than the upper. This Willow leaf (Fig. 12) is *narrowly lanceolate*. The leaf of the Lombardy Poplar, or of the White Birch (Fig. 13), is so broad at the base as to form a three-sided figure, like the Greek letter (Δ) delta. Hence it is a *deltoid* leaf.

---

14. What is the figure of the Quince leaf?
15. Describe the figure of the leaf of Flowering Almond; of the Weeping Willow; of the Lombardy Poplar, &c.

16. In the next four kinds of leaves you will notice that the broadest place is midway between the base and apex. Thus the *orbicular* (Fig. 14), or rounded, leaf is about as broad as it is long. The *oval* leaf (Fig. 15) is about one-third longer than broad. This Plum leaf is an example. The *elliptical* (Fig. 16) is about twice longer than broad, and the *oblong* (Fig. 17) is three or four times longer than broad. Here are examples.

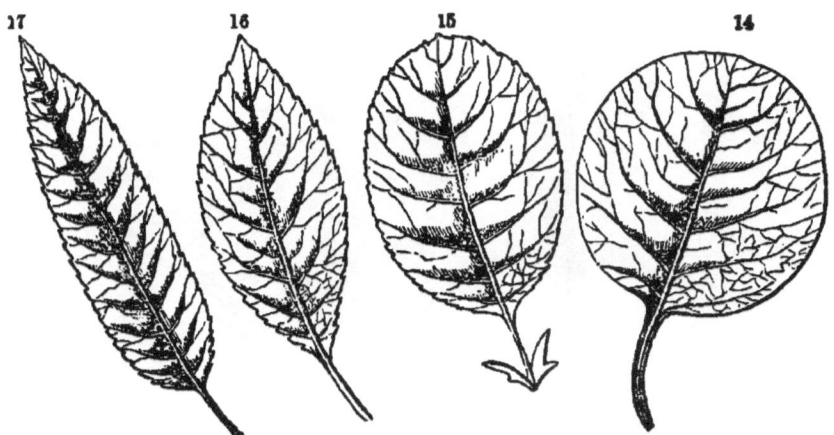

Fig. 14. Orbicular leaf of Winter-green (*Pyrola*).
Fig. 15. Oval leaf of the Plum-tree.
Fig. 16. Elliptical leaf of Black Haw.
Fig. 17. Oblong leaf of a Willow.

17. We next have four varieties of forms which are broader towards the apex than base. First, the *obovate* (Fig. 18), like this leaf of the Smoke-tree. Its outline is like that of

16. When is the figure of a leaf called orbicular? Will you show me specimens? Describe an oval leaf, and give specimens. Describe an elliptical leaf, and give examples. Describe an oblong leaf, and give examples.
17. When will the figure of a leaf become obovate? Give examples of

the egg inverted. A similar form, but narrower, is the *ob-lanceolate;* that is, the inverted lance-shaped, like the leaf of Papaw, or Fig. 19.   Next, still narrower, is the *spatulate,* a figure compared to the surgeon's spatula (Fig. 21); and lastly, the wedge-shaped, or *cuneate,* tapering from a broad apex to a slender base, as in Fig. 20.

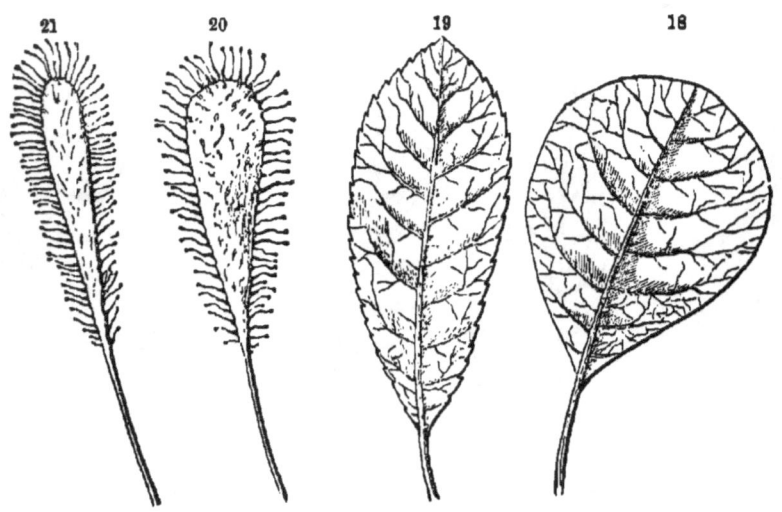

Fig. 18. Obovate leaf of the Smoke-tree (*Rhus cotula*).
Fig. 19. Oblanceolate leaf of Muhlenburg's Willow.
Fig. 20. Cuneate leaf of a Sundew (*Drosera longifolia*).
Fig. 21. Spatulate leaf of a Sundew (variety of *D. longifolia*).

18. Thus we have arranged these twelve forms of feather-veined leaves into three classes.

such leaves. Oblanceolate? Show us examples. Spatulate? Show us examples. Wedge-shaped, or cuneate? Give examples.

18. Repeat the names of the four leaf-forms broadest at base;—the four broadest in the middle;—the four broadest towards apex.

LESSON IV.

FORMS AND FIGURES OF LEAVES.

19. In many kinds of leaves we find the parts at the base more or less enlarged backwards, as you see in this picture

Fig. 22. The Morning-glory.

of the Morning-glory leaf (Fig. 22). This is the heart-shaped, or, more properly, the *cordate* leaf. It is truly an elegant figure in this and in the Lilac, &c. But sometimes this peculiar enlargement at base becomes excessive, and the figures more curious than elegant. Such is the arrow-shaped figure, called *sagittate*, having long-pointed base lobes, as seen in the Arrow-head (Fig. 47), the Scratch Knot-grass, &c. (Fig. 26.)

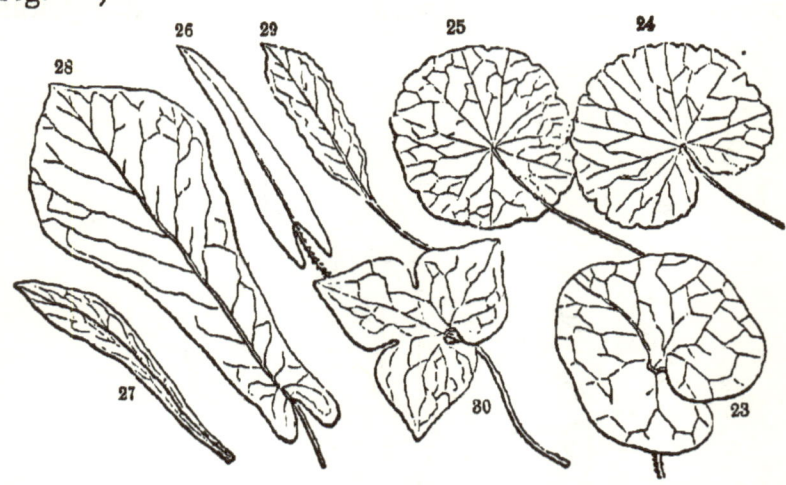

Fig. 23. Reniform leaf of Wild Ginger.
Fig. 24. Reniform leaf of Pennywort.
Fig. 25. Peltate leaf of Pennywort.
Fig. 26. Arrow-shaped leaf of Scratch Knot-grass.
Fig. 27. Spatulate leaf of Silene Virginica.

Fig. 28. Fraser's Magnolia: obovate-spatulate, auriculate at base.
Fig. 29. Oblong leaf of the Toothed Arabis.
Fig. 30. Three-lobed leaf of Liverwort.

20. In the common Sorrel leaf, and in Fraser's Magnolia leaf (Fig. 28), these base lobes remind one of *ears*, and such leaves are said to be *auriculate* (from the Latin *auricula*, an

19. Describe the cordate leaf, and give examples. The sagittate, and examples.

20. Describe the auriculate form, and give examples. The reniform. Examples.

ear). In some leaves these lobes are very broad and rounded, giving to them a kidney-shaped form, that is, *reniform*,
as you see in this Wild Ginger leaf (Fig. 23), and in the
Pennywort (Fig. 24). The *peltate*, or shield-shaped leaf (Fig.
25—another Pennywort) has its base lobes united, and its petiole fixed to the under side. See, also, Nasturtion leaves.

21. We will next study a class of forms with deeply lobed
or cleft blades, not well filled up between the veinlets.

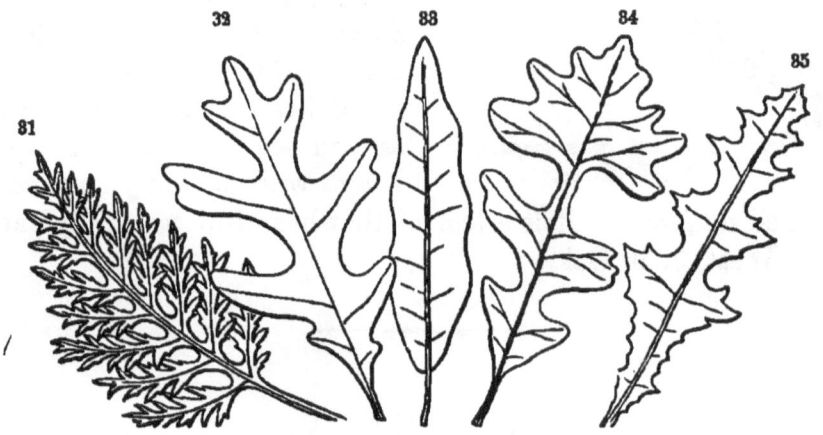

Fig. 31. Bi-pinnatifid leaf of Pig-weed.
Fig. 32. Sinuate-lobed leaf of White Oak.
Fig. 33. Undulate-lobed leaf of Jack Oak.
Fig. 34. Lyrate leaf of Moss-cup Oak.
Fig. 35. Lobed leaf of *Mulgedium* (Blue Milkweed).

First, look at this Liverwort leaf (Fig. 30). It is cleft in
two places, rendering it three-lobed. The Sweet-gum leaf
(Fig. 4 *a*) is five-lobed. Oak leaves are lobed in many patterns, according to the kind. The White Oak has a *sinuate-*

21. What is the figure of the Liverwort leaf? What the figure of the
Maple leaf? What kind of venation have these last two? Define the figure of the White Oak leaf. Of the Mossy-cup Oak.

lobed leaf (Fig. 32), the Mossy-cup Oak has a *lyrate* leaf, having its terminal lobe larger than any other (Fig. 34).

22. Fig. 35 is the leaf of a kind of Milkweed, called *Mul-*

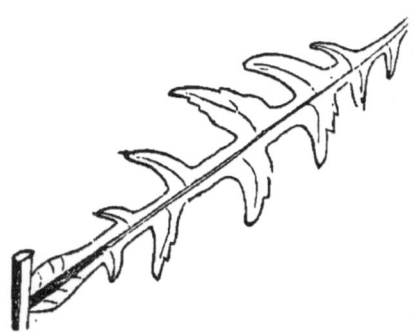

Fig. 36. Leaf of *Lactuca elongata*, or Wild Lettuce.

gedium, with sharp lobes projecting at *right angles* to the midvein ; and Fig. 36 is of the Wild Lettuce, with lobes pointing or hooking backwards. Such leaves are called *runcinate*. The Dandelion has also runcinate leaves. When a leaf has only shallow lobes, as you see in Fig. 33, it appears with a wavy outline, called *undulate*. It is a leaf of that beautiful tree called at the West, Jack Oak.

LESSON V.

OTHER FORMS AND FIGURES.

23. It is now time to learn the difference between a simple and a compound leaf. The *simple* leaf has but one blade, as the Quince leaf, and all the leaves which we have hitherto noticed. We have now before us a *compound* leaf, one plucked from a Rose-bush (Fig. 39), consisting of several distinct blades on one petiole. It has also one pair of stip-

---

22. What of the figure called runcinate? Describe the undulate leaf. What example? What kind of venation have the last four forms ?
23 What is a simple leaf? A compound leaf?

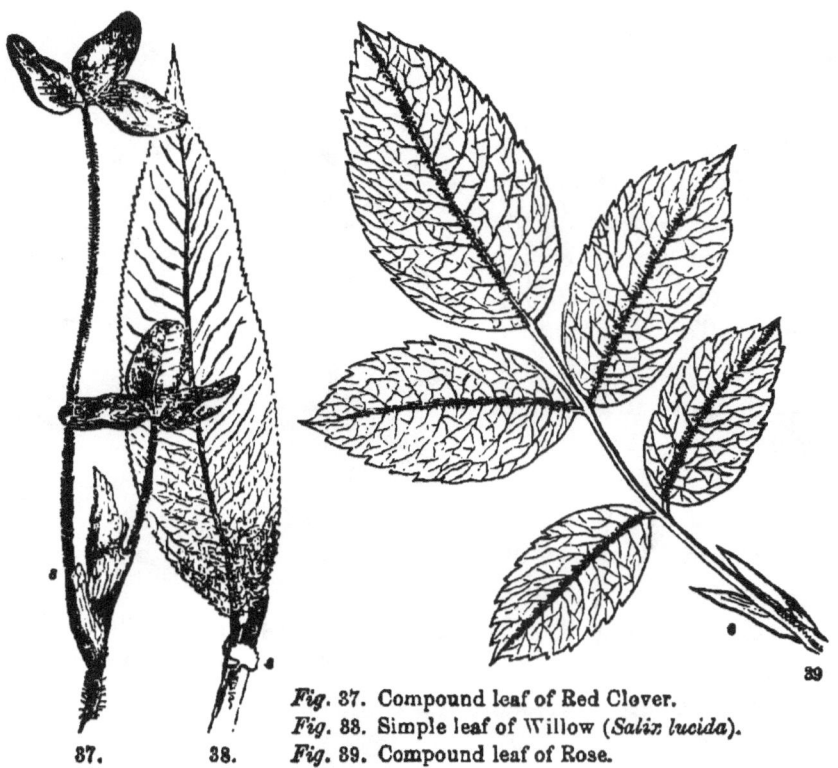

Fig. 37. Compound leaf of Red Clover.
Fig. 38. Simple leaf of Willow (*Salix lucida*).
Fig. 89. Compound leaf of Rose.

ules, like a simple leaf. This Clover leaf is also compound (Fig. 37), having stipules (*s*), as well as this simple leaf of the Shining Willow (Fig. 38).

24. But here is a leaf, the Celandine (Fig. 40), which is almost, but not quite, compound. The blade is feather-veined, and deeply divided into several parts, called *segments*. Such a leaf is called *pinnatifid*. In Fig. 31, the leaf of a garden weed (Ambrosia), you observe that the segments are themselves pinnatifid, so that the leaf is *twice* or *bi-pinnatifid*.

24. Please define the pinnatifid leaf. The bi-pinnatifid.

25. But what form of leaf is this (Fig. 41) of the Fennel-flower, with such a multitude of narrow segments? You may call it *pin-nat-i-sect*, if the long word does not try your short memory too much. The Thistle leaf is also pinnatisect, although quite different in form.

26. Fig. 42 represents a *pedate* leaf of a Passion-flower. Observe its palmate venation, each of its veins bearing a seg-ment, and each lower

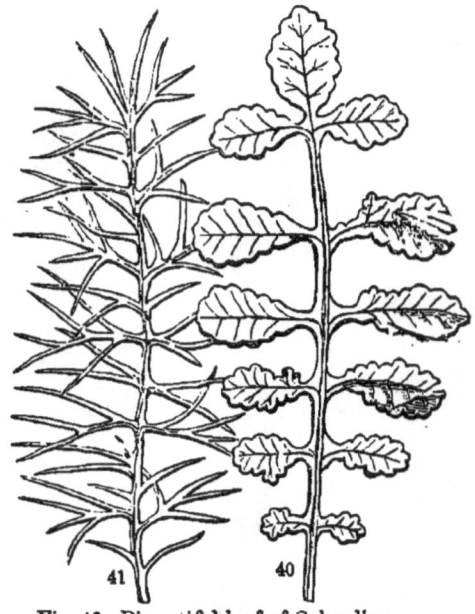

Fig. 40. Pinnatifid leaf of Celandine.
Fig. 41. Pinnatisect leaf of Fennel-flower.

segment double, so resembling a bird's foot. *Pedate* means foot-shaped.

Fig. 42. Pedate leaf of Passion-flower.

Fig. 43. Laciniate leaf of Monk's-hood.

25. What do you call such leaves as those of the Fennel-flower?

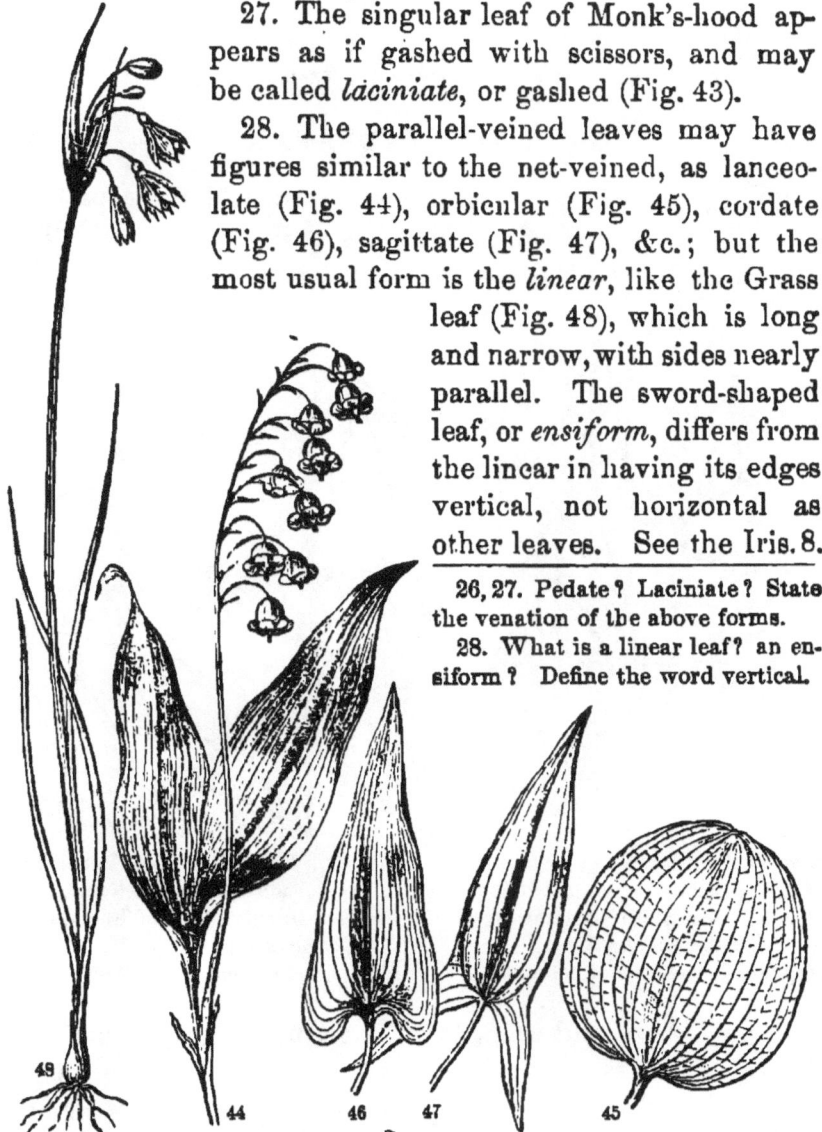

27. The singular leaf of Monk's-hood appears as if gashed with scissors, and may be called *laciniate*, or gashed (Fig. 43).

28. The parallel-veined leaves may have figures similar to the net-veined, as lanceolate (Fig. 44), orbicular (Fig. 45), cordate (Fig. 46), sagittate (Fig. 47), &c.; but the most usual form is the *linear*, like the Grass leaf (Fig. 48), which is long and narrow, with sides nearly parallel. The sword-shaped leaf, or *ensiform*, differs from the linear in having its edges vertical, not horizontal as other leaves. See the Iris. 8.

26, 27. Pedate? Laciniate? State the venation of the above forms.

28. What is a linear leaf? an ensiform? Define the word vertical.

Fig. 44. Lanceolate,—Lily of the Valley.   *Fig. 46.* Cordate leaf of Pond-weed.
Fig. 45. Orbicular,—Round-leaved Orchis.   *Fig. 47.* Sagittate leaf of Arrow-head.
Fig. 48. Linear leaves of Blue-eyed Grass (*Sisyrinchium*).

2

LESSON VI.

MARGIN AND APEX.

29. In describing a leaf we are to consider the patterns of its border, or margin, which are quite various, and often elegant. Some of the leaves heretofore noticed have the

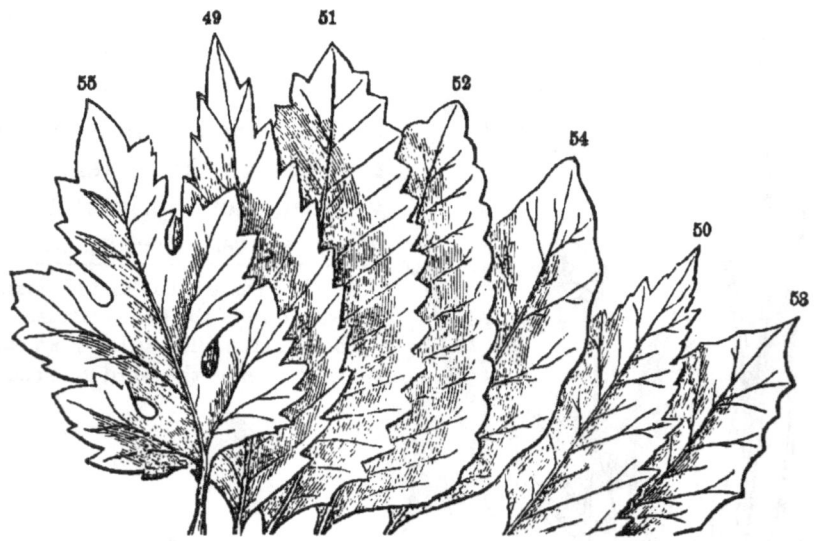

Fig. 49. Serrate leaf of Chestnut.
Fig. 50. Doubly serrate leaf of Elm.
Fig. 51. Dentate leaf of Arrow-wood (*Viburnum dentatum*).
Fig. 52. Crenate leaf of Catmint.

Fig. 53. Repand leaf of Enchanter's Night-shade (*Circæa Lutetiana*).
Fig. 54. Undulate leaf of Shingle Oak (*Q. imbricaria*).
Fig. 55. Lobed leaf of *Chrysanthemum*.

margins *entire* and even, as in the Quince leaf, or the Lily. But most leaves are notched in various ways. For example,

30. This Willow leaf (Fig. 3) is notched in the margin

---

29. When is the margin said to be entire?

like a saw, with the teeth projecting forward. Such a margin is said to be *serrate*, or, if the teeth are quite small, *serrulate*. When the teeth point neither forward nor backward, but *outward*, we call the margin *dentate*, or toothed; and if the teeth are quite small, *denticulate*. See Figs. 49, 50, 51, &c.

31. Some leaves are margined with rounded and blunt teeth, and we call them *crenate* (Fig. 52); or, if such teeth are very small, *crenulate*.

32. In Figs. 13 and 50, you see that the teeth themselves are again toothed, an arrangement called *doubly dentate*. So we may find leaves doubly serrate or doubly crenate. Thus we have described seven modes or styles of bordering. Several other modes are found described in the larger botanies.

APEX.

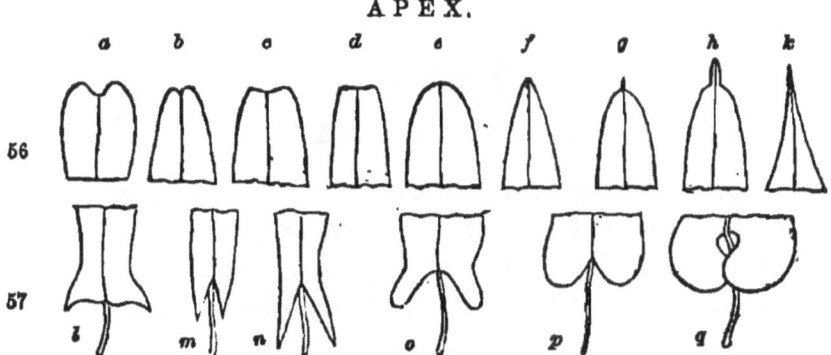

Fig. 56. Apex of leaves: *a*, obcordate; *b*, emarginate; *c*, retuse; *d*, truncate; *e*, obtuse; *f*, acute; *g*, mucronate; *h*, cuspidate; *k*, acuminate.

Fig. 57. Bases of leaves: *l*, hastate; *m, n*, sagittate; *o*, auriculate; *p*, cordate; *q*, reniform.

30. When is the margin serrate? When serrulate? How does the dentate differ from the serrate?

31. What sort of teeth does the crenate imply? Crenulate?

32. Explain doubly dentate, &c.

33. It is also necessary to be acquainted with the various forms of the apex of leaves. This diagram (Fig. 56) will assist the memory. The apex may be *acuminate*, ending in a long, tapering point; or *cuspidate*, suddenly contracted to a sharp, slender point; *mucronate*, tipped with a spiny point; *acute*, simply ending with an angle; *obtuse*, blunt.

34. Or the leaf may end without a point, being *truncate*, as if cut square off; *retuse*, with a rounded and slightly depressed end where the point should be; *emarginate*, having a small notch at the end; *obcordate*, having a deep indentation at the end. See also, and explain, the diagram of the bases of leaves (Fig. 57).

LESSON VII.

COMPOUND LEAVES.

35. A COMPOUND leaf consists of several distinct blades borne on one petiole. (See Lesson V., first paragraph.) These separate blades are called *leaflets*. You notice that in Fig. 39 each of the five leaflets has its own foot-stalk, called *petiolule*, and its own midvein, &c.

36. The Rose leaf (Fig. 58) is pinnately compound, or sim-

---

33. What does the term acuminate imply? What sort of apex is cuspidate? mucronate? acute? obtuse?

34. When may we call the apex truncate? retuse? emarginate? obcordate? Please name these several forms of the bases of leaves.

35. Define a compound leaf. What is a leaflet? What do you call the foot-stalk of the leaflet?

Fig. 58. Leaf of the Rose.

ply *pinnate*, having several leaflets arranged along both sides of the common stalk. This common stalk, answering to the midvein of a simple leaf, is called the *rachis*.

37. Among pinnate leaves, there are, at least, three important distinctions. Observe the Figs. 59, 60, and 61. One of them ends with an odd leaflet, and is called *odd-pinnate*. Another ends with a pair of leaflets, and is *equally pinnate*. Another still has its alternate leaflets smaller, and is *interruptedly pinnate*.

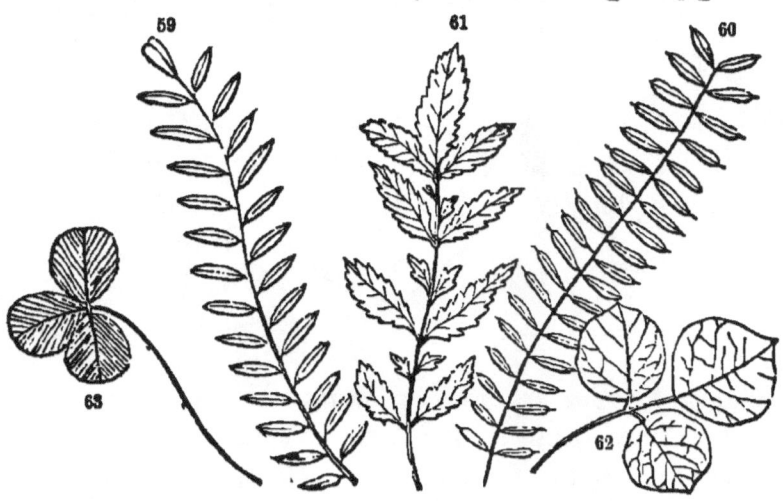

Fig. 59. Odd-pinnate leaf (*Tephrosia*). *Fig.* 61. Interruptedly pinnate (Agrimony).
Fig. 60. Equally pinnate leaf (*Cassia*). *Fig.* 62. Pinnately ternate (*Desmodium*).
Fig. 63. Palmately ternate (Clover).

36. Define the pinnate leaf. What is the rachis?

37. Give the distinction between odd-pinnate and equally pinnate. What leaf is interruptedly pinnate?

38. Every one knows that the number of leaflets in the Clover is three; also in the Bean, and in this figure (62) of the Desmodium leaf. Such leaves are called *ternate*. But here the pupil will notice another important distinction. In this Desmodium leaf, the odd leaflet is stalked, and is said to be *pinnately* ternate; in Clover, the odd leaflet is nearly sessile, like the other leaflets; this is *palmately* ternate.

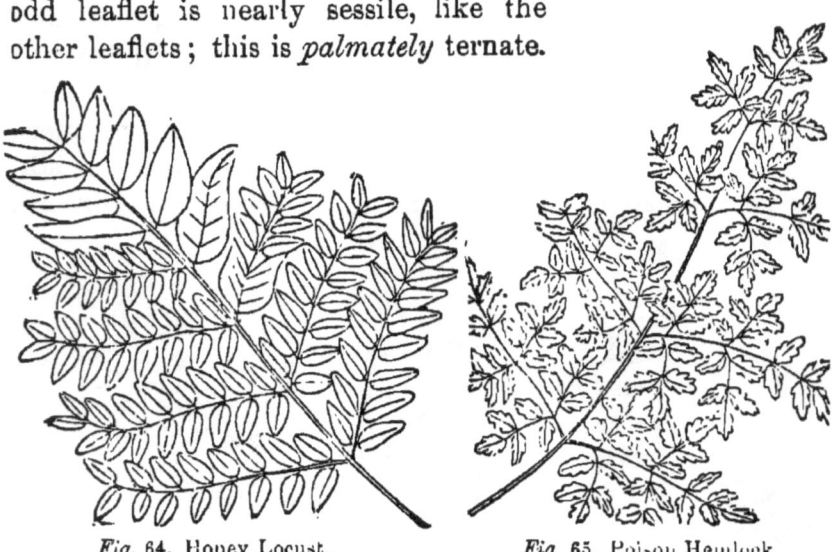

Fig. 64. Honey Locust.　　　　　*Fig.* 65. Poison Hemlock.

39. Fig. 64 represents a *bi-pinnate* (that is, *twice* pinnate) leaf of Honey Locust. The simple leaflets seem to have each become itself a pinnate leaf. And still more compound is this Poison Hemlock leaf, being *tri-pinnate*, or *thrice* pinnate (Fig. 65). In the same manner, we have *bi-ternate* and *tri-ternate*.

---

38. How many leaflets in a ternate leaf? Difference between the pinnately and the palmately ternate?

39. Can you define a bi-pinnate leaf? Tri-pinnate? What is a bi-ternate leaf? A tri-ternate?

40. All the above forms of compound leaves, except the Clover, are founded on the pinnate venation; but the palmate venation gives us the palmately ternate (Clover, already described); the *quinate*, with five leaflets; the *septinate*, with seven leaflets, &c. See the leaves of Horse-chestnut, of Hemp, and of this Lupine (Fig. 66).

Fig. 66. A leaf of Lupine.

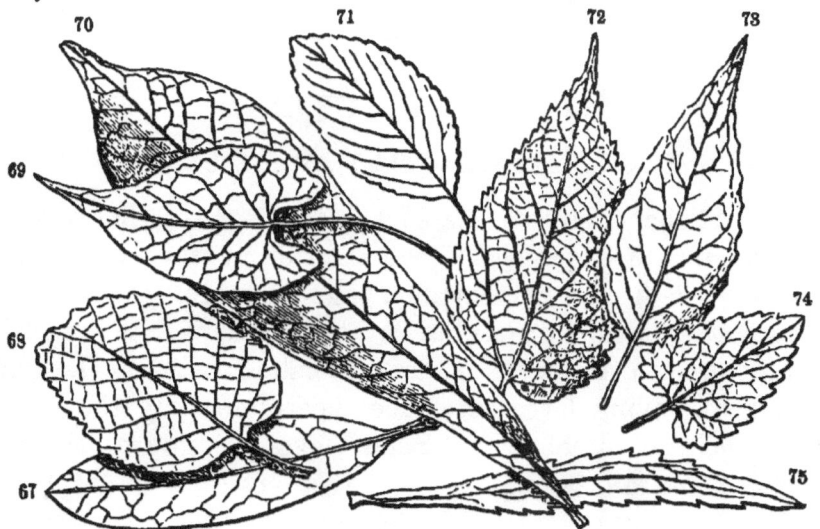

Fig. 67. Rose-bay (*Rhododendron*).
Fig. 68. Alder (*Alnus glauca*).
Fig. 69. Knot-grass (*Polygonun sagittatum*).
Fig. 70. Papaw (*Asimina triloba*).
Fig. 71. Touch-me-not (*Impatiens fulva*).

Fig. 72. Sugar-berry (*Celtis Americana*).
Fig. 73. Enchanter's Night-shade (*Circœa lutetiana*).
Fig. 74. Catmint (*Nepeta Glechoma*).
Fig. 75. Goldenrod (*Solidago Canadensis*), a triple-veined leaf.

The pupils should be required to describe the leaves in this cut, as to venation, figure, margin, apex, and base.

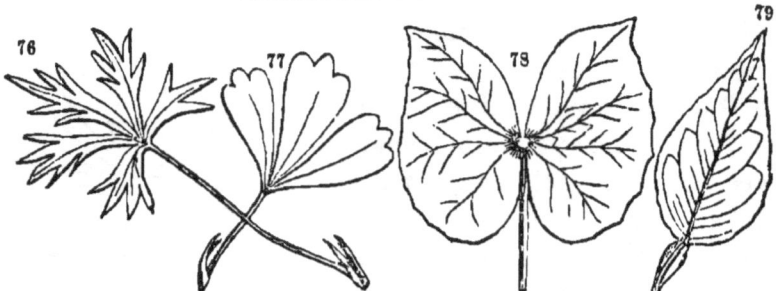

Fig. 76. Potentilla anserina; leaf with five cut lobes, almost quinate.
Fig. 77. Potentilla tridentata; ternate, with palmate, three-toothed leaflets
Fig. 78. Jeffersonia diphylla; a binate leaf.
Fig. 79. Lemon; a simple leaf jointed to the petiole.

LESSON VIII.

SESSILE LEAVES—FORMS OF STIPULES.

41. WE have already stated (Lesson I., § 5) that many leaves are without petioles (foot-stalks), or, in other words, are sessile. The figures presented on page 33 exhibit some of the modes of attachment peculiar to sessile leaves. In Fig. 80 (an Aster) you see leaves of the form called spatulate (Lesson III., § 5), having large base lobes nearly clasping the stem at the point of attachment. Such leaves are said to be *amplexicaul* (stem-clasping).

42. In the next figure (81, Bellwort) the leaves are elliptical, parallel-veined, and not only clasp the stem at base, but the lobes there grow together on the opposite side, appearing as if the stem passed through the leaf; that is, *perfoliate.*

40. What kind of venation have all these forms? On the palmate venation what forms are founded?

41. When are leaves said to be sessile? Define an amplexicaul leaf.

42. Can you define a perfoliate leaf?

Fig. 80. Amplexicaul leaves of Aster laevis.
Fig. 81. Perfoliate leaves of Bellwort (*Uvularia perfoliata*).
Fig. 82. Connate leaves of Honeysuckle (*Lonicera sempervirens*).

43. In Fig. 82 (Trumpet Honeysuckle) the leaves placed opposite are joined together by pairs, base to base. Such are *connate* leaves.

44. The forms of the petiole, when the petiole exists, are also various. Generally, it is merely a rounded, slender stem, but you will often find it flattened. Have you ever noticed the structure of the Aspen (Poplar) leaf, which so easily flutters in the gentlest breeze? Its petiole is flattened vertically, so that its edges turn sky-ward and earth-ward. Such a form of leaf-stalk is called *compressed*, and it must be very nicely balanced in order to hold the blade at rest.

---

43. When are leaves said to be connate?

44. What is the usual form of the petiole? Carefully describe the petiole of the Aspen.

2*

45. A *winged* petiole is flattened horizontally. A *sheathing* petiole embraces the stem with its winged edges like a sheath. You can find plenty of examples of these forms.

Fig. 83. Rose,—stipules adnate.　　*Fig.* 84. Violet (*V. tricolor*),—gashed stipules.

46. Let us now study more particularly the varying forms of the stipules. We have already defined them. (See Lesson I., § 4.) Here is seen the leaf of a Rose and of a Pansy (Figs. 83, 84), both with quite showy stipules. The former

Fig. 85. Leaf of Conioselinum,—tri-pinnate, with sheathing petiole.
Fig. 86. Leaf of Polygonum Pennsylvanicum, with its (*s*) ochrea.
Fig. 87. Stem of Grass, with joint (*j*), leaf (*l*), ligule (*s*).
Fig. 88. Leaf of Pear-tree, with slender stipules.

45. What difference between a winged and compressed petiole? Can you describe a sheathing petiole? Give examples of these three forms.

has its stipules *adnate;* that is, growing to the petiole. The Pansy has large stipules deeply cleft into many segments.

47. Figs. 85–88 are very instructive. Fig. 88 is a Pear leaf, with an ovate blade, a slender, cylindric petiole, and a pair of small, narrow stipules (*s*). Fig. 86 is a Knot-grass leaf, with an *ochrea* (*s*); that is, a pair of stipules so joined at the edges as to form a sheath around the stem Fig. 87 is a Grass leaf, linear, with a *ligule* (*s*) supposed to be the top of a doubled stipule. Fig. 85 is a very compound leaf of Conioselinum, having a broadly winged, sheathing petiole.

LESSON IX.

ARRANGEMENT OF LEAVES AND BUDS.

48. If you carefully notice how the leaves are distributed over any plant,—the Corn plant, for example,—you will soon admire their order and exactness in this respect. At first view, we might suppose their positions all accidental; but it is not so, and much of the peculiar aspect of the plant depends upon this circumstance.

49. In the Corn plant, or in this figure of Lady's-slipper (89), we find the leaves *alternate,*—that is, one on this side, the next one higher and on that side, and so on. So it is in

46. Stipules; can you repeat the definition? Describe the stipules of the Rose. Describe the stipules of the Pansy.

47. Describe the stipules of the Pear. Stipules of Knot-grass—what called? Stipules of Grass—what called?

48. Are the positions of the leaves on the plant accidental?

49. Can you describe the alternate arrangement? How is this arrangement more accurately described?

the Elm, Cherry, Willow, and many other plants. But it would be more accurate to say that the arrangement, in all these cases, is *spiral*. (See Class Book, § 224.)

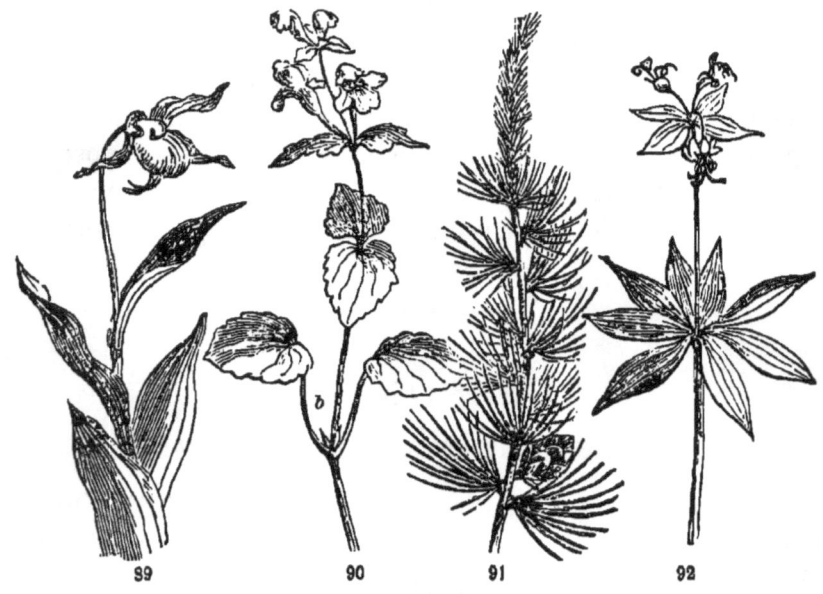

Fig. 89. Lady's-slipper (*Cypripedium*),—leaves alternate.
Fig. 90. Synandra,—leaves opposite.
Fig. 91. Larch (*Larix Americana*),—leaves fasciculate.
Fig. 92. Indian Cucumber (*Medeola*),—leaves whorled.

50. In the Maple, Lilac, Phlox, and in this figure of the Synandra (90), a wild western plant, the leaves are *opposite* that is, two opposite ones stand at each joint. The Meadow Lily, and this Medeola (Fig. 92) of the New England woods have *whorled* or *verticillate* leaves; that is, several in a circle at each joint. Again, look at this Larch (Fig. 91), the Pines, &c., whose leaves, gathered in little tufts or bundles, are *fasciculate*.

50. Define the opposite arrangement. The whorled; fasciculate.

51. In early spring, before the leaves are expanded, we find them folded up in the buds. This is called the *vernation* of the leaves (from the Latin *vernus*, spring). In this condition the young leaves are closely packed in many curious modes, which are described in the Class Book, §§ 209–214.

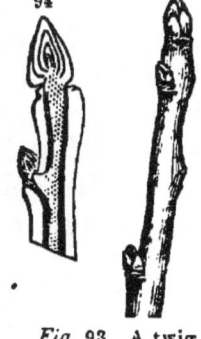

52. If we dissect and carefully examine a swelling leaf-bud in early spring, we observe in the midst of it a tender point of a growing pith, bearing and covered by many circles of little leaves and scales, packed as close as possible. Fig. 94 shows a twig with two buds as if split through the axis, exhibiting the pith, growing point, young leaves, and scales.

Fig. 93. A twig, with two lateral and one terminal bud.

Fig. 94. Same, split through the two buds.

53. According to this figure and the next (94), buds are either *terminal* (*t*), situated at the end of the stem or twig, or *lateral* (*a*), situated on the side. But we must more carefully define the position of the lateral buds. Should we tell you that they are *axillary*, or located in the *axil* of a leaf, you would not understand, until knowing that the *axil* of a leaf is the upper angle between the leaf-stalk and the stem. (See *b*, Fig. 90.) Now, remember this rule, which you may soon confirm by your own observation, that *there is a bud at the termination of every stem or branch, and in the axil of every leaf.*

---

51. What is the meaning of the term *vernation?*
52. Give a careful definition of a leaf-bud.
53. What is a terminal bud? What an axillary? Where are buds always found?

LESSON X.

APPENDAGES, ETC.

54. The *tendril* is a very common appendage. You have seen it in the Grape-vine, the Pea-vine, the Greenbrier, &c It is like a stout, green thread, reaching out its curved poin like a finger, until it touches some object; then it quickly entwines itself around it, and soon acquires a firm hold. We do not find tendrils on any plants except such as, like vines, are too weak to stand without support.

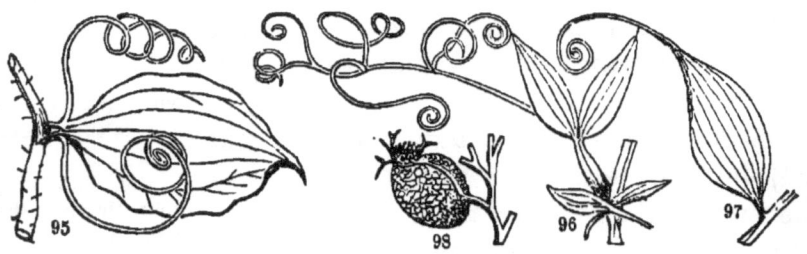

Fig. 95. Leaf of Greenbrier, with tendrils in place of stipules.
Fig. 96. Leaf of Everlasting Pea,—tendrils at end of rachis.
Fig. 97. Leaf of Gloriosa,—apex ends in a tendril.
Fig. 98. Air-bladder of Horn-pondweed.

55. But tendrils are quite various in habit. Those of the Pea (Fig. 96) grow from the extremity of the rachis. Those of the Greenbrier (Fig. 95), from the base of the leaf-stalk, in the place of stipules; those of the Grape (Fig. 000) are opposite the leaves, in the place of clusters.

56. Many plants are armed with sharp thorns, spines, or

54. What is the first appendage mentioned? Please describe the form and use of the tendril.

55. State the habit of the tendril of the Pea · Greenbrier; Grape-vine.

prickles, as if in self defence. See the Thorn-bush (Fig. 99), where the long straight *thorns* come from the axils of the leaves, and are woody. The terrible thorns of the Honey Locust (Fig. 100) are branched. Those of the common Locust are in the place of stipules. Those last mentioned, and all others which originate with the leaves (as in Berberis, Thistle, &c.), are more properly called *spines*.

57. As for the Rose and Bramble, they are armed with *prickles*, which are horny in substance, connected with the bark only, not with the wood. (See Fig. 101.)

Thorns.—*Fig.* 99. Cratægus parvifolia (thorns axillary). *Fig.* 100. Honey Locust (branched thorns).

58. *Glands* are little wart-like bodies which secrete the peculiar fluids of the plant, sometimes imbedded in the leaf or the rind of the fruit, as in the Lemon, where it is filled with a fragrant volatile oil; sometimes raised on a hair (Figs. 102, 103), as in Sundew, exuding a clammy liquid.

59. *Stings* are piercing hairs, having a bag at the base filled with an acrid fluid. When touched the tip breaks off, the hair penetrates the skin, and the poison is injected into the wound. (See Fig. 106.)

---

56. What is the habit of the thorns of the Thorn-bush? of the Honey Locust? of the common Locust? What of the habit of spines?

57. What of prickles?

58. Describe glands, the two kinds.

59. What is the structure and action of stings?

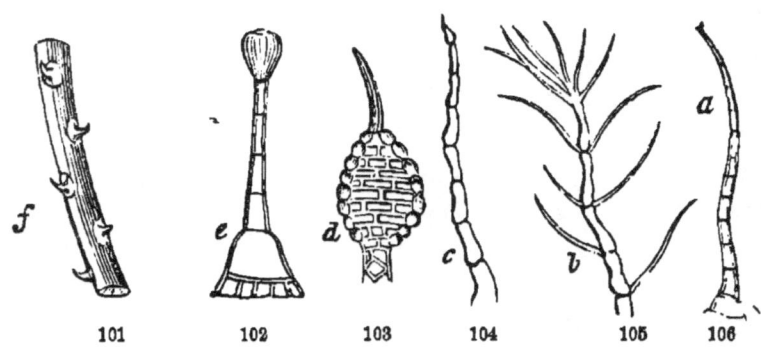

101   102   103   104   105   106

What do these figures represent ?—*Fig.* 105 represents a branched hair as it appears under a strong magnifier; *Fig.* 104, an unbranched or simple hair; *Fig.* 102 is a hair with a gland on it; *Fig.* 103, also, is a gland on the top of a hair; *Fig.* 101 represents the hooked prickles of a Rose-bush, not magnified; *Fig.* 106 represents a sting of a Nettle, much magnified.

60. *Hairs* of various kinds (Figs. 104, 105) are found on the leaves and other parts of plants. By this clothing peculiar qualities are given to the surface, named and described as follows.

61. A dense coat of hairs will make the surface *pubescent* when the hairs are short and soft; *villous*, when rather long and weak; *sericious*, or silky; *tomentous*, when matted like felt, &c.

62. But thinly scattered hairs make the surface *hirsute* when they are long; *pilous*, when short and soft; *hispid*, when short and stiff, &c.

---

60. How are plants clothed?
61. Define the term pubescent; villous, &c.
62. Define the term hirsute; hispid, &c.

LESSON XI.

ORGANS OF THE FLOWER.

63. To-day, we commence the study of the beautiful flower. We have before us the Meadow Lily (Fig. 107), whose organs are large and perfectly distinct. Observe, in the first place, that its brighter colors form a striking contrast with the soft green of the leaves. The coloring, the structure,

107   109   108   110

Fig. 107. Meadow Lily (*Lilium Canadensis*).
Fig. 108. Wake-robin (*Trillium erectum*).
Fig. 109. Stamens (*s, s*) and pistil (*p*) of the Lily.
Fig. 110. Stamens (*s, s*) and pistils (*p*) of the Trillium.

and the fragrance of the flower are all worthy of its Infinite Creator, and remind us of his wisdom and goodness.

64. As to the structure of the flower, it is always compound, being composed of several or many pieces nicely adapted to each other. In this Lily, for example, you may count thirteen pieces, or organs, attached in a close order to the summit of the flower-stalk (Fig. 113, *a*). You may call the flower-stalk the *peduncle*, and the point of attachment (*r*) the *torus*, or receptacle. The former is the better name.

65. Two circles of leaf-like organs form the envelopes of this flower, and each circle consists of three pieces. The outer circle is the *calyx*, and the three pieces which compose it are called *sepals* (*s, s, s*). The inner circle is the *corolla*, and the three pieces which compose it are called *petals* (*p, p, p*). In the Lily and some other flowers the calyx is colored like the corolla. But it is not so generally. In the Rose, Strawberry, Pink, and in this Trillium (Fig. 108), the calyx is green, while the corolla is almost always distinguished by some brighter color.

66. Now, taking both calyx and corolla together as a whole, we call them the *perianth* (a Greek word, meaning *around the flower*). This name is very convenient when we speak of such flowers as this, where the calyx and corolla are not much different.

---

63. What is the subject of to-day's lesson? What do you notice as to the color of the flower?

64. What is said of the compound nature of the flower? Of how many pieces is the flower of the Lily composed? What is the peduncle? What is the torus?

65. Will you point out and define the calyx? sepals? Will you point out and define the corolla? petals? What of the colors of these organs?

66. What is the use of the word perianth? Will you point out and define the stamens? What of their number? What is the pistil? How many?

67. Next within the perianth of the Lily we find six long, slender organs of peculiar form and color, called *stamens*. In the Rose you find a larger number (perhaps one hundred) of stamens, while in the Speedwell you find but two. But the most common number is five. Count them in the Morning-glory, the Bellwort, Primrose, &c.

68. Lastly, this central, club-shaped body (*p*), here as long as the stamens, but of totally different structure, is the *pistil*. Other flowers have more than one pistil, as the Pink, which has two; the Rose, which has many.

69. Thus, we have learned that the flower—at least *this* flower—is compounded of four kinds of organs, those of each kind being arranged in a circle by themselves. The outer circle, of sepals, constituting the calyx; the second circle, of petals, constituting the corolla; the third circle, the stamens; the fourth circle, the pistils.

LESSON XII.

MORE ABOUT THE CALYX AND COROLLA.

70. Let us examine the flower of the Pink (Fig. 112), the Strawberry (Fig. 111), the Crowfoot, the Single Rose. In either you observe five green sepals, and the same number of colored petals. Notice also the positions of those organs, —how the petals stand alternating with the sepals, and that they are all distinct and separate. This is the general rule, but there are many exceptions.

67. Lastly, review the whole arrangement.

70. What is the rule as to the number of petals and sepals? What is the rule as to their relative position, &c.?

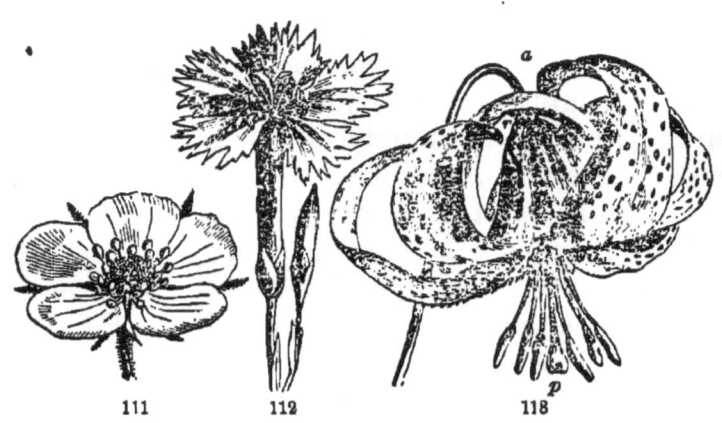

Fig. 111. Flower of the Strawberry.　　　*Fig.* 112. Flower of the Pink.
Fig. 113. Flower of the Lily.

71. Often in the petal, and sometimes in the sepal, you can distinguish two parts,—namely, the broad, expanded part above, called the *lamina*, and the narrow part at base by which it is attached to the torus; this is the *claw* (Fig. 116, *c*).　The petal of the Pink has a long claw; of the Rose or Buttercup (Fig. 119), a short one.

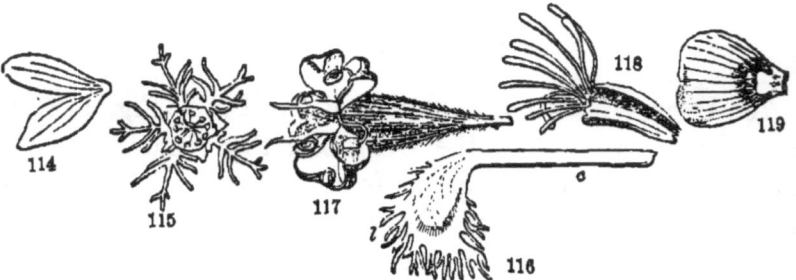

72. The forms of the petal are almost infinite in variety, like the leaf; as ovate, orbicular, oblong, &c., and some-

71. Will you define the lamina? the claw?
72. Please mention some of the forms of petals.

times very singular. See these figures. Fig. 114 is the form of the bifid petal of a Cerastium; Fig. 115, the flower of Mitella, with five pinnatifid petals; Fig. 117, the flower of Sweet Cicely, with five petals inflected at the point; Fig. 116, fringed, long-clawed petal of Silene stellaria; Fig. 118, many-cleft petal of Mignonette; Fig. 119, rounded, short-clawed petal of Crowfoot, showing its honey scale, or nectary, at base.

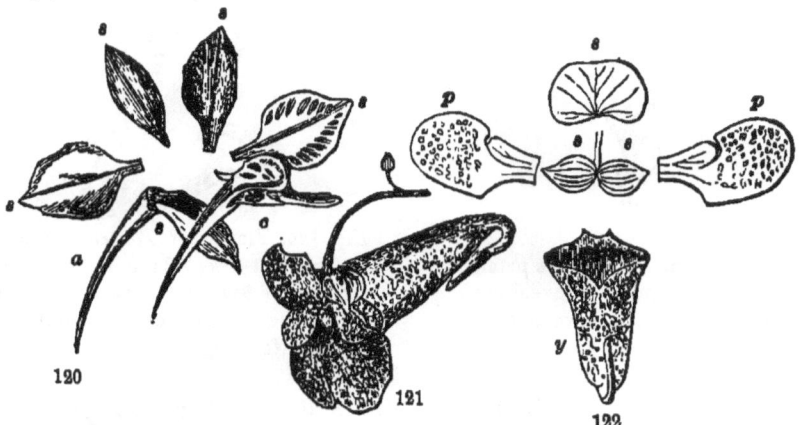

120
121
122

Fig. 120. Larkspur, its petals and sepals separated: *s, s, s, s, s,* sepals; *a,* the upper sepal spurred; *c,* the petals all united into one, and produced backwards into a spur which is sheathed in the spur of the calyx.

Fig. 121. Touch-me-not. *Fig.* 122. Its petals and sepals displayed: *p, p,* the two double petals; *s, s, s, y,* the four sepals, *y* being in the form of a sack, with a spur.

73. A nectary is found also in the petals of Columbine (Fig. 361), Larkspur (Fig. 120), Touch-me-not (Fig. 121), &c., distorting them into grotesque shapes, called *spurs*.

74. Before us now is the flower of Pink (Fig. 123). The calyx (*c*) appears as a green tube, with five notches or teeth at the top. It is evident that this is made up of five sepals

73. What is a nectary? What is a spur? Examples.

cohering (united) by their edges. The Convolvulus (Figs. 128, 144), the Phlox (Fig. 126), the Pink-root (Fig. 127), &c., show a similar cohesion (union) of their petals into a tube more or less complete.

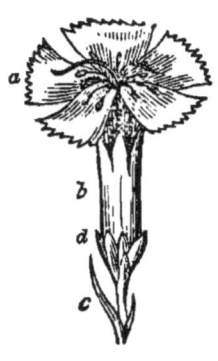

Fig. 123. Pink: *a*, the five petals; *b*, the calyx, composed of five united sepals, *c*, a bract; *d*, several bractlets.

Fig. 124. Flower of *Tecoma radicans* (the Trumpet-creeper): *c* is the calyx, composed of five united sepals; *t*, the tube; *s*, the *segments* of the corolla or the petals, forming the *border*.

75. The calyx with united sepals is called *monosepalous*, and the corolla of united petals *monopetalous* (from the Greek *monos*, one), from the mistaken idea that this calyx consisted of only one sepal, &c. *Gamopetalous* and *gamosepalous* are similar words, used in the same sense. Opposed to these terms are *polysepalous* and *polypetalous* (Greek *polys*, many).

76. The gamosepalous calyx or gamopetalous corolla, although composed of several pieces, is described as a single organ, and its lower part, formed by the united claws, whether long or short, is the *tube* (Fig. 124, *t*); the upper

74. Describe the calyx of Pink; corolla of Phlox.
75. Meaning of the terms monopetalous, &c.?
76. Define the limb of a monopetalous corolla; the tube; the throat.

part, composed of the united laminæ, is the *limb* (Fig. 128, *s*);
the opening of the tube above is the *throat.*

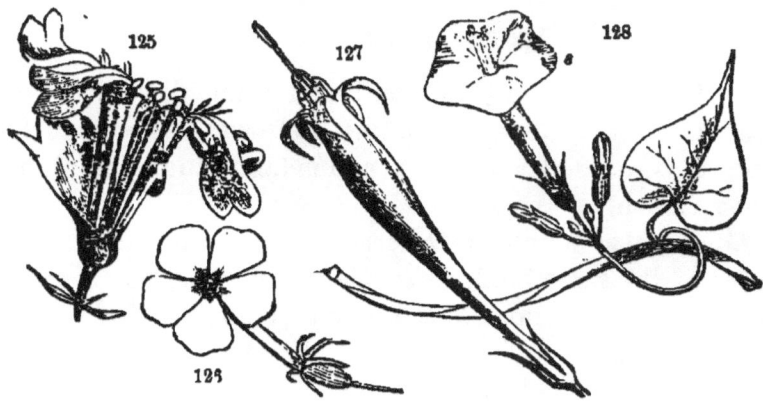

Fig. 125. Flower of Saponaria (Bouncing-Bet); petals and claws quite distinct.
Fig. 126. Phlox; claws united, with laminæ distinct.
Fig. 127. Spigelia (Pink-root); petals still further united.
Fig. 128. Quamoclit coccinea; petals united throughout.

77. In the Figs. 125–128, you may see how the petals in
different flowers are distinct, or in various degrees united.
In the Bouncing-Bet, the petals, with their long claws, are
entirely *distinct.* In Phlox, the claws unite in a tube, while
the laminæ are distinct. In Pink-root, only the narrow tips
of the laminæ are distinct; and in Quamoclit, the laminæ
also are wholly united.

77. What is the condition of the petals in Pink Soapwort? What their
condition in Phlox? What their degree of cohesion in Pink-root? What
n Quamoclit?

LESSON XIII.

ABOUT ADHESIONS.

78. WE fear that the pupil will find some difficulties in this lesson. Yet if he bring to the task *eyes* determined to see, and a *mind* determined to understand, the difficulties will soon vanish.

79. *Cohesion*, as taught in the last lesson, implies the union of organs of the same kind, as sepals with sepals, petals with petals; but *adhesion* implies the union of one kind of organ with another kind.

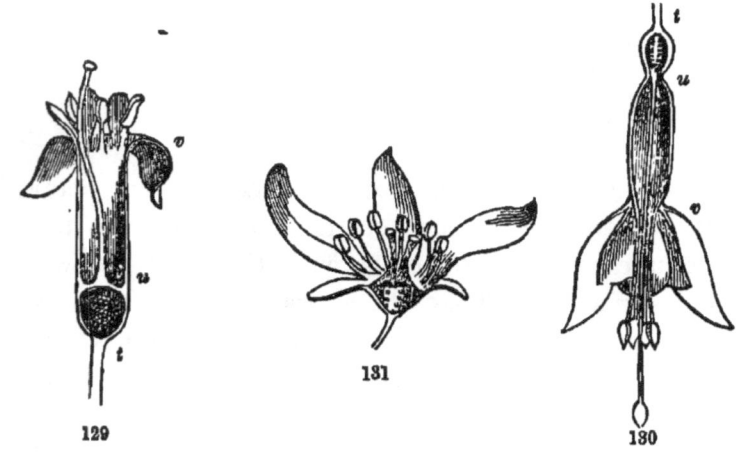

Fig. 129. Section of the flower of the Golden Currant, showing its parts.
Fig. 130. Section of the flower of Fuchsia.   *Fig.* 181. Of Early Saxifrage.

80. For example, split a flower of Phlox, and you will see the five stamens adhering to the inner side of the corolla tube, appearing as if inserted into it.

79. Can you state how adhesion differs from cohesion?

81. Now we take it for granted that all the organs of the flower have their starting-point or origin at the same one point, namely, at the torus (*t*, Fig. 129), hence in this figure of the Golden Currant, it is understood that from *t* to *u* the calyx, corolla, stamens, and pistil, adhere together; from *u* to *v*, the calyx, corolla, and stamens, adhere; and at *v*, all the organs are separate, that is, *free.* Observe the same structure in the Ear-drop (Fig. 130).

82. In this and like cases, the calyx is said to be *superior*, because it seems to stand upon the pistil (ovary) and fruit, but the more correct term is, calyx *adherent*.

132.

133.

Fig. 132. Section of the flower of Yellow Violet: *t*, the torus. The stamens are hypogynous.

Fig. 133. Section of the flower of Pear: *o, o,* sepals; *p, p,* petals; *s, s,* stamens, —perigynous; *o,* ovary,—inferior or adherent.

83. There are two other terms used in similar cases, which, although hard to pronounce, you may as well become acquainted with now. When the stamens adhere to the calyx

81. What do we take for granted? Please show the adhesions in the Golden Currant.

82. When is the calyx adherent? When superior?

3

or corolla they are said to be *perigynous* (a Greek word,
meaning "around the pistil"). Otherwise, when *free*, they
are said to be *hypógynous*, meaning "under the pistil."

84. Now study attentively these figures, or rather, the
flowers themselves. The figures are sections, *i. e.*, show the
flowers as if split. Fig. 132 (the Violet) shows the stamens
hypogynous and the organs all free. Fig. 133 (the Pear)
shows the stamens perigynous, adhering to the calyx. Fig.
131 (the Saxifrage) shows the stamens perigynous and the
calyx half adherent. Do not fail to examine many flowers
until these troublesome terms become familiar, for these
distinctions are very important.

LESSON XIV.

FORMS OF PERIANTH.

85. While all flowers agree in certain general characteris-
tics, so that you are never at a loss to recognize any one of
them as a *flower*, yet in form and fashion they appear in
infinite variety, each form endowed with its own peculiar
grace. It is impossible to describe or name every form, but
we will endeavor to reduce them to a few classes of forms.

86. Notice first that all forms are either polypetalous or
gamopetalous, as already described (§ 75). Again, they are
either *regular* or *irregular*. Compare the flower of Flax

---

83. When are the stamens said to be perigynous? When hypogynous?
84. How are they in Saxifrage? in Pear? in the Rose? the Violet?
86. What is the first division of the corolla forms? What is the second
division? When is a flower said to be regular? irregular?

185. 136. 134. 187.

Polypetalous corollas.—*Fig.* 134, Wild Apple (*Pyrus coronaria*),—rosaceous.
Fig. 185. Wall-flower,—cruciform. *Fig.* 136. Scarlet Catchfly,—caryophyllaceous.
Fig. 137. Atamasco Lily,—liliaceous.

and Pea. The former is equally and similarly developed all around, and each petal is like all the other petals. It is a *regular* flower. The Pea flower (Fig. 138) is unequally developed, some of the petals differing in form and size from the others, as shown in Fig. 139; therefore it is *irregular*.

87. The figures at the head of this page represent four different styles of corollas which are polypetalous and regular. Fig. 134 (Wild Apple) is a *rosaceous* corolla, that is, rose-like, having five short-clawed petals. Fig. 135 (Wall-flower) is a *cruciform* (cross-shaped) corolla, with four long clawed petals.

88. Fig. 136 (Scarlet Catchfly) is a *caryophyllaceous* corolla,

---

87. Name the four forms of polypetalous, regular flowers. Can you describe the rosaceous corolla? What sort of corolla is the Wall-flower Describe it.

88. Please describe the Catchfly or Pink. What sort is it? The Lily please describe. What sort of corolla is it?

pink-like; a form with five long-clawed petals. Fig. 137 (Atamasco Lily) is a *liliaceous* corolla, having a six-leaved perianth, made up of three sepals and three petals, all colored alike.

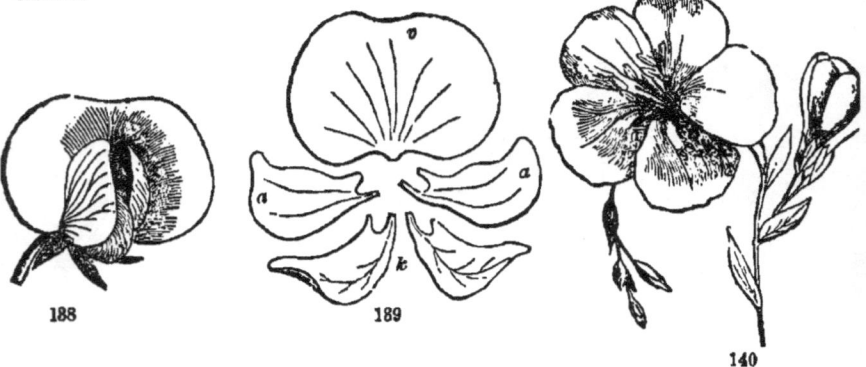

188   189   140

Fig. 138. Pea,—an irregular flower.   *Fig.* 189. Its five petals shown separate, *viz.*, *v*, the banner; *a, a*, the wings; *c, c*, the keel-petals.
Fig. 140. Flax (*Linum grandiflorum*),—a regular flower.

89. Fig. 145 is the flower of Sweet Pea, an irregular corolla, called *papilionaceous*, or butterfly-shaped, consisting of five petals, as displayed in Fig. 139, *viz.*, one odd petal above, very large, called the *banner*, two smaller petals below (*k*), called the *keel*, and two lateral petals (*a, a*), called the *wings*.

90. We next propose to examine the principal forms of gamopetalous corollas. Here we have a beautiful array of them. Among the regular forms is, first, the *Rotate*, wheel-shaped or star-shaped, having a very short tube, and a flat, spreading border; as Fig. 141 (Campanula Americana).

91. *Campanulate*, bell-shaped, having a wide tube and

89. Can you describe the papilionaceous corolla?
90. Of monopetalous corollas, describe the rotate.   91. The campanulate.

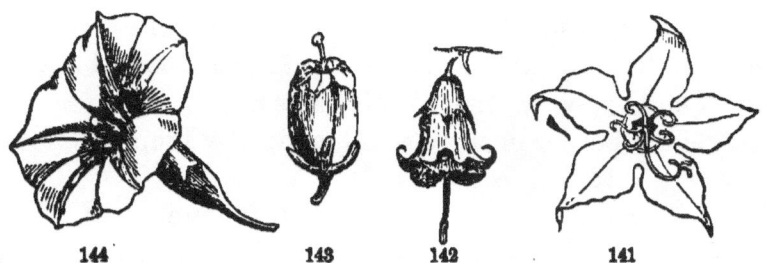

144   143   142   141

Gamopetalous corollas.—*Fig.* 141. Campanula Americana,—wheel-shaped. *Fig.* 142. Campanula divaricata,—campanulate, or bell-shaped. *Fig.* 143. Andromeda,—urceolate. *Fig.* 144. Field Bindweed (*Convolvulus*),—a funnel-form corolla.

narrow border, as in the Bell-flower (Fig. 142), and in Canterbury Bells.

92. *Urceolate*, urn-shaped, an oblong or globular corolla with a narrow opening, as the Whortleberry, Heath (Fig. 143).

93. *Funnel-form*, narrowly tubular below, gradually enlarging to the border, as Morning-glory (Figs. 22, 144).

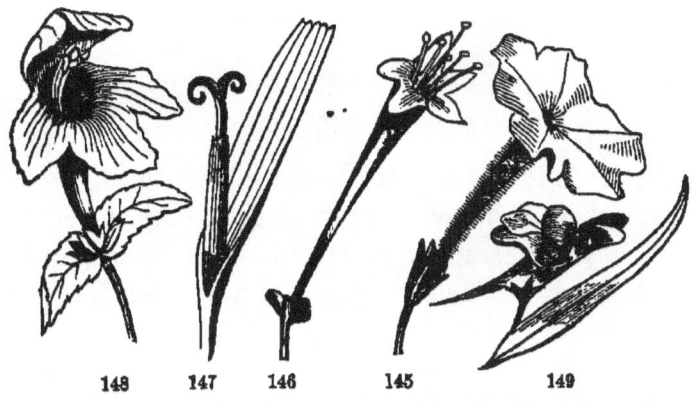

148   147   146   145   149

Fig. 145. Petunia,—salver-form. *Fig.* 147. Dàndelion,—ligulate.
Fig. 146. Honeysuckle,—tubular. *Fig.* 148. Synandra,—labiate.
Fig. 149. Toad-flax,—labiate-personate.

92. The urceolate.   93. The funnel-form.

94. *Salver-form*, the tube suddenly spreading out in a horizontal border, as in Phlox, Petunia (Figs. 126, 145).

95. *Tubular*, when the corolla is nearly all a slender tube with a small border or none at all, as in the Trumpet Honeysuckle (Fig. 146).

96. *Ligulate* (from the Latin *ligula*, tongue), as if formed by splitting the tubular on one side. The notches at the end plainly indicate the number of united petals which compose it, as also do the parallel seams. See the flowers of the Dandelion (Fig. 147), also of Cichory.

97. *Labiate* (Latin *labium*, lip), resembles the mouth of an animal. It is a very irregular corolla, having the petals of dissimilar shape and dissimilarly united. See (Fig. 148) a flower of Synandra, or Catmint, or Catalpa. In Fig. 149 (Snap-dragon), the mouth is closed and said to be *personate*, which means *masked*.

———•••———

LESSON XV.

CONCERNING THE STAMENS.

98. SAFELY infolded within the perianth, we find a number of delicate, thread-shaped organs, quite unlike the sepals and petals. They are arranged in one or more circles, and called the *essential organs*, because they are absolutely necessary to the perfection of the seed.

---

94. Describe the salver-form.　　95. The tubular.　　96. Ligulate.

97. Labiate. Now repeat the regular forms. Repeat the names of the irregular forms.

98. Where do we find the essential organs? How arranged? Why are they so called?

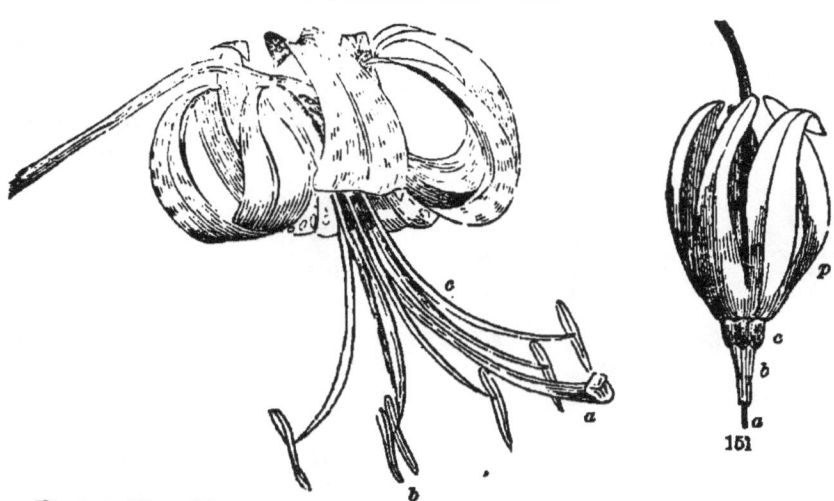

Fig. 150. Tiger Lily.

Fig. 151. Flower (enlarged) of Dodecatheon : *a*, pistil; *b*, anthers; *c*, filaments; *p*, petals.

99. Let us look at this picture of the Lily (Fig. 150), or at some real flower. The slender organs marked *a*, *b*, *c*, are the essential organs of which we are speaking; and you see at once that there are two kinds of them. Those which stand in the outer row next to the petals are the stamens.

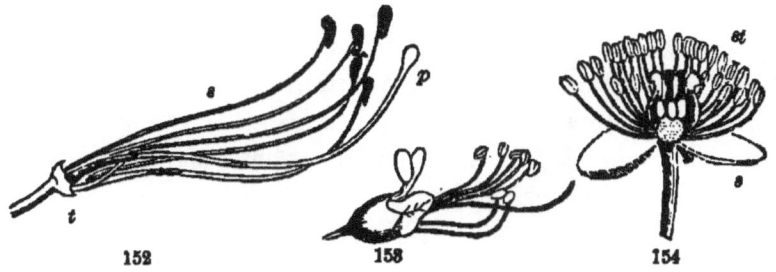

Fig. 152. Rhododendron ; only the torus (*t*), the five stamens (*s*), and the pistil (*p*).

Fig. 153. Buckeye, whole flower ; 7 stamens, 1 pistil, 3 petals.

Fig. 154. Hydrastis, split through the centre (a section), showing the torus, 2 sepals (*s*), many hypogynous stamens (*st*), and several pistils in the midst.

The central organ (or organs) is the pistil. We now propose to notice the form of the stamens.

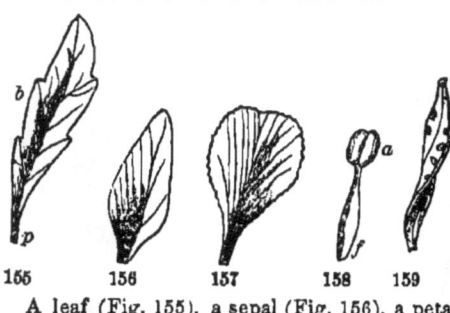

155 156 157 158 159

A leaf (Fig. 155), a sepal (Fig. 156), a petal (Fig. 157), a stamen (Fig. 158), and a pod (pistil, *Fig.* 159) of Draba arabizana, placed side by side for comparison.

100. The stamen may be compared to the leaf. Its slender, thread-like stalk is the filament, answering to the petiole of the leaf (*f*, *p*). Its head (*a*) is the *anther*, answering to the blade. Moreover, the anther contains within its cells many dust-like particles called *pollen*. When the cells burst the pollen escapes. Thus it appears that the stamen consists of three members. See them illustrated in this figure (161) of a stamen of the Morning-lory.

101. The filament is usually of a thread-like form (as its name, from the Latin *filum*, a thread, implies), longer than the anther, and more or less elastic. But the filament is no more necessary than the stem of a leaf, and is often wanting.

102. The anther is an oblong body at the top of the filament, consisting of two hollow lobes joined to each other and to the filament by the *connectile* (*c*), which answers to the midvein of the leaf. The two lobes are usually marked along their outer edge by a *seam*, which at length opens into the cells. This opening, however it takes place, is called the *dehiscence*. If there be no filament, the anther is *sessile*.

99. How many kinds? Situations of the two kinds respectively?

100. How does the stamen compare with the leaf? Specify the three members of the stamen.

101. Describe the filament. 102. The anther; the dehiscence.

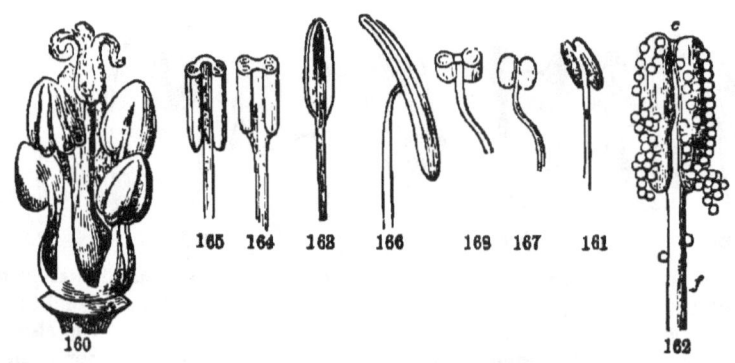

165 164 168 166   169 167   161

160   162

Fig. 160. Frankenia, showing the five stamens (around the one style, which has three stigmas at top).

Fig. 161. Stamen (adnate) of Morning-glory.

Fig. 162. Same, enlarged, with pollen-grains discharged: *f*, filament; *a*, anther, —two-lobed; *c*, top of connectile.

Fig. 163. Buttercup. Fig. 164. Same, cut across.

Fig. 165. Iris, cut across (extrorse).

Fig. 166. Amaryllis,—versatile. Figs. 167, 168. Larkspnr,—innate.

103. But dehiscence takes place very variously. When all regular, it is a chink running lengthwise along the outer edge, as you see in this stamen of a Buttercup (Fig. 163). But here, in this stamen of Iris (Fig. 165), it appears on the back of the anther (looking towards the petals), and we say that the anthers are *extrorse*, that is, turned outwards. A term of opposite meaning is *introrse*, denoting that the lines of dehiscence turn inwards towards the pistil, or at least do not turn outwards. For example, the anthers of the Violet (Fig. 173).

104. Moreover, other modes of dehiscence besides chinks are occasionally found. The anthers of Berberis, Sassafras, &c. (see Figs. 171, 172), open by lids hinged at the top. The

103. When is the anther said to be extrorse? introrse?

104. Can you distinguish the *opercular* and *porous* dehiscence?

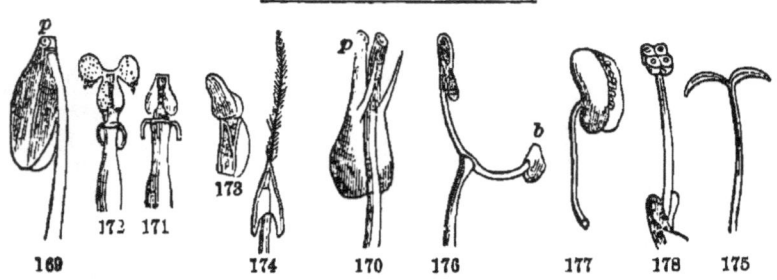

Peculiar forms of stamens.—*Fig.* 169. A stamen of Pyrola rotundifolia: *p*, two openings (pores) at top where the pollen escapes. *Fig.* 170. Stamen of Bilberry (*Vaccininium uliginosum*): *p*, its pores at the top of two horns; it has also two spurs. *Fig.* 171. Berberis aquifolium, anther closed. *Fig.* 172. Anther open by two lids upwards. *Fig.* 173. Anther of Violet with an appendage at top. *Fig.* 174. Oleander,—an arrow-shaped anther appendaged at top. *Fig.* 175. Catalpa,—lobes of anther separated. *Fig.* 176. Sage,—lobes of anther widely separated on stipes; *b*, barren lobe without pollen. *Fig.* 177. Mallows,—anther one-celled. *Fig.* 178. Ephedra,—anther four-celled.

anthers of Huckleberry, Blueberry, Wintergreen, and others of the Heath family, open through two little tubes at the top. The former is *opercular* dehiscence, the latter *porous*. (See Figs. 169, 170.)

105. It is also interesting to notice how the anther is attached to the filament in various ways. Generally, it is *innate*, that is, seeming to stand erect on the top of the filament. Again, it is *adnate*, which means, attached by its back to the side of the filament, as in Buttercups. And thirdly, it is joined by a single point in its back to the slender tip of the filament, as if lightly balanced upon it. This is the *versatile* anther, common in the Grasses (Figs. 150, 166).

105. What three distinctions in the attachment of the anther? Describe that of the stamens of the Pink; the stamens of Buttercups; of the Grasses.

LESSON XVI.

MORE ABOUT THE STAMENS.

106. THE careful student will find a great and interesting variety in the number, arrangement, and form of the stamens. In regard to number, as we have already seen, the Lily has six stamens, the Pink has ten, the Speedwell two, the Indian Shot only one. Some flowers have numerous stamens, as the Rose with forty, fifty, or one hundred, and the Cactus with

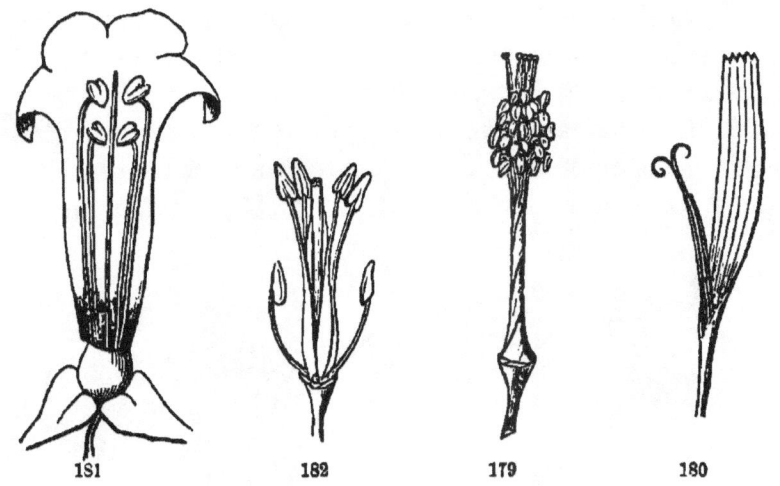

181   182   179   180

Fig. 179. Stamens and pistils of Mallow; the filaments (*f*) are united into a tube sheathing the styles.

Fig. 180. Floret of Dandelion,—anthers (*a*) united into a tube.

Fig. 181. Corolla of Lophospermum, split open to show the four stamens (didyn'a-mous) and the one style.

Fig. 182. Cardamine,—stamens six, tetradyn'amous.

106. What number of stamens in Pink? Speedwell? Indian Shot? What in the Rose? Cactus? Apple? or in *these* flowers? Define "stamens definite;" "stamens indefinite."

two hundred. Let us learn how to distinguish between flowers with *definite* and with *indefinite* stamens. Definite, when they are not more than ten, indefinite, when more than ten, or not readily counted.

107. The stamens are usually separate and distinct, as in the Lily, Rhododendron, &c. (Figs. 150, 152), while in the Mallow (Fig. 179), Pea, and other flowers, they grow together, forming a tube around the pistil; in other words, they are *monadelphous* (Greek, *monos*, one, *adelphos*, brotherhood). The Pea, or Dielytra, is *diadelphous*,—the stamens in two sets; and the St. Johnswort, *polyadelphous*,—in three or more sets. Another mode of cohesion is seen in the floret of Dandelion (Fig. 180), where the *anthers* cohere while the filaments are distinct, *i. e.*, *syngenecious*.

108. In two cases we may definitely mark the relative length of the stamens. *Didyn'amous* stamens (as seen in the Mint tribe, and in the Figworts, Fig. 181) are four in num-

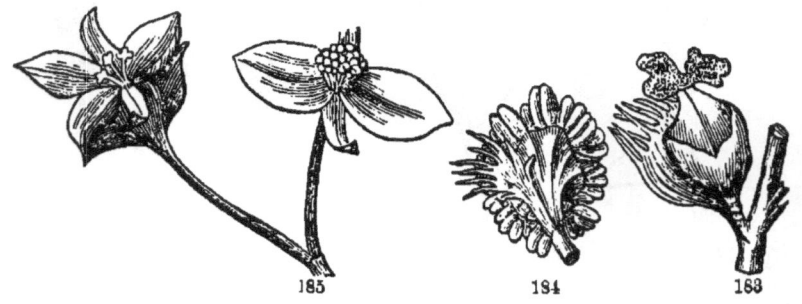

185   184   183

Fig. 183. Pistillate flower of Balm of Gilead.
Fig. 184. Staminate flower of the same.
Fig. 185. Begonia: *a*, staminate flower; *b*, pistillate flower.

107. Define "stamens monadelphous." Give examples. Diadelphous. Give examples. Polyadelphous. Example. Define "stamens syngenecious." Mention examples.

ber, two long and two short. *Tetradyn'amous* stamens are six in number, four long and two short (as in the Mustard tribe, Fig. 182). Again, *hypógynous* stamens may be seen in the Crowfoot tribe and in Fig. 132; and *perigynous* stamens in the Rose tribe and Fig. 133. What is the difference? You need not be told the meaning of these words (§§ 83, 84).

109. Some plants have their essential organs separated, so that the stamens are all found in one sort of flowers, the *sterile*, and the pistils are all in another sort, the *fertile*. So

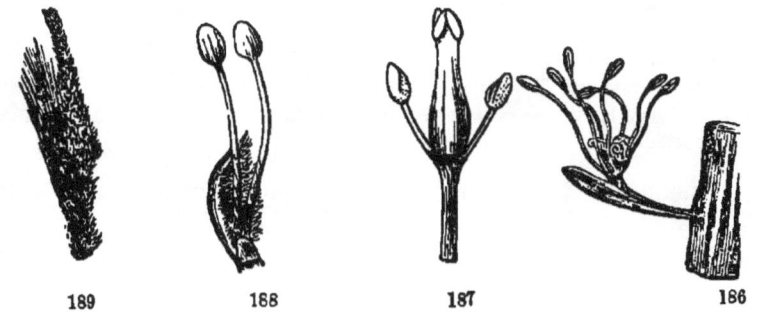

189   188   187   186

Fig. 186. Flower of Lizard-tail (*Saururus*); it is perfect, but *naked*, *i. e.*, with no floral envelopes; stamens seven, pistils three.

Fig. 187. Flower of Ash (*Fraxinus*),—naked, with two stamens and one pistil.

Fig. 188. Staminate flower of Willow,—made up of two stamens and a bract.

Fig. 189. Pistillate flower of the same,—merely one pistil and a bract.

it is in the Begonias (Fig. 185), and in the Willows (Figs. 188, 189). All such flowers are called *imperfect*, and only the fertile bear fruit.

110. A *perfect* flower is one that has both stamens and

108. In what two cases do we mark the length of stamens? Define "stamens didynamous;" "stamens tetradynamous;" "stamens hypogynous;" "stamens perigynous."

109. What do you understand by "sterile flowers?" by "fertile flowers?"

110. What is a perfect flower? complete? imperfect?

pistils. A *complete* flower has *all* the organs, *viz.*, sepals, petals, stamens, pistils. A *naked* flower lacks both the calyx and corolla.

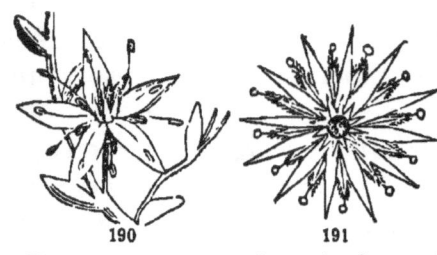

111. A *symmetrical* flower has each of these several organs in an equal number or, at least, the same number of pieces in each *circle* of organs. For example, the Flax flower is symmetrical, having sepals five, petals five, stamens five, and pistils five. The Lily is also symmetrical, having three sepals, three petals, six stamens (in two equal circles), and three pistils (which are combined in one).

Fig. 190. A symmetrical, regular flower of Iceland Moss (*Sedum acre*); it has five sepals, five petals, twice five (ten) stamens, and five pistils,—all separate and distinct.

Fig. 191. House-leek (*Sedum sempervivum*), —twelve-parted.

LESSON XVII.

THE PLAN OF THE FLOWER.

112. It is very instructive and delightful to study the symmetry of flowers in the way mentioned in the last lesson. We are thus led to the discovery of a truth in the science of botany at once beautiful and sublime,—worthy of the wisdom of the Infinite Creator. That truth or *principle* is, that *all flowers, though infinitely various in form and fashion,*

---

111. What a symmetrical flower? How is the Lily symmetrical?

112. Please state the principle learned from studying the symmetry of the flower.

are built upon one only plan, and that plan founded in the science of numbers.

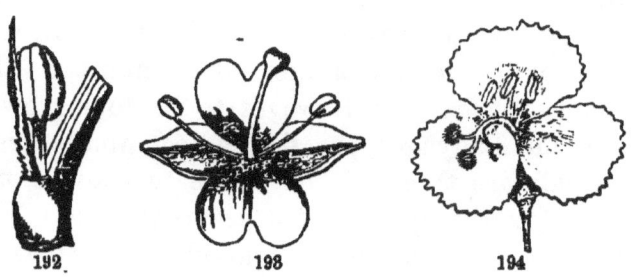

Fig. 192. Flower of Hippuris,—one-parted.
Fig. 193. Circæa Lutetiana; flower two-parted.
Fig. 194. Yellow-eyed Grass (*Xyris*); flower three-parted.

113. Let us, then, examine the Flax. Here all the organs are in fives. The Circe has them all in twos; the Iris, in threes. And every plant is distinguished in this way by some *number* which we call the *radical number*, according to which its organs are parted. Now in the Mock Orange, or Philadelphus, although the stamens seem to be indefinite, still the radical number is four. The stamens occur in many circles, with four in each circle, so that these are also in fours. As for the pistils, they are evidently four, but so united as to form apparently but one. Examine also the Bloodroot. Its stamens will be found in fours, the radical number, and the stamens of the Apple will be found in fives. So the petals of Bloodroot are twice four (8), and of the Magnolia twice three (6), or three times three (9).

114. It is therefore a general law, that when any organ is

113. Can you define the radical number of the flower? What is it in Circe? Iris? Flax? How is it in Philadelphus? How in Bloodroot?
114. State the law of multiplied organs.

multiplied, its new number is only a repetition of the radical number.

115. Also, when any organ is diminished in number, we find generally that the deficiency is only apparent, and does not interfere with the law of the radical number. Thus in Philadelphus, the *one* pistil proves to be four growing together. In the Lady's-slipper, the radical number is three, and the sepals are three, although the two lower ones are united *almost* to the tip into what seems but one. Thus the true number is often curiously disguised *by cohesions.*

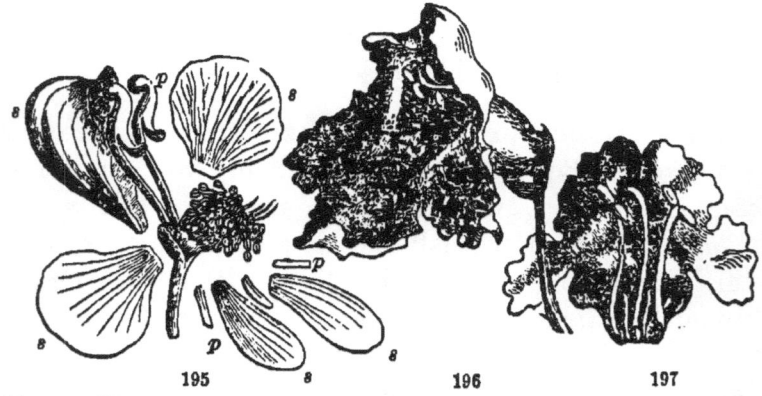

195   *s*   196   197

Fig. 195. Flower of Aconitum Napellus displayed ; *s, s, s, s, s,* the five sepals, the upper one hooded ; *p, p, p,* the five petals, of which the two upper are nectaries covered by the hood, and the three lower very minute.

Fig. 196. Flower of Catalpa,—two-lipped, five-lobed.

Fig. 197. Corolla laid open, showing the perfect stamens and rudimentary.

116. Again, the five petals of Monk's-hood (Fig. 195) are apparently but two, while three of them are so very small as to be overlooked. In the Mint tribe, as Peppermint, Cat-

---

115. How does cohesion interfere with the radical number in Philadelphus How in the sepals of the Moccasin flower ?

116. How does suppression interfere in Monk's-hood ? In the Mint tribe Catalpa ? Mustard ? What tendency do you see in all these cases ?

mint, while five-parted, the flowers have generally but four stamens; but on close observation we often find a small rudiment of the fifth stamen in its proper place, *as if its growth had been early stopped.* And in Monarda and Catalpa, only two stamens grow up to maturity, while three are *mere rudiments* (Fig. 202). Nevertheless, such flowers are said to be unsymmetrical. So the flowers of the Mustard tribe. The stamens are in two rows of four in each; but of the outer row (or circle) two were checked in growth (or *suppressed,* as the botanists say) at the outset. (See Fig. 97.) The tendency to symmetry is manifest in all these cases.

117. We must carefully distinguish between the terms *unsymmetrical* and *irregular.* The former refers to number only, the latter to form and size (Less. XIV.). The Mustard flower is unsymmetrical, but not irregular. The Orchis is irregular, but not unsymmetrical. Snap-dragon is both irregular and unsymmetrical.

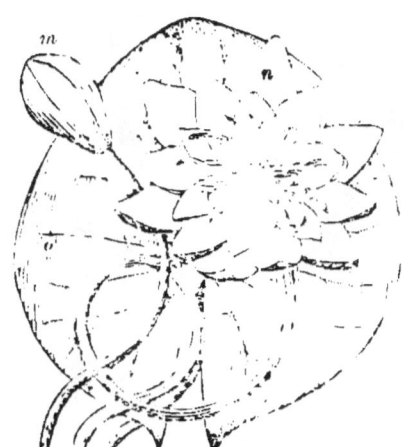

Fig. 198. Nymphæa odorata.
Fig. 199. Petals gradually passing into stamens.

118. Here is a figure of the Water Lily (198), and a separate view of its sepals, petals, and stamens. Observe

117. What difference between unsymmetrical and irregular? Examples.

how the form of the slender stamen gradually changes to the broad petal, the anther becoming smaller and smaller. One can scarcely say where it ceases to be a stamen and begins to be a petal. So, also, the petals gradually pass into sepals, and in other plants, Peony for instance, the sepals just as gradually pass into leaves. (See Class Book of Botany, § 113.)

119. This transformation of one sort of organ into another (always from stamen *back* towards the leaf) is quite common among cultivated plants. It is in this manner that the Rose, Carnation, Peony, &c., become *double, viz.*, by the stamens, and often the pistils too, becoming petals: for in the wild state these flowers have but five petals.

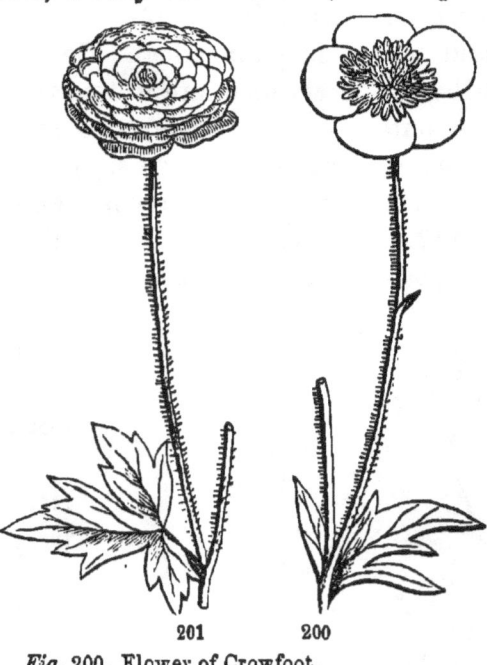

201   200

Fig. 200. Flower of Crowfoot.

Fig. 201. Double flower of the same; the stamens and pistils have become petals.

120. From these examples and others like them, we conclude that the different organs of the flower, and the leaf also, although commonly very different, *have all one common nature and origin;* or, in other words, the organs of the flower may all be considered as *transformed leaves.*

118. Show the graduation of organs in Water Lily.
119. How do the Rose, Peony, &c., become double?
120. What great principle is derived from these facts?

LESSON XVIII.

OF THE PISTILS.

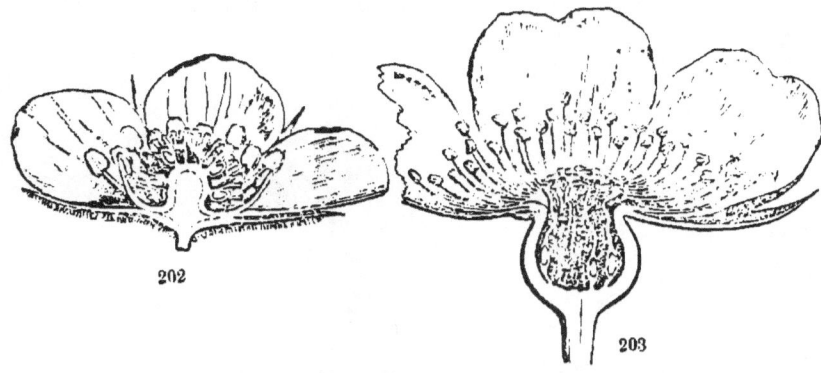

Fig. 202. Section of flower of Strawberry,—ovaries many, on a raised torus.
Fig. 203. Section of a Rose,—ovaries sunk into a hollow torus.

121. THE pistils occupy the centre of the flower, at the end or centre of the torus. Their number varies in different plants from one to one hundred, or more. When they are several they stand arranged in a circle like the other organs. When they are many they are commonly heaped together in a spiral manner, and raised on the conical torus, as in Buttercup, Strawberry, or sunk into the cavity of a hollow one, as in Rose. (See Figs. 202, 203.)

122. The pistil consists, plainly, of three parts, as may be seen in Fig. 204. At the top is the *stigma* (*s*), at base is the *ovary* (*o*), and between them is the *style* (*sty*). The style

121. In what part of the flower are the pistils situated? What is their number? their arrangement? How situated in the Rose?

122. Please describe the pistil and each of its parts. In what case is the stigma sessile?

being a mere stalk, like the filament of a stamen or the petiole of a leaf, may, like them also, be wanting, without loss to the pistil. In this case the stigma is sessile (sitting) upon the ovary, as in the pistils of Anemone (Fig. 207), and of Trillium (Fig. 206).

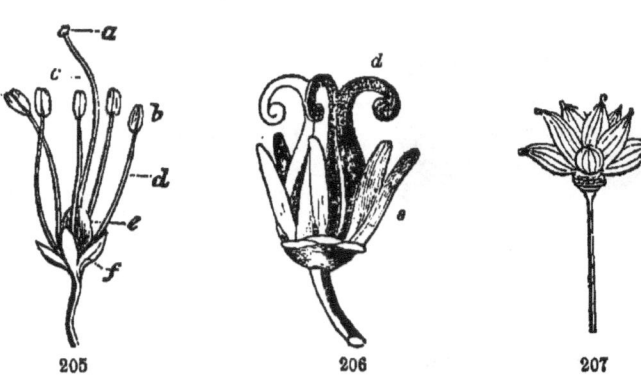

Fig. 204. Pistil of Tobacco.
Fig. 205. Pistil, stamens, and calyx of Azalea.
Fig. 206. Trillium,—stigmas (*d*) and anthers (*s*) nearly sessile.
Fig. 207. Pistils of Rue Anemone (*A. thalictroides*),—stigmas sessile.

123. The ovary is a kind of sac or case, enclosing the *ovules* (see Fig. 215, where there is but one, or in Fig. 209, where there are five, and Fig. 202, where there are many ovaries.) When full grown, the ovary becomes the fruit, and the ovules the seeds.

124. It is very important to distinguish between the *simple* and the *compound* pistil, for when there are several in the same flower they often grow together, forming a single body with members more or less distinct. As the petals grow

123. Describe the ovary and the ovules.
124. Name an important distinction in ovaries. When is the ovary or pistil compound?

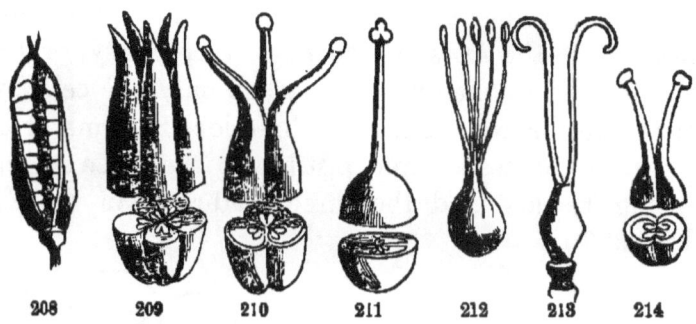

208   209   210   211   212   213   214

Fig. 208. Simple pistil of Larkspur.
Fig. 209. The five simple pistils of Columbine, all distinct.
Fig. 210. The three pistils of a St. Johnswort,—ovaries united but styles distinct.
Fig. 211. Compound pistil of another St. Johnswort, the three pistils entirely united.
Fig. 212. Flax,—the five ovaries united but the styles distinct.
Fig. 213. Pink,—the two ovaries united, styles distinct.
Fig. 214. Saxifrage,—the two pistils slightly united.

together, forming a monopetalous corolla, so the pistils may combine into a *compound pistil.* The parts of such a pistil are conveniently called *carpels.*

125. As to the extent of this union of the pistils, it is found in all possible degrees, always beginning at base and proceeding upwards. For example, in Columbine (Fig. 209), the five carpels (pistils) are entirely distinct; in Early Saxifrage (Fig. 214), the two carpels are united at the base; in Pink (Fig. 213), the two unite to the top of the ovary, leaving the styles distinct; so also in Flax (Fig. 212); in Evening Primrose, the four pistils cohere to the top of the style, leaving the stigmas distinct; and finally, in the Lily, the three carpels are united throughout. (See Figs. 209–214.)

126. We may know the number of carpels in a compound

125. As to the cohesion or union of pistils,—how is it in Columbine? in Pink? in Early Saxifrage? Evening Primrose? Lily?

pistil by the number of separate styles, or by the separate stigmas, or by the *lobes* of the stigma or ovary, or by the number of cells in the ovary, or (when only one cell) by the number of seed-rows. Thus the three-lobed stigma or ovary of the Lily indicates a triple pistil, also the three stigmas of the Spring Beauty, and the three seed-rows in the Violet. (See Fig. 229.)

215   216

Fig. 215. Section of the flower of Alchemilla, showing the stamens perigynous, the style single, simple, and lateral.

Fig. 216. Section of flower of Jeffersonia,—stamens hypogynous, pistil single, simple, with one seed-row.

127. But when the pistils remain separate and distinct we call each one a simple pistil. Thus in Columbine (Fig. 209) there are five simple pistils; in Anemone (Fig. 207), and in Buttercups, many; while in Cherry, Peach, Bean, Alchemilla, and Jeffersonia, there is just one simple pistil in each flower. Such a pistil is usually of an irregular form, with its style lateral (on one side), and only one seed or seed-row. (See Figs. 215, 216.)

126. Please tell us how you detect the *number* of *carpels* in the compound ovary of Spring Beauty; of Lily; of Violet.

127. What peculiarity in the form of a simple pistil?

LESSON XIX.

HOW THE LEAVES ARE FOLDED IN THE BUD

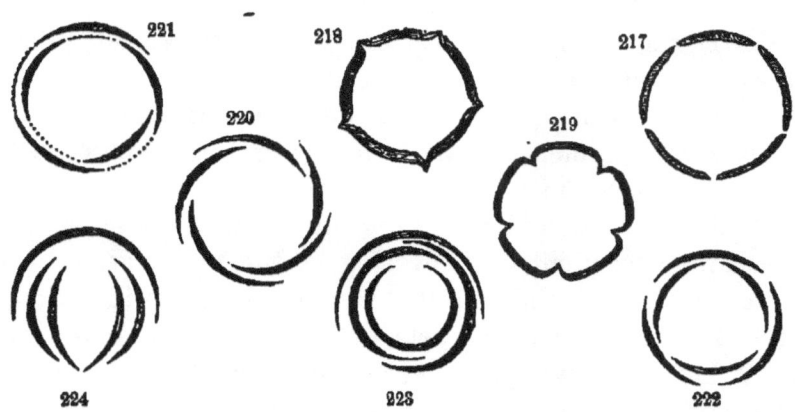

Æstivation.—*Fig.* 217. Valvate calyx, as of Mallow. *Fig.* 218. Sepals of Holly-hock,—valvate-reduplicate. *Fig.* 219. Sepals of Clematis,—valvate-induplicate. *Fig.* 220. Petals of Flax,—contorted. *Fig.* 221. Petals of Wild Rose,—quincuntial. *Fig.* 222. Petals and sepals of Lily or Tulip. *Fig.* 223. Petals of Wall-flower,—convolute. *Fig.* 224. Petals of Pea,—vexillary.

128. THERE is the *leaf-bud*, consisting of many scales and young leaves, folded up in such a manner as to occupy as little space as possible; and the *flower-bud*, consisting of the organs of the flower in their early state, also closely packed. Now if you study the arrangement of the pieces composing the bud of either sort, you will be surprised and delighted with its variety and elegance. As each species of plant has the same invariable mode of folding in all its buds, this study well becomes a matter of science.

LESS. XIX.—What is the subject of this Lesson?
128. Two kinds of buds; please describe each.

129. With a sharp knife let us make a cross-section (that is, a cut square across) of a flower-bud just ready to open ; we may thus obtain some such views as are here drawn. For example, in Fig. 217, we have the *valvate* arrangement. Here the pieces composing the circle barely touch each other by the edges, as in the sepals of Mallows, petals of Lilac, valves of a seed-vessel. (See, also, Figs. 218, 219.)

130. In the Phlox, Flax, Oleander, we find a twisted or *contorted* arrangement of the petals (Fig. 220), where each piece overlaps the next, all in one direction.

131. The bud is said to be *imbricated*, when some of the pieces are wholly outside, covering by the two edges others which are wholly inside. But this may take place in various ways. See how it is in the petals of the Eglantine, or Apple (Fig. 221). Here two petals are outside, two inside, and *one* partly both. In the Tulip, *one* sepal is outside, *one* inside, and *one* partly both. And just so with its three petals (Fig. 222).

132. The bud is *convolute* when each leaf wholly involves all that are within it, as do the petals of Magnolia and Wall-flower (Fig. 223) ; and it is *vexillary* in the Pea tribe, where only the outside petal, larger than the rest, infolds them all (Fig. 224).

133. The *plicate* arrangement is found in monopetalous flowers, as in Thornapple, Potato, where the corolla is folded in a manner somewhat like a fan.

129. How do we prepare a bud for examination ? What do you understand by a cross-section ? Define the valvate arrangement, with examples.

130. What æstivation do we find in Flax, Phlox, &c. ?

131. What is the imbricated æstivation ? Describe it in the petals of Tulip ; Apple ; Eglantine.

132. How are the petals arranged in the bud of Wall-flower ?

133. How in the flower of Thornapple ? or Potato ?

134. The pupil should make himself well acquainted with these seven modes of *æstivation* (so the botanists call it). Other modes are described in larger works. (Class Book of Botany, p. 79.)

135. Also in the leaf-bud we find similar modes of leaf-folding (here called *vernation*, from the Latin *vernus*, spring, as *æstivation* is from *æstivus*, summer). The figures following represent cross-sections of various sorts of leaf-buds. In the bud of Sycamore the infolding scales are imbricate, but the young leaves within are somewhat plicate.

136. In the leaf-bud of Cherry (Fig. 230) we find the convolute vernation, similar to the *æstivation* of Wall-flower. The leaf-bud of Lilac (Fig. 231) gives us another form of imbricate.

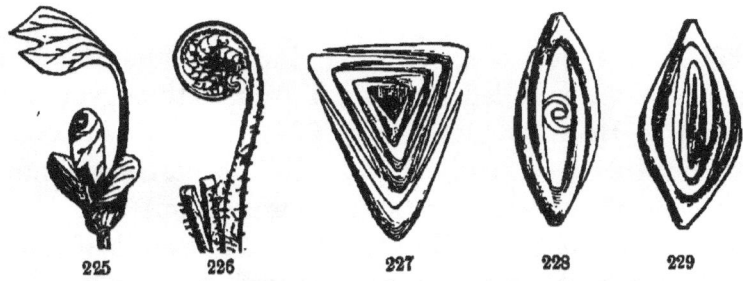

Vernation.—*Fig.* 225. Unfolding leaf-bud of Tulip-tree,—reclinate. *Fig.* 226. Fern leaf-bud,—circinate. *Fig.* 227. Sedge,—equitant. *Fig.* 228. Sage,—obvolute. *Fig.* 229. Iris,—equitant.

137. Fig. 229 represents the vernation of Iris, and Fig. 227, of a Sedge-grass. Both are *equitant* (which means, in Latin,

---

134. What is the meaning of the word *æstivation?*

135. What is the meaning of the word *vernation?* Please describe the vernation in Sycamore leaf-bud.

136. In the leaf-bud of Cherry; Lilac.

137. What of the equitant vernation?

riding horseback). Each leaf, first on this corner, then on that, infolds or overlays all that is within it.

138. *Obvolute* vernation appears in the leaf-bud of Sage (Fig. 228), where each leaf infolds only half of the blade of its opposite leaf.

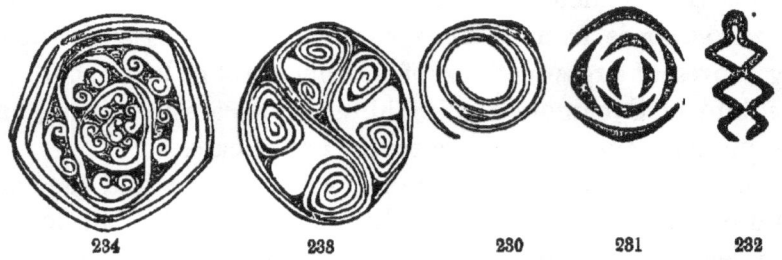

234   233   230   231   232

Vernation.—*Fig.* 230. Cherry leaf-bud,—convolute. *Fig.* 231. Lilac,—imbricate. *Fig.* 232. Birch leaf,—plicate. *Fig.* 233. Dock,—revolute. *Fig.* 234. Balm-of-Gilead,—involute.

139. In the bud of Dock (Fig. 233) we find the young leaves *revolute*, or rolled backwards from both edges; but in the bud of Balm-of-Gilead (Fig. 234) they are *involute*, or rolled inwards from both edges. This is best seen under a microscope of one lens, *i. e.*, a single microscope.

140. In the bud of Tulip-tree (Fig. 225) each leaf is *reclinate*, being bent over forward and infolding all within it; and in the Fern (Fig. 226) it is *circinate*, or coiled from the top downwards, like a watch-spring.

138. The obvolute?
139. What the vernation of Dock? of Balm-of-Gilead?
140. Please describe the reclinate; the circinate.

LESSON XX.

HOW THE FLOWERS ARE ARRANGED ON THE PLANT.

141. WE may now devote one or two lessons to the arrange
ment and position of the flowers upon the plant, a subject to
which botanists give the name of *inflorescence.*

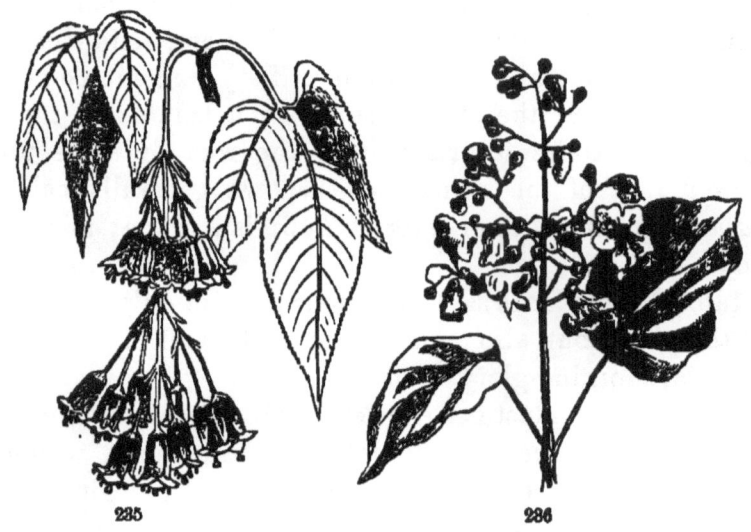

235   236

Fig. 235. Staphylea trifolia; a pendulous, paniculate cyme.
Fig. 236. Catalpa; a panicle.

142. Every one has observed such facts as the following, namely, that flowers are sometimes alone, and often in
clusters; that they are sometimes raised on stalks, and some

141. What is the meaning of the word *inflorescence?*
142. What common facts in inflorescence does everybody notice?

times sessile (or without stalks); and that they may arise
from terminal buds, or from axillary. With the meaning of
the words terminal and axillary you were made acquainted
in Lesson IX.

143. The stalk which supports the flower, or the cluster of
flowers, we call *peduncle*. Now the peduncle may be either
simple, bearing a single flower, or divided into branches and
bearing a cluster of flowers. In the latter case, the branches
or branchlets are called *pedicels*.

144. When the peduncle arises from terminal buds it
seems like a continuation of the main stem, as in Foxglove,
Horse-chestnut; and when from axillary buds, it comes out
from the side of the stem just above a leaf, as in the Currant. Sometimes it arises from the root or some underground part of the stem, and then we generally call it a
scape. Thus the flower-stalk of Tulip is a scape; also of the
Dodecatheon.

145. The flower is said to be *solitary*, not only when alone
on the plant, but also when alone in the axil of a leaf, as in
Fuchsia, Morning-glory, Petunia.

146. Among clustered flowers, you will often meet with
the following twelve varieties of inflorescence, which we
must now try to represent and describe. We begin with the
spike, such a cluster as we see in the Plantain, Mullen, &c.
We may define it thus: A long peduncle (called *rachis*),
having sessile flowers arranged along its sides. But before
we go further with inflorescence, we must examine the *bracts*
which accompany it.

---

143. Please define *peduncle;* also *pedicel*.
144. When are the flowers terminal? axillary? Define *scape*.
145. Why is the flower called solitary in Fuchsia, Petunia, &c. ?
146. Define a spike. Explain to us the *rachis*.

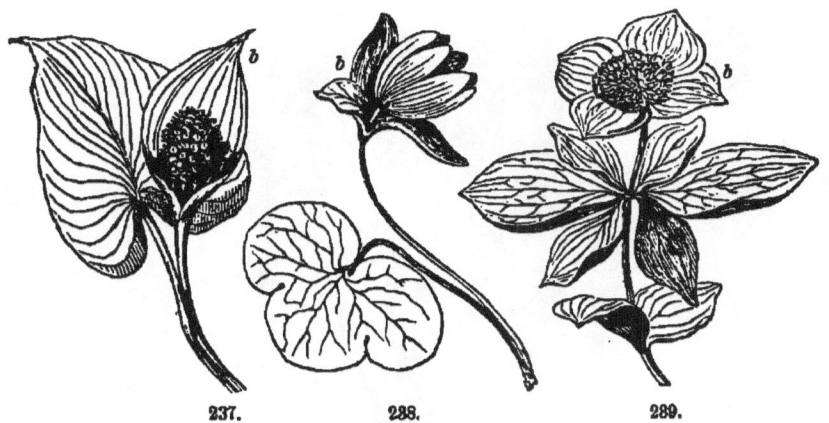

237.   238.   289.

Bracts (*b, b, b*). *Fig.* 237. Cornus Canadensis, with an involucre of four colored bracts. *Fig.* 238. Hepatica triloba, with an involucre of three green bracts. *Fig.* 239. Calla palustris, with a colored spathe of one bract, inclosing the spadix.

147. *Bracts* are evidently of the same nature as leaves, differing only in their diminished size, and in their position on the flower-stalks, or near the flowers. They are sometimes colored as brightly as flowers, as in Painted-cup, or in Balm. When several bracts are arranged in a whorl at the base of the cluster of flowers, an *involucre* is formed, such as we find in Carrot, and most of the Umbelworts (Fig. 244). In the Flowering Dogwood the large involucre is colored white.

148. Next in resemblance to the spike is the *spadix*, an inflorescence seen in the Calla (Fig. 237), Golden-club (Fig. 241), and Cat-tail. It may be defined as a thickened, club-shaped spike, often with a large bract (called *spathe*) at base, as in Jack-in-the-pulpit, or without a spathe, as in Fig. 241.

147. What sort of leaves grow on the peduncles, if any? Define bracts What is an involucre? How is it in Cornus?

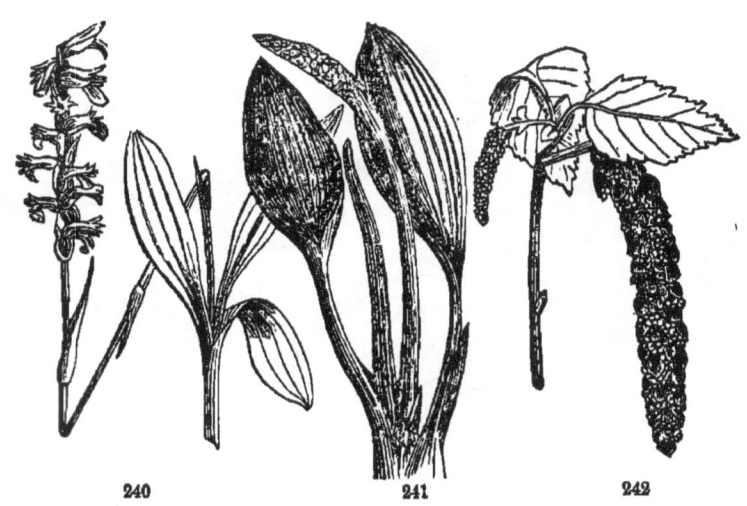

240   241   242

Fig. 240. Lady's-tresses (*Spiranthes*),—flowers in a twisted spike.
Fig. 241. Golden-club (*Orontium*),—flowers in a spadix with no spathe.
Fig. 242. Birch (*Betula*),—flowers in aments.

149. An *ament*, called also *catkin*, is a more slender and
delicate spike, filled with colored scales and flowers, and all
falling together without separating, such as adorn the Birches
(Fig. 242), Willows, and Poplars in early spring. The Hop
also bears aments.

LESSON XXI.

THE INFLORESCENCE, CONTINUED.

150. The flowers of the Black Cherry, Currant, Foxglove,
Locust, and Moth-mullen are in *racemes*. The raceme, then,
is a rachis bearing its flowers on distinct, simple pedicels

148, 149. Can you define the spike? ament? spadix? and spathe?
150. Please name and describe the inflorescence of Black Cherry.

(not sessile, as in the spike). It is often pendulous, often erect.

242*a*   243

Fig. 242. Secund (one-sided) raceme of Andromeda racemosa.
Fig. 243. Pendulous raceme of Currant.

151. The *corymb* differs from the raceme in having the lower pedicels lengthened so as to elevate all the flowers to about the same level, as in the Yarrow or Wild Thorn.

152. The *umbel* appears in Milkweed, Onion, Ginseng, &c. It consists of several pedicels of similar length, all arising from the same point at top of the peduncle. But in Caraway, Carrot, and most of the Umbelworts (Fig. 244), the *umbels* are *compound*, as if each of the pedicels had become

151. How does a corymb differ from a raceme?
152. Please name and describe the inflorescence of the Milkweed. How

itself an umbel. These secondary umbels we call *umbellets*. At the base of the umbel there is usually a whorl of bracts forming an involucre (*a*), and often also at the base of each umbellet (*b*), when we call it an *involucel*.

153. The fine flowers of the Catalpa are in *panicles* (Fig. 235), also the flowers of Oats. We may describe a panicle as if a raceme should have its pedicels irregularly branched.

154. A cluster resembling a panicle, but more compact, such as you see in Lilac, is called a *thyrse*.

155. A *head* of flowers, such as we see in Clover or the Button-bush,

Fig. 244. Compound umbel of Sweet Cicely (*Osmorhiza*).

hardly needs description. We might say that the head is a reduced umbel, having its flowers all sessile at the top of the peduncle.

156. The great family of the Asterworts has all its flowers in heads, so dense and so nicely arranged as to be easily mistaken for a single flower. But if you carefully examine such a head, say of an Aster, or especially of a Sunflower, you will see that it is composed of many little flowers or florets. The florets of the outer row are enlarged and open, so as to

---

does that of Carrot differ? What is an umbellet? What the whorl of bracts at the base of the umbellets?

153. Please describe the panicle. 154. The thyrse. 155. The head.

156. What the inflorescence of the Asterworts? How is the head of Aster made to resemble a single flower? What the florets of the ray? What the florets of the disk?

resemble the petals of one corolla, and the involucre, formed of many imbricated scales, resembles a calyx. This head is often called a *compound flower*. The outer florets are the *florets of the ray*, the interior are the *florets of the disk*. See all this illustrated in Figs. 245–250.

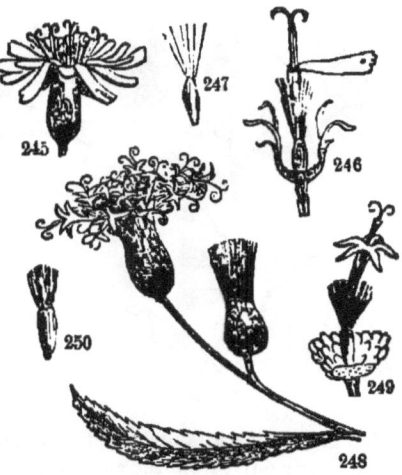

157. The forms of inflorescence heretofore described result from axillary buds; but the three following come from terminal buds. *Cyme* is the general name given to all the forms of terminal inflorescence. You may recognize them by the order in which the flowers open. Thus, in

Fig. 245. Head of Blue Milkweed (*Mulgedium*); all its florets are ligulate. *Fig.* 246. A view of one of them remaining on the receptacle. *Fig.* 247. A fruit crowned with its pappus.

Fig. 248. Heads of Ironweed (*Vernonia*); all its florets are tubular. *Fig.* 249. One of them remaining on the receptacle. *Fig.* 250. Fruit.

the cyme, the terminal and central flowers open first, but in the forms before mentioned, the lower and outer flowers first.

158. When the cyme is spreading and level-topped, we call it a *cymous corymb*, as in the common Elder; and when not level topped, it may become a *cymous panicle*, as in Chickweed (Fig. 251), Spergula, and Staff-tree (Fig. 234).

159. The *scorpoid cyme* is a very remarkable form of in-

157. Are the forms hitherto described terminal or axillary? Please define the cyme.

158. Cymose corymb; Cymose panicle.

159. What is a scorpoid cyme? Name and describe the inflorescence of Bunch Pink; the inflorescence of Catmint.

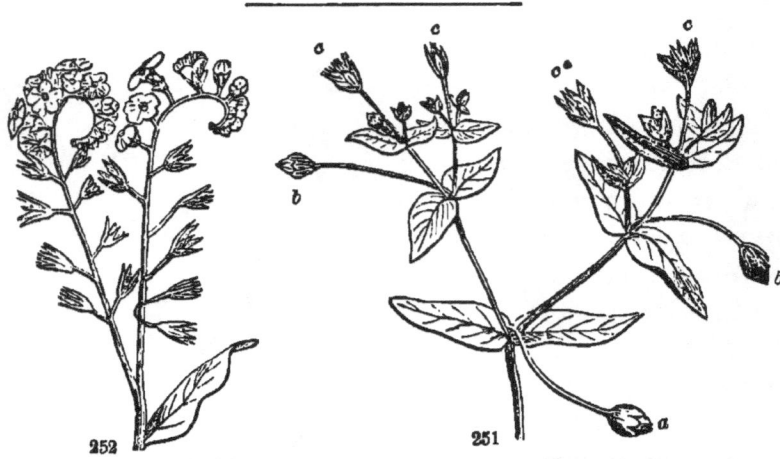

Fig. 251. Cyme of Chickweed (*Stellaria media*). First, the terminal flower (*a*) opened; secondly, from the axils of its highest leaves arose two branches, and terminated in the flowers *b*, *b*; thirdly, from their highest axils arose the flowers *c, c, c, c*, from whose axils a fourth set is seen to start, and so on.

Fig. 252. Scorpoid cyme of Forget-me-not (*Myosotis palustris*).

florescence, as shown in Pink-root and Forget-me-not (Fig. 252). Before flowering it is coiled from the tip downwards, and it uncoils as it blossoms. In its nature it is a half-cyme. The *fascicle* is a densely packed cyme, as seen

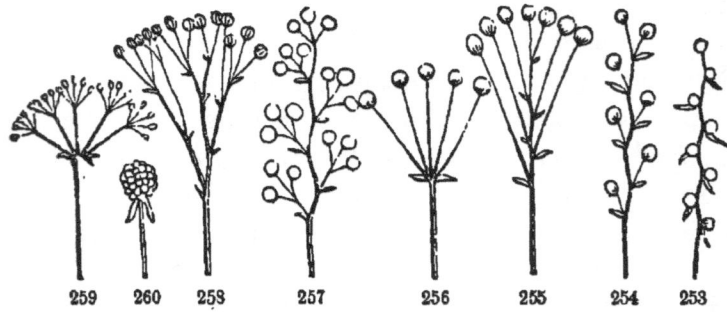

Diagrams of the forms of axillary inflorescence, showing how they gradually pass into each other. *Fig.* 253. Spike. *Fig.* 254. Raceme. *Fig.* 255. Corymb. *Fig.* 256. Umbel. *Fig.* 257. Panicle. *Fig.* 258. Compound corymb. *Fig.* 260. Head. *Fig.* 259. Compound umbel.

in Bunch Pink or Pycnanthemum. The *glomerule* is a small, dense cyme appearing in the axils of the leaves, as in Catmint and the Mint tribe generally.

160. The preceding diagrams may be carefully studied. They will convey a general idea of all these forms of inflorescence, and how they are related to each other.

LESSON XXII.

CONCERNING THE FRUIT.

161. The flower is of short duration. After a few hours or a few days of blooming beauty, it fades and disappears.

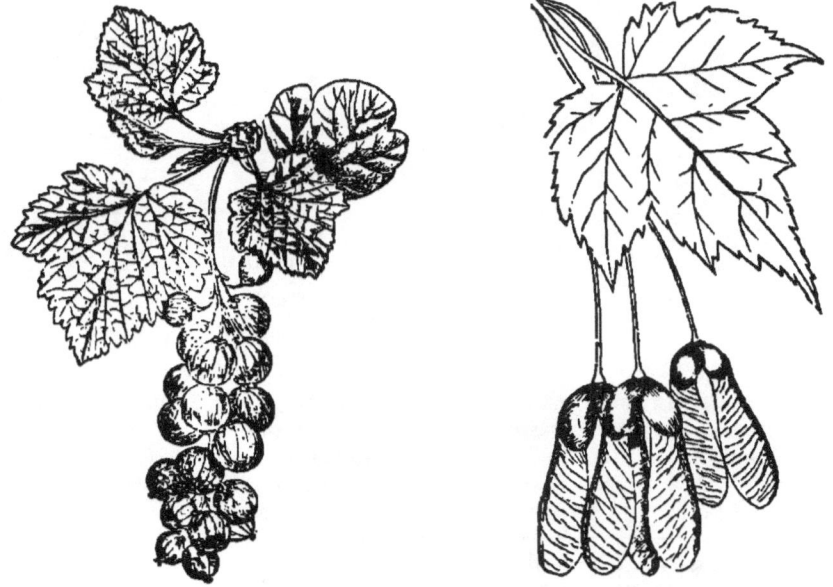

Fig. 261. Fruit of Currant,—a berry.   *Fig.* 262. Fruit of Maple,—samara.

160. Please explain the diagrams 253–260.

The stamens and petals have accomplished their work, and are dead. The sepals also, when colored like petals, are dead. But the pistil, especially the ovary, yet remains in its place, living and growing until the seeds which it contains are perfect.

162. Thus the *fruit* is the ovary or pistils brought to perfection.

163. During the growth and ripening of the pistil, great and manifold changes occur, so that at last the fruit is very different in form, size, substance, and color. The little pistil in the flower of the Cherry must undergo a great alteration in becoming a plump Ox-heart!

Fig. 263. An umbel of Cherry blossoms,—namely, a bud, an entire flower, and a section showing the *one* pistil and the perigynous stamens.

Fig. 264. The drupe, cut through to show the stone and one seed.

Fig. 265. A corymb of Strawberry,—flower and fruit. The achenia are seen on the surface of the fruit, which is only the overgrown torus.

161. Can you tell us what parts of the flower perish? What parts remain in place and still grow?

162. How do you define the fruit?

163. Mention some of the changes occurring from ovary to fruit.

164. In the fruit we see the end and aim of plant-life accomplished, according to the wise and good design of the great Creator. While it serves to reproduce and keep alive the plants upon the earth, it also serves as food for animals and for man.

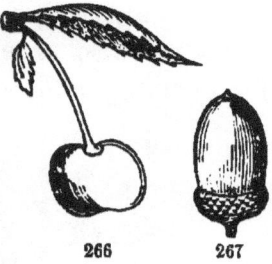

165. It is curious to observe how different are the parts of the fruit which in different plants become food. In the Apple, we eat the calyx which here

Fig. 266. Drupe,—a ripe Cherry.

Fig. 267. Tryma,—acorn of Red Oak.

adheres to the ovary, and in ripening was thickened and enlarged by the nutritious substance. In the Strawberry, we eat the enlarged, pulpy torus, which bears, all over its surface, the little dry, seed-like fruit. In Peach, the luscious morsel is the outer coats of the ovary itself; and in the Orange, it is the inner coat. In the Nut, Pea, Wheat, and most plants, the nourishing

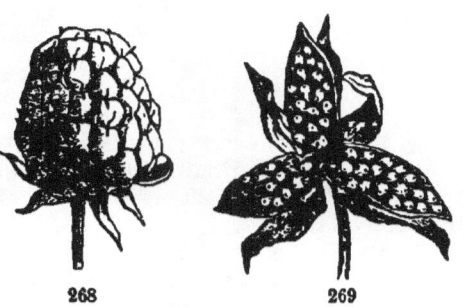

Fig. 268. Etærio,—a Blackberry.
Fig. 269. Capsule of Violet, open.

matter is laid up in the seeds, while the carpels ripen into a dry fruit.

166. The fruit consists of the seeds and the seed-vessels. The word *pericarp* means the same as seed-vessel. When

164. Mention some of the uses of the fruit.

165. Can you tell us what part of the Apple is eaten? What part of the Strawberry is the eatable part? What part of the Peach? the Orange? In what part is the nutritious matter deposited in the Pea? Wheat? Almond?

the pericarp is ripe, it may open in some special manner of itself and discharge the seeds; or it may have no provision for opening, and remain closed until it grows or decays

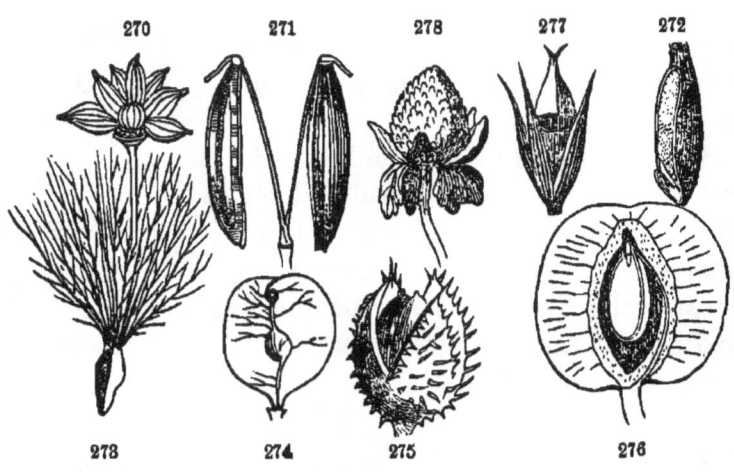

Fig. 270. Achenia of Rue Anemone, in a head.

Fig. 271. Fruit of Caraway, consisting of two achenia.

Fig. 272. Kernel of Wheat,—a sort of achenium called *cariopsis.*

Fig. 273. Fruit of Thistle,—another sort of achenium, crowned with a pappus which serves as wings.

Fig. 274. Fruit of Elm,—a samara, or winged achenium.

Fig. 275. Fruit of Beech,—two nuts, inclosed in the burr.

Fig. 276. The Peach (a drupe),—cut open, showing the seed inclosed in its stone, and the stone in the thick pulp.

Fig. 277. Fruit of Pigweed,—a one-seeded pericarp called *utricle.*

Fig. 278. Fruit of India Strawberry,—a fleshy torus bearing the achenia outside.

Fruits that open we will call *dehiscent* fruits, and those which do not open, *indehiscent.* We will first study some of the forms of indehiscent fruits, arranged as follows:

167. First Division: Fruits indehiscent, one-seeded, dry; namely, ACHENIUM, SAMARA, GLANS.

---

166. Of what two parts does the fruit consist? What is a dehiscent fruit? indehiscent?

Second Division: Fruits indehiscent, one-seeded, fleshy; namely, DRUPE, TRYMA, ETÆRIO.

Third Division: Fruits indehiscent, several-seeded; namely, BERRY, PEPO, POME.

168. The *achenium* is such a fruit as we find in Buttercups, Anemone, Sage. Usually there are several produced together from one flower. We must not mistake them for seeds. They are pericarps, each inclosing one seed, as you see in the figures. The grain of Wheat or Corn (called cariopsis) is much the same, but the one seed cannot be separated from the pericarp.

169. The *samara* is merely an achenium with a wing, as in Ash, Elm, Maple. The latter fruit is a double samara.

170. A *glans* (or nut) is such a fruit as Acorn, Chestnut, Hazelnut, much like achenium, but larger, and seated in a cup or involucre.

171. A *drupe* is such a fleshy fruit as the Cherry or Peach. It is well called a stone-fruit. The *stone* incloses the one seed, and is itself inclosed in a juicy pulp.

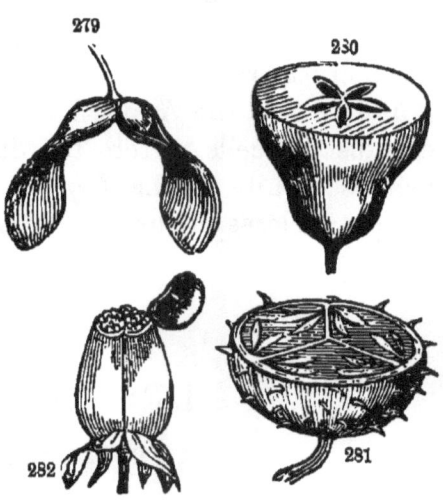

Fig. 279. Maple,—a double samara.
Fig. 280. Pear,—a pepo.
Fig. 281. Gooseberry, cut across; an enlarged view, showing the seeds lying in the pulp.
Fig. 282. Fruit of Henbane,—a pyxis with its lid open.

---

167. Please define our first division of fruits. What special fruits belong to it? the second, &c.; the third, &c.

172. *Tryma* is the name for such fruits as Walnut, Cocoanut. Like the drupe, it has a stony seed-shell, but its outer coat is rather woody than pulpy.

173. Such fruit as the Raspberry or Blackberry we call *etærio*. It consists of many little fleshy drupes growing fast together or to the torus. In the Blackberry they grow to the torus (Fig. 268).

174. The *berry* is a thin-skinned, pulpy fruit, holding its several seeds loose in the pulp, as Currant, Grape (Fig. 261). The Orange, &c., is much like a berry, but on account of its thick rind has been called by another name (hesperidium).

175. *Pepo* is such a fruit as Squash, many-seeded, with a hard, crusty rind.

176. *Pome*, the Apple, Pear, Haw, a fleshy fruit with several distinct cells. Here the fleshy calyx grows fast to the ovaries; while in the *Hip*, or Rose-fruit, the fleshy calyx merely iucloses the ovaries, as seen in Fig. 203.

LESSON XXIII.

FRUITS, CONTINUED.

177. The dehiscent pericarp,—that is, those which open to discharge the seeds,—are generally dry fruits, known as pods. The various forms have the following names: PYXIS, FOLLICLE, LEGUME, SILIQUE, CAPSULE.

168–176. The student will now please define and name the fruit of Buttercups, Corn, Ash, Maple, Oak, Hazel, Plum, Walnut, Raspberry, Grape, Orange, Squash, Pear, Haw, and Rose.

177. Please give the names of the dehiscent pericarps.

178. The *pyxis* is the most curious and singular of all pods. It opens crosswise by a lid, like a snuff-box. Fig. 284 is the likeness of the pyxis of Rheumatism-root, common in Ohio. It is formed of *one carpel* only. Fig. 282 is the pyxis of Henbane, formed of two carpels. So the pyxis of Poor-man's-weather-glass (Anagallis, Fig. 344) is formed of several carpels.

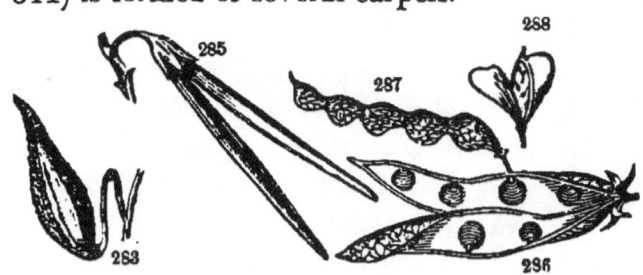

Fig. 283. A follicle of Milkweed (*Asclepias*).
Fig. 284. A pyxis,—fruit of Jeffersonia, the Rheumatism-root.
Fig. 285. A pair of follicles,—the fruit of the Dogbane (*Apocynum*).
Fig. 286. A legume, open,—fruit of the Pea-plant.
Fig. 287. A jointed legume, or loment,—fruit of Desmodium.
Fig. 288. A silicle,—fruit of Shepherd's-purse.

179. *Follicle* is the name of such pods as those of Columbine (Fig. 208), Milkweed (Fig. 283), and of Dogbane (Fig. 285). They are formed of a single carpel, and open lengthwise, on one side only. It is easy to see the resemblance between the follicle and a leaf, the leaf being folded so as to bring its two margins together. (See Fig. 207.)

180. *Legume* is the proper name of the Pea pod, Bean pod, &c., of one carpel, one cell, one row of seeds, and commonly

178. Give the character of the pyxis. How does the pyxis of Henbane differ from that of Jeffersonia?

179. Can you describe and name the fruit of Columbine? How is its leafy character seen?

180. Describe and name the Pea pod What is a loment?

opening by two valves (Fig. 286). Such a pod is sometimes divided crosswise by joints (as in Fig. 287, Desmodium); we then call it a *loment.*

181. *Silique* is a two-carpeled pod, such as we find in Mustard. It has two cells, separated by a thin partition, and two rows of seeds (Fig. 291). A short silique, or one not much longer than wide, such as we find in Pepper-grass or Shepherd's-purse (Fig. 288), is called a *silicle.* (See Fig. 290).

182. *Capsule* (the word means casket). This name is applied to all other forms of dry, compound fruits, formed of several unit-ed carpels. In opening, they commonly split into several valves, as in Iris;

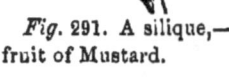

Fig. 291. A silique,— fruit of Mustard.

Fig. 290. Silicle of Draba (en-larged).

or divide into several parts (carpels) like so many follicles, as in St. Johnswort; or they open by small pores, as in Poppy.

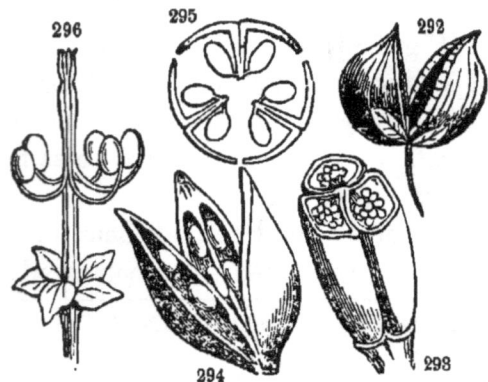

296   295   292 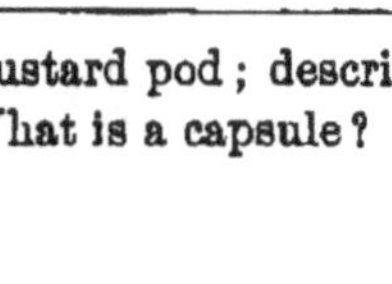   294 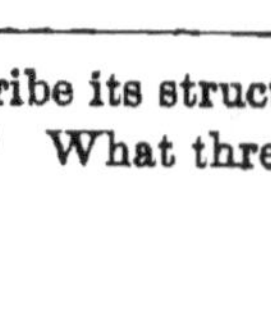   293

Fig. 292. A capsule,—fruit of Scrophularia ; it is two-celled, two-carpeled, or two-valved.

Fig. 293. A three-celled cap-sule of Colchicum; it opens *be-tween* the carpels.

Fig. 294. Capsule of Iris, open-ing *into* the carpels.

Fig. 295. Cross-section of the same, showing how it opens.

Fig. 296. Fruit of Geranium; its five carpels separate, and are carried up on the curving styles (called a *regma*).

181. Mustard pod ; describe its structure and name. What is a silicle ?
182. What is a capsule ? What three modes of opening are m‹

183. We should not omit altogether to notice the aggre-gated fruits, such as the Pine-cone (Fig. 300), Pine-apple, &c. These fruits are composed not merely of the pistil, but of the entire flower, or even of the whole inflorescence, bracts and all, grown thick, and consolidated into one fleshy mass. This is evidently the nature of the Pine-apple and of the Mulberry.

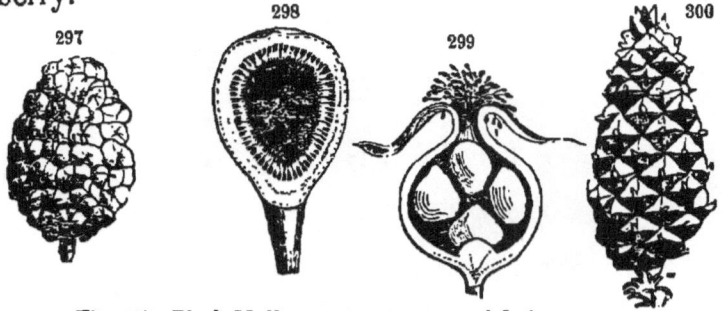

Fig. 297. Black Mulberry,—an aggregated fruit.
Fig. 298. Fig, cut open, showing the little flowers within.
Fig. 299. Hip of a Rose, cut open, showing the achenia within.
Fig. 300. Pine-cone, composed of thick scales.

184. As for the Fig, it is a great hollow torus, having its innumerable flowers within the cavity, growing from the walls, and all together become a sweet, pulpy mass.

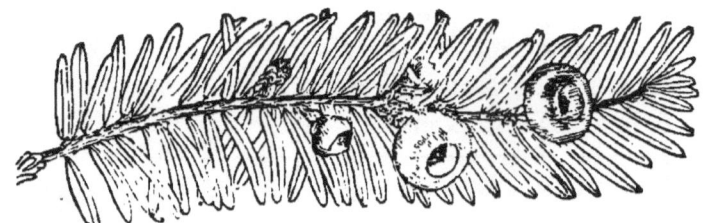

Fig. 301. A branchlet of the Canada Yew, showing the fruit.

183, 184. Mention some examples of aggregated fruits. Can you describe a Pine-apple? a Fig?

185. But there are some kinds of fruit almost or quite destitute of a pericarp, consisting of naked seeds. On the preceding page is a figure (301) of the Canada Yew, a trailing shrub of New England and Canada. The fruit is a single naked black seed, seated in a fleshy, coralline-red cup. The *cone* (of Pine, Fir, &c.) is made up of thick woody bracts, each covering in their axils two or more winged seeds (Fig. 300).

LESSON XXIV.

CONCERNING THE SEEDS.

186. LAST and most important is the *seed*, the perfected ovule, containing the germ of a new plant like its parent plant. The seed consists of a kernel and its shell. Place a bean in water, and soon its softened shell or skin is easily separated from the kernel.

187. The *shell* of a seed may be of any color, as white, black, yellow, red, &c.; may be polished and shining, or dull and rough; may be of any shape, as round, or oval, or egg-shaped; may be winged, as in Catalpa, or may be clothed with long hairs, called *coma*. The silk of Silk-grass (Asclepias) is the coma of the seed, and cotton is the coma of Cotton seed. The seed of Poplar (cotton-wood) or Willow is also furnished with coma.

---

185. What plants have no pericarps? Please describe a cone of Pine; fruit of Yew.

186. What is the seed, and what does it contain? Of what two parts does it consist?

187. What do you remember concerning the color and shape? Describe the coma of a seed.

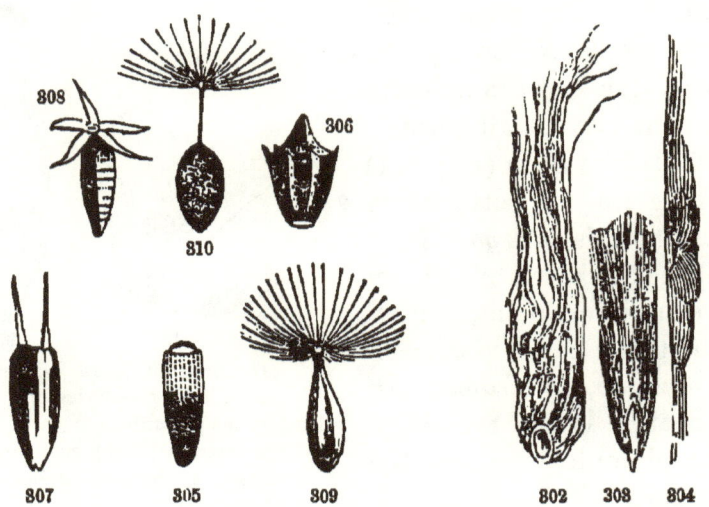

Fig. 302. A seed of the Cotton-plant, with its tuft of coma, or cotton.
Fig. 303. A seed of the Cotton-tree (*Populus*), with its silky coma.
Fig. 304. A winged seed of the Catalpa.
Fig. 305. Achenium of Eclipta; it has no pappus.
Fig. 306. Achenium of Horseweed; scarcely any pappus.
Fig. 307. Achenium of Sunflower; has two awns for pappus.
Fig. 308. Achenium of Ageratum; has five sepals for pappus.
Fig. 309. Achenium of Blue Milkweed; has abundant pappus.
Fig. 310. Achenium of Wild Lettuce; with pappus raised on a beak

188. The learner must distinguish between the *coma* of a seed and the pappus of a fruit. The down of Thistle or Dandelion is *pappus*, for the little fruit on which it grows is not merely a seed, but a pericarp (achenium), also containing one seed. In a word, the *seed* may be fledged with a coma, but the *fruit* is fledged with a pappus, both intended as wings to bear away the seed to distant places. (See Class Book of Botany, § 485.)

188. What is the distinction between coma and pappus?

189. As to the seed-kernel, it may consist of two parts, namely, the *germ* and *albumen*, or it may be all germ.

190. In the Bean (Fig. 311) it is all germ. A better name for the germ is *embryo*. Now in all seeds, the embryo is, in fact, a miniature plant, consisting of three parts, viz., *radicle, plumule, cotyledons*. In this Bean, *r* is the radicle, *p* is the plumule, *c, c*, are the cotyledons.

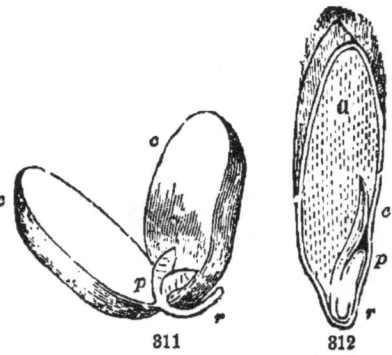

Fig. 311. Seed of Bean, without its shell: *c, c* are the two cotyledons; *r*, the radicle; *p*, the plumule.

Fig. 312. Seed of Wheat, cut open: *a* is the albumen; *c*, the *one* cotyledon; *p*, plumule; *r*, radicle.

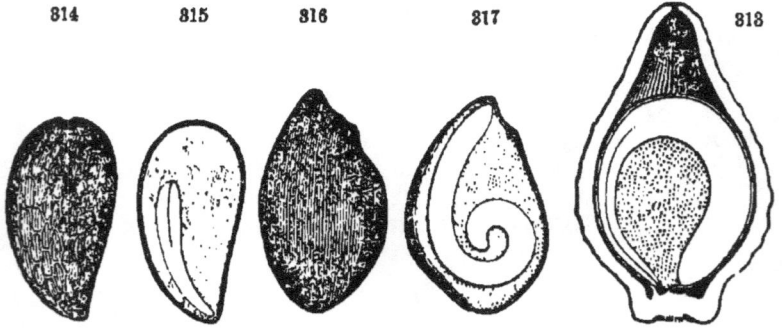

Fig. 313. Seed of Four-o'clock; embryo two-cotyledoned, coiled; *a*, albumen.

Fig. 314. Seed of Heather. *Fig.* 315. A section of the same, showing the curved embryo, with two cotyledous, lying in albumen.

Fig. 316 Seed of Onion. *Fig.* 317. Section of the same, showing the coiled embryo, one cotyledon, in albumen.

191. The radicle is the part destined to grow downwards

---

189. Of what two parts may the seed-kernel consist?
190. Describe the parts of the seed of bean.

and become root. The plumule is the young bud destined to expand upwards and become stem and leaves. The cotyledons are two young leaves, thick and bulky, full of starchy matter to feed the embryo when it shall awake and begin to grow.

192. In the Wheat-seed (Fig. 312) we find, besides the embryo, a white, mealy mass (*a*), well known when ground into flour. This mass is evidently intended to answer the same purpose as the starchy cotyledons of the Bean—to nourish the embryo. The radicle (*r*), the plumule (*p*), the cotyledon (*c*). and the albumen (*a*), are clearly shown. Fig. 313 (seed of Four-o'clock) also shows albumen; here the embryo is coiled into a ring around the albumen. Thus we see that the food of the young plantlet is laid up somewhere in every seed, either in the bulky cotyledons of the embryo itself, or in the albumen outside the embryo.

193. We have, then, seeds albuminous, and seeds exalbuminous; seeds two-cotyledoned, and seeds one-cotyledoned.

LESSON XXV.

THE SEED BECOMING A PLANT.

194. WE have seen that the ripened seed is a miniature plant, living, but sleeping; packed and sealed up for transportation. It may continue to sleep, perhaps, for years, if

191. Describe the nature and destiny of the radicle; of the plumule; of the cotyledons.

192. Of what does the Wheat-seed consist? What is the intention of the albumen? the position of it in Wheat? in Four-o'clock?

193. What seeds are albuminous? exalbuminous? What seeds are two-cotyledoned? one-cotyledoned?

kept dry; but if exposed to moisture, it soon awakes and commences its wonderful course of development.

195. In the Spring of the year the melting snows or the warm rains supply the proper moisture to the seeds which have fallen to the ground, and they may be seen everywhere swelling, bursting, and growing. The young botanist must not fail to watch their development.

196. Beneath some Oak, for example, buried in the old leaves, we find acorns in all stages of growth, showing at one view all the steps in the process of *germination*. Here is an acorn with its shell softened and its kernel a little swollen. We divide it lengthwise with a sharp knife, and the section (Fig. 318) shows the two thick cotyledons (*c c*) and the radicle (*r*).

197. In another acorn (Fig. 319) the cotyledons have absorbed yet more water, and enlarged so much as to burst the shell, and the radicle growing, has come forth, a little root, directing its course downwards.

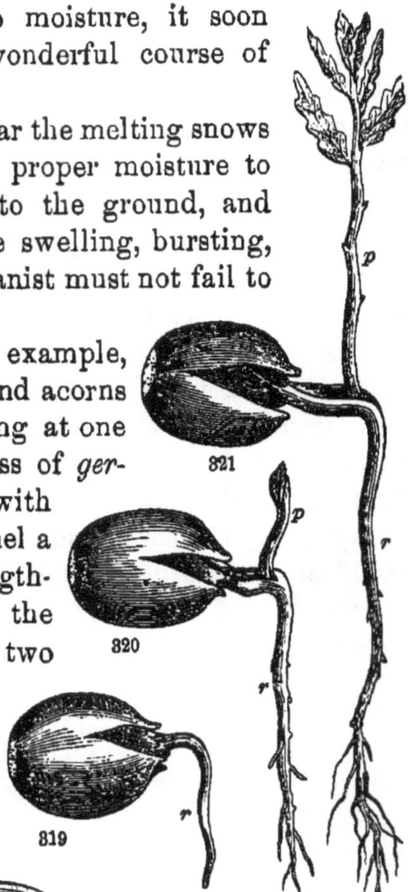

321

320

319

318

Fig. 318. Acorn, seed of the Oak, cut open,—showing *c, c,* the cotyledons; *r*, the radicle.

Figs. 319, 320, 321. Show the progress of germination: *r*, radicle; *p*, plumule.

194. Please tell us again what a seed is. In what condition is a seed? When will it awake?

195. Condition of the seeds generally in Spring?

198. In the next stage of growth (Fig. 320) the two stalks of the cotyledons (*s*, petioles, Less. I.), make their appearance, and from between them, at the top of the rootlet, the plumule shoots forth, a little stem with a bud at the top, directing its course upwards. The rootlet, meanwhile, has grown longer, entered the soil, and divided itself into branches and fibres all covered over with fine white hairs. These hairs, called fibrils, may be seen under a microscope, as in Fig. 322, which represents the end of a fibre of Maple with its fibrils much magnified.

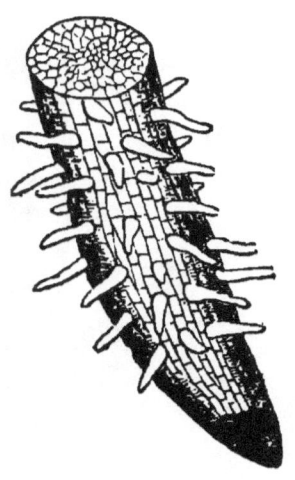

Fig. 822. The extreme end of a rootlet of Maple, greatly magnified under a lens, showing the fibrils.

199. Up to this stage, the growing rootlet and bud have drawn all their nourishment from the store of food laid up beforehand in the thick cotyledons for this very purpose; but now the rootlet has reached the soil, and by means of its numerous fibrils, which are so many little mouths, it is beginning to draw its nourishment from the earth.

200. Another acorn, or the same one a few days later (Fig. 321), shows root and stem well organized, and the young Oak fairly started on its grand journey of growth and life. The root has descended deeper and spread its branches wider

196. What is the meaning of the word *germination?* Describe the section of an acorn in Fig. 318.

197. Describe that stage of growth seen in Fig. 819.

198. Describe the third stage, as represented in Fig. 320.

199. The first source of food for the embryo? the second?

200. How does the plant appear in Fig. 321?

in the soil, while the bud has mounted higher, unfolding itself into stem and leaves, and spreading itself in the air and light.

201. The young plant has now become independent of the seed, which will soon wither and perish. The cotyledons, in this case, are never able to throw off the shell, but perish

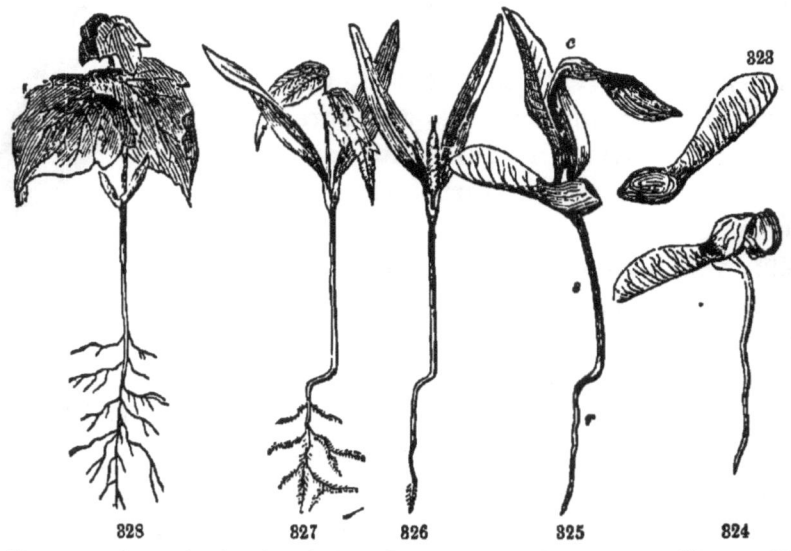

Progress of germination in Maple.—*Fig.* 323. A seed (samara). *Fig.* 324. The same, just beginning to grow; the rootlet descends, the cotyledons have burst the shell. *Fig.* 325. The leaf-like cotyledons (*c*) nearly open, the stem (*s*) and root (*r*) lengthening. *Fig.* 326. The terminal bud appears. *Fig.* 327. The first pair of true leaves expanded. *Fig.* 328. The second pair appear, &c.

together with it. In other plants, however, as in Maple (Fig. 325), the two cotyledons escape from the shell, change color, and become leaves,—the first pair on the plant (*c*).

202. The bud, which we called plumule, is still seen at the

201. When does the seed perish? Cotyledons of Oak and Maple—how differ in development?

top, arising higher and higher, as it unfolds its axis into the joints (called *nodes* and *internodes*) of the stem, its outer scales into leaves, and is itself continually renewed from within. Thus the ascending stem, or *axis*, is always terminated by a bud.

829   830

Fig. 829. Bud of Currant unfolding,—the scales (*s*) gradually becoming leaves.
Fig. 830. Bud of Tulip-tree,—the scales unfolding into stipules (*s*).

203. Soon other buds appear. There is one in the axil of each leaf. So long as the terminal bud only is developed, the plant grows up a simple stem. But by the growth of these axillary buds, if they grow at all, branches are produced; and these branches, from *their* axillary buds, produce branchlets, and so on.

---

202. What do you understand by the *nodes* and *internodes?* How is the axis always terminated?
203. In what case will the stem be simple? How are branches produced

LESSON XXVI.

LIFE OF THE PLANT, OR ITS BIOGRAPHY.

204. The water which the plant imbibes by its roots becomes *sap* in the stem, and circulates in every part as the blood circulates in the animal frame. The leaves, by their broad, thin forms, serve as lungs, to bring all the sap which passes through them into contact with the air and light.

205. By this means the sap is changed into a nourishing food, fitted to sustain the growth of the plant in every part. Thus the leaves are designed, not only as an ornamental robe, but as organs of breathing and digestion.

206. In the second stage of growth, when the plant depends no longer upon the seed for nourishment, it goes on increasing in stature and multiplying its leaves and branches. It now consists of three parts, namely, root, stem, and leaves. These are called the *organs of vegetation*.

207. The third stage of plant-life is the period of flowering. Before this period, all its activity was devoted to its own nourishment and growth. Now it begins to live and act for the continuance of its own kind after it upon the earth, according to the Divine decree in Genesis, i., 11. Some of its buds undergo a striking change, and open each a flower instead of a leafy branch.

208. A flower is therefore a leafy branch transformed (as

---

204. What becomes of the water which the roots imbibe? What part do the leaves act?

205. What change takes place in the sap?

206. What is the second stage of plant-life?

207. The third stage? Whence come the flowers?

shown in the Class Book, p. 23), having its axis undeveloped, its leaves in crowded circles, moulded into more delicate forms and tinged with brighter colors, not only to adorn the face of nature, but to prepare the way for fruit.

209. The fourth stage of plant-life is the period of its fruit bearing. The flowers have gradually faded and disappeared, but the pistil, having received the quickening pollen (see Class Book of Botany, p. 148), remains in its place, holds fast all the nourishing matter which continues to flow into it through the flower-stem, grows, and finally ripens into the perfected fruit and seed.

210. The fifth and last stage in the biography of the plant is its *hibernation* (winter's sleep), or its death. If the event of flowering and fruit-bearing occur within the first or second year of the life of the plant, it is generally followed by its speedy death. In all other cases it is followed by a state of needful repose, wherein it is commonly stripped of its leaves, and gives few, if any, indications of life, until awaked, with renewed vigor, in the following Spring.

211. According to their different terms of life, we distinguish plants as annuals, biennials, and perennials. An *annual herb* completes its whole history in one year. In the Spring it germinates; in Summer it grows, blooms, bears fruit; and in Autumn its work and life are ended. The Mustard, Maize, and Morning-glory are such.

212. A *biennial herb* lives two years. During the first it

---

208. Please state the nature of the flower.

209. Please describe the fourth stage of plant-life.

210. The fifth stage.

211. In regard to their term of life, how are plants divided? Describe an annual herb.

212. A biennial herb.

germinates, grows, and bears leaves only ; and in its second year it blossoms, bears fruit, and dies. Such are the Beet and Radish.

213. A *perennial* plant survives several or many years. There are *herbaceous* perennials and *woody* perennials. The herbaceous perennials, or *perennial herbs*, are such as survive the winter only by their roots or their parts which grow underground. These in Spring send up leaves, flowers, and and often stems, all of which perish in Autumn, leaving only the parts underground alive as before. Such are the Hop, Asters, Violets.

214. *Woody perennials* survive the winter by their stems as well as roots, and usually grow several years before flowering, and thence flower annually during their existence. According to their size, such plants are trees, shrubs, undershrubs. A *tree* is the largest among plants, having a permanent, woody stem, usually unbranched *below*, and dividing into branches *above*. The Oaks, Elms, and Pines are familiar examples.

215. A *shrub* is smaller than a tree, usually growing in clusters from one underground mass of roots. The Lilacs, Roses, Alders, are *shrubs*. Small shrubs, about of our own stature, as the Currants, Brambles, we call *bushes*. Very low shrubs, as the Blueberries, Box, &c., are *undershrubs*.

---

213. Describe a perennial plant. Of what two sorts? Describe a perennial herb.

215. A tree, a shrub, bush, undershrub,—how distinguished? To which of the above-mentioned sorts does the Cabbage belong? To which the Hollyhock? the Balsamine? Four-o'clock? To which the Tulip? Golden-rod? Lily? Pink? Quince? &c.

LESSON XXVII.

CONCERNING THE AXIS OF THE PLANT.

216. THE term *axis* expresses the central column or body of the plant around which the branches and other organs are arranged. As we have already noticed, the axis grows and extends in two directions, —upwards and downwards. The ascending part is the stem, the descending part is the root. The former loves and seeks the air and light, the latter the dark, damp bosom of the earth.

217. The ROOT serves the twofold purpose of fixing the plant firmly in its place, and of imbibing the necessary food from the soil. The food when thus imbibed is never in a solid

Fig. 331. An entire plant (Shepherd's-purse), showing the axis (*a* to *r*). The part from *c* to *r* is the descending axis, or root; from *c* to *a* the ascending axis, or stem; *b, b,* branches, bearing racemes of flowers and fruit.

831

state, but dissolved in water, and consists of certain earths, alkalies, and gases. (See Part II., Chap. 7, Class Book of Botany.)

218. It is the nature of the root to divide itself into branches, and the only organs which properly belong to it are branches, fibres, and fibrils. It puts forth no buds nor leaves unless the plant be in some unnatural state.

219. The roots of woody plants, especially, are *branching roots*. Year after year they multiply and extend in branches and branchlets beneath the

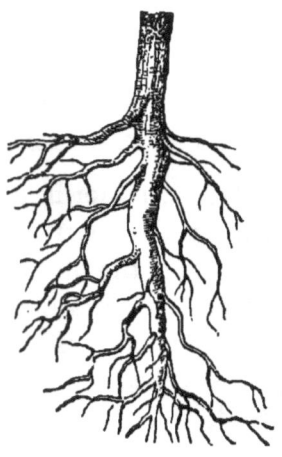

Fig. 332 Branching root of a young tree.

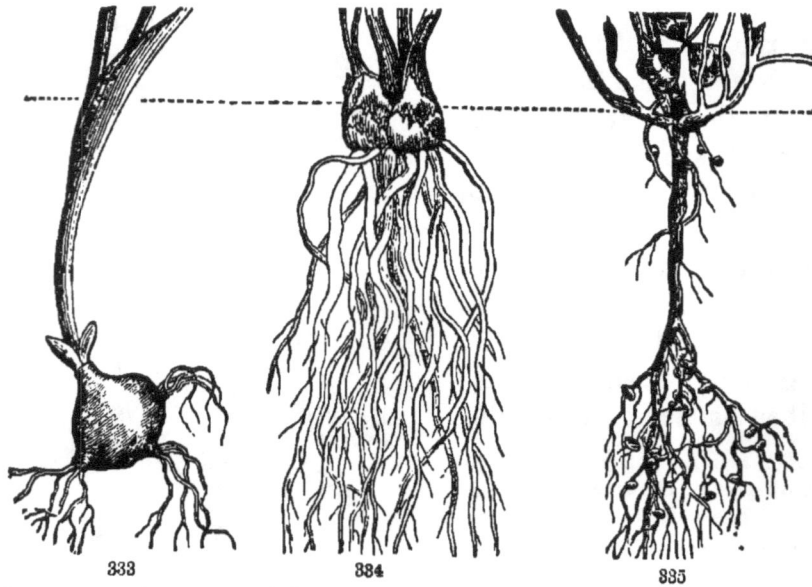

Fig. 333. A tuberous root (Erigenia).   *Fig.* 334. Fibrous roots (Buttercups).
Fig. 335. Branching root (White Clover), with tubercles.

ground, in proportion to the growth of the branches and twigs of the stem above. The axis itself may not descend to any great depth, and after a few years may be found far exceeded in growth by its own branches which extend horizontally in a better soil. The greater the growth of the root, the more firm will be its hold upon the ground, and the greater its capacity for drinking in liquid nourishment for the tree.

220. The roots of herbaceous plants take a great variety of forms. Some are tuberous, some fibrous. The *tuberous* are such as consist of a large axis or body, with small branches; as in the Beet, Ground-nut, Spring Beauty, and many other biennial plants.

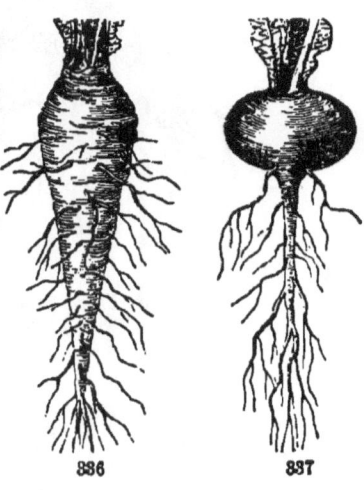

221. The *fibrous* are such as consist mostly of fibres, with scarcely any axis; as in Buttercups, Grasses. In such cases the axis ceased to grow immediately after germination, and long thread-like branches supplied its place.

886   887

Fig. 886. Tuberous and *fusiform* root of Beet.

Fig. 887. Tuberous and *napiform* root of Turnip.

222. The *fibro-tuberous* roots are such as have some of their fibres thickened and fleshy, as seen in the Peony, Dahlia,

---

216. Please explain the meaning of the term axis. In what two directions does it grow?

217. What is the twofold purpose of the root? What does it imbibe? In what state is this food when imbibed?

218. What is said of the nature of the root? What are its only proper organs? What is said of leaves or buds?

219. Describe the roots of woody plants, and their growth.

220–222. Describe tuberous roots; fibrous; fibro-tuberous; tubercular.

Spiræa. If little tubers here and there are attached to the fibres, the root is *tubercular*, as in Squirrel-corn.

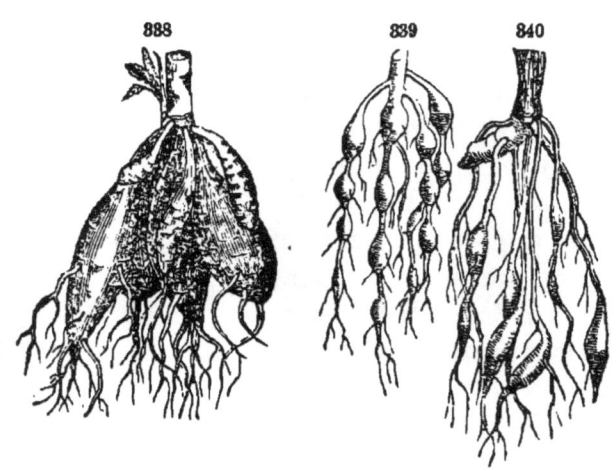

Fig. 338. Fibro-tuberous root of Peony.
Fig. 339. Fibro-tuberous root of Spiræa filipnndula.
Fig. 840. Fibro-tuberous root ef Mourning Geranium.

223. All these fleshy forms, whether tuberous or fibro-tuberous, are filled with *starch*, deposited there in store, for use in the future growth of the plant. Many other forms of roots are described in larger works.

LESSON XXVIII.

OF THE STEM OR ASCENDING AXIS.

224. The stem tending upward in its growth is often called the ascending axis. It does not in all cases continue to arise

223. What purpose do fleshy roots serve?
Less. XXVIII. What is the subject of this lesson?

Fig. 341. Spotted Prince's Pine, entire plant: the stem is a "leaf-stem."
Fig. 342. Diclytra (*D. cucullaria*), whole plant; it has a "scale-stem."

in growing, but often becomes oblique or horizontal. Therefore we have, besides *erect stems*, stems *prostrate, procumbent,
trailing*, when running along flat on the ground, or over
bushes, as the Partridge-berry, White Wintergreen (Fig. 343);
and, also, stems *decumbent*, first arising and afterwards reclining on the ground, as the Poor-man's-weather-glass (Fig.
344).

225. There are, also, *subterranean* stems, never arising

224. What of the direction of the growth of stems? How does the stem
of White Wintergreen grow? How the stem of Anagallis?

Fig. 343. Tne White Wintergreen (*Chiogenes*); it has a procumbent stem

above the ground at all, but only sending up leaves and
flowers with their stalks, as the Tulip.

226. It is the nature of the stem to produce buds, as it is of
the root to produce none. At first the stem is itself a bud,
and as it grows it bears this bud always at the summit and
produces a new bud in the axil of every new leaf.

Fig. 344. Poor-man's-weather-glass (*Anagallis*); it has a decumbent stem.

227. The stem has *nodes* and *internodes*. The joints where
the leaves severally come out are the *nodes*, and the portion
of stem between, the *internodes*. In the bud the internodes
are quite undeveloped, and the nodes close together; but as
it develops into a regular leaf-stem, the internodes grow, and
the nodes with their respective leaves are separated.

228. But in some plants, the nodes only are developed, and
the axis never extends itself above ground, and covers itself

225. What of the stem of Tulip?
226. What the nature of the stem with respect to buds?
227. Please tell us what are nodes and internodes.

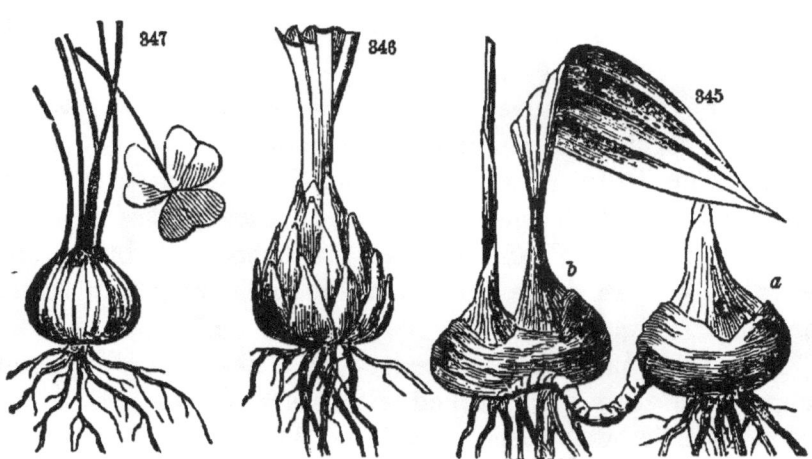

Fig. 345. Corms of Putty-root (*Aplectrum*): *a*, of last year; *b*, of the present year.
Fig. 346. Scale bulb of White Lily.
Fig. 347. Scale bulb of Violet Sorrel (*Oxalis violacca*).

with scales instead of leaves. Thus we have two classes of stems; namely, *leaf-stems* and *scale-stems*. These figures, one of the delicate Diclytra and the other (Fig. 341) of the

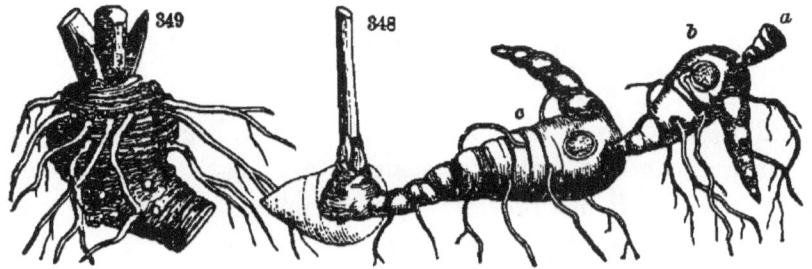

Fig. 348. Rhizome of Solomon's Seal: *a*, fragment of the first year's growth; *b*, the second year's growth; *v*, the third year's growth, bearing *d*, the stem of the present year, which will leave a scar (the seal), like that of the others.
Fig. 349. *Premorse* rhizome of Trillium.

228. What two classes of stems have we to consider? What is the difference between them? To which class does Diclytra belong? Prince's Pine?

Prince's Pine, make a fine contrast of the two kinds of stems.

229. Several varieties of scale-stems must be distinguished; as, bulb, corm, rhizome, creeper, tuber, &c.

230. The Tulip, Hyacinth, Onion, Lily, have *bulbs;* you see (Figs. 346, 347), they consist of roundish masses of thick scales with a small axis—in fact, an overgrown bud. The *corm* is like it in shape, but has a thick axis with thin scales or none. (Fig. 345.)

231. The *rhizome*, or root-stock, is a fleshy, underground stem, often scaly and marked with scars, as you see in the Bloodroot, Solomon's Seal (Figs. 348, 349).

Fig. 350. Creeper of "Nimble Will," or Witch-grass: *a,* bud; *b, b,* bases of the stems which rise above-ground.

232. The *creeper* is more slender, much branched, many-jointed and many-scaled, as seen in this figure of the Witch-grass. It sends out rootlets from its joints, and is very tenacious of life, binding the soil into turf wherever it abounds.

233. The *tuber*, such as grows on the underground stems of the Potato-plant, is evidently a *stem* (not a root), for it always produces buds.

229. Name five sorts of scale-stems.
230. Describe the bulb; the corm. 231. The rhizome.
232. The creeper. 233. The tuber.

234. Of the leaf-stem class we must describe three kinds, the trunk, caulis, and vine. *Trunk* is the name given to the stems of woody, erect plants, especially of trees. They are the representatives of loftiness and strength, in poetic phrase, lifting their summits to the skies and doing battle with the storm. There are, indeed, few objects in nature possessed of a truer grandeur than the White Pine's trunks of the Northern forests.

235. *Caulis*, is the general name given by botanists to the

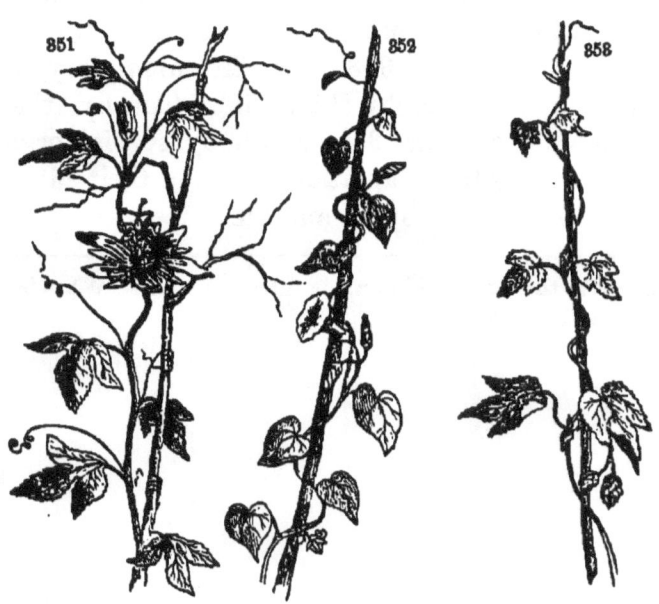

Vines. *Fig.* 351. Passion-flower (*Passiflora lutea*), climbing by tendrils. *Fig* 352. Morning-glory, twining from left to right. *Fig.* 353. Hop, twining from right to left.

234. Name, next, three kinds of leaf-stems. Describe the trunk.
235. The caulis. Meaning and use of caulescent? acaulescent? Give examples of each.

stems of herbs.   From this word come two adjectives much
used and quite convenient, viz., *caulescent* and *acaulescent ;*
the former denoting the presence of stems above-ground, the
latter of only underground stems.   Thus the Buttercup is
caulescent, while the Pitcher-plant is acaulescent ; the Garden
Violet or Pansy is caulescent, while the wild Blue Violet is
acaulescent.

236.   *Vine,* as every one knows, denotes a slender stem, too
weak to stand alone, and supporting itself by the aid of other
plants or objects.   Some vines are woody, some herbaceous.
The Hop twines itself around its supporter, turning from
right to left, as in Fig. 353.   The Morning-glory, also, but
it turns from left to right (Fig. 352).   Thirdly, the Grape
and Passion-flower (Fig. 351) climb by special organs, *the
tendrils,* of wonderful adaptation, showing their Maker's design more truly than if by an audible voice.

---

236. Describe the vine.   What their three varieties ?   What is the read-
mirable in the tendril ?

LESSON XXIX.

PLANTS TO BE ARRANGED IN CLASSES.

Fig. 854.

Clover

and

Grasses.

PLANT may be studied by itself, as an individual, separate from other plants or objects; or it may be considered in its relations to other plants, as constituting a part of a system. In this latter view we discover one vast design embracing the innumerable millions of plants as one kingdom, leading us to adore the wisdom and goodness of him who planned and created the world. For we see that he has not only made each plant with so much loveliness and perfection in itself, but has assigned to each its proper rank in the system, and endowed it with just that nature, habit, and style of beauty, which adapts it to that rank.

238. To study plants as constituting a system, as we now propose to do, is useful in two ways: first, it gives us a larger and truer conception of the Vegetable Kingdom; and

---

237. What two modes of studying the plant are mentioned? In the second mode what discovery is made?

238. In the systematic study of plants what two other advantages?

secondly, it teaches us how to recognize by name the plants with which we meet, so as to avail ourselves of all that has been recorded concerning the same by botanists before us.

239. Suppose the pupil, in his study, has dropped a single Flax-seed on a lock of cotton floating in water in a bulb-glass. It grows, filling the clear water with its silvery radicles, while its stem shoots upwards covered with leaves and finally blooming with flowers. This is an individual plant. He studies its organs, colors, portrait, and carefully writes its history.

240. Meanwhile, other Flax-seeds, by thousands, have been sown in the fields, and from each, also, a plant has arisen. The student finds them in flower, tinging all the plain in ocean blue. Now, shall he, as a botanist, repeat his study over each of all these millions? Certainly not. He finds himself already acquainted with them, for each bears an exact resemblance to that which he has already described. His knowledge of one individual Flax-plant, therefore, avails him for each and all the myriads of Flax-plants growing everywhere.

241. In this manner we obtain the idea of a SPECIES. Thus, a species of plants consists of many individuals of the *same kind,* having descended from a common stock, and resembling each other and their common parent in every feature.

242. The common Blue Flax, of which linen is made, is *a species;* the wild Yellow Flax is another; and the Purple Flax of the gardens is another. The White Clover is a spe-

239. Can you give us an idea of an *individual plant?*
240. Having studied *one* individual Flax-plant, why do we not need to udy the others?
241. Please state your idea of a species.
242. Please illustrate your idea of a species.

cies with its progeny of millions of plants; the Red Clover is another; the Yellow Clover another; the Buffalo Clover another. In like manner all the plants of the globe are grouped into species, and this is the first step in classification.

243. The second step carries us to the genus, which we may thus define : A GENUS is an assemblage of species which are much alike; especially in their flowers and fruit. Thus, FLAX is a genus made up of the several species mentioned above, and other similar species. CLOVER is a genus composed of 150 species, some of which we have just mentioned. Every one notices the resemblance between White Clover, Red Clover, &c. PINE is a genus, embracing as species White Pine, Yellow Pine, Pitch Pine, Long leaved Pine, and many others.

244. Individuals of the same species may differ somewhat among themselves, and these differences constitute *varieties*. Thus Apple-trees differ in their fruit, and there are hundreds of varieties although only one species. Roses differ in their form, color, and fragrance of their flowers, forming many varieties under each species. Probably no two plants of any species were ever *exactly* alike. Sameness, or monotony, is not a characteristic of Nature.

LESSON XXX.

THE NATURAL SYSTEM.

245. IN attempting to classify and arrange the genera of plants, according to their natural resemblances and differ-

---

243. Can you now define a genus? Please illustrate your idea of a genus.
244. What is a variety? Illustrate your meaning.

ences, botanists have formed a system called the Natural System. Let us now briefly notice this system of classification.

246. We have already stated that the plants of the globe are all created in *species*, and that this is the first step in classification. Then, in the second place, the species are grouped into *genera*. Now the number of species of plants already known is about 100,000, and the genera 20,000.

247. The third step in our system carries us forward to the NATURAL ORDERS. These are made up of genera. As we associate similar species to form a genus, so we associate similar genera to form the natural orders. The number of orders described in the Natural System is about three hundred. For example, the natural order Crucifiræ, or the Crucifers, embraces such genera as Mustard, Cress, Cabbage, Turnip, Radish, Wall-flower, which every one sees to bear resemblance to each other in many respects.

248. How then shall we define a natural order? It is a group of similar genera; or, a group of genera closely related to each other. Therefore, individuals form species; species form genera; genera form orders. But how shall we classify these three hundred orders?

249. Suppose we take an excursion into the mountains. We walk beneath the forest trees, and among the shrubs. We tread upon the lesser herbs, the matted grasses, and the mosses and lichens which cover the rocks. Everywhere we see plants, and behold the domain of the vegetable kingdom.

245. What is the subject of this Lesson XXX.?

246, 247. What is the first step in classification? the second? the third? What number of species known in all the vegetable world? What number of genera? of orders? (Ans. 303.)

248. Define a natural order. Please review these three steps.

249. Show how we may divide the vegetable kingdom.

Now viewing this as one grand whole, we want to divide it into two subkingdoms. How shall we do it?

250. Every attentive observer has noticed that some of these plants produce no flowers; as, *e. g.*, the Ferns and Mosses. Let us then take all such plants and consider them as forming one sub-kingdom, viz., the FLOWERLESS PLANTS All other plants will of course constitute the other sub-king dom, *viz.*, the FLOWERING PLANTS. Botanists call the latter the Phænogamia, and the former, the Cryptogamia (Greek words of the same import).

251. Now these two sub-kingdoms have other distinctions besides *flowering* and *not-flowering*. See the fruit-dots growing on the back of Fern leaves. The microscope shows them to be clusters of hollow cases, and each case filled with a fine yellow dust. But this dust is not seeds, with embryo, radicle, &c. (Less. 24), but little sacs, containing a fluid, similar to the pollen grains (Less. 15). We call them SPORES. See, also, the Mushrooms having no leaves, and the Lichens

Some of the Cryptogams.—*Fig*. 355. A Fern, showing the fruit dots. *Figs*. 356, 857 858, are Lichens, some appearing to have stems, and some with no appearance of any

often, also, without stems. Hence we may say of the Cryptogams that they are not only *flowerless*, but *seedless*, and often *leafless* and *stemless*.

252. We will now dismiss the Cryptogams for the present, and consider the Flowering Plants (Phænogams), as one sub-kingdom;—how shall *this* be divided? Every one notices a striking difference between plants with parallel-veined leaves and those with net-veined leaves. The former have their flowers three-parted, while the latter have their flowers two, four, or five-parted, &c.,—the former have seeds one-lobed (monocotyledoned, Less. 15), the latter, two-lobed (dicotyledoned, Less. 15). Let ns, then, divide the Phænogamia into two provinces; as Nature has already done.

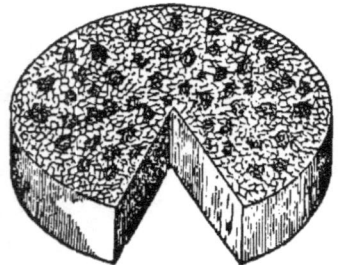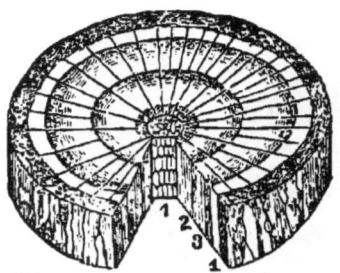

Fig. 859. Cross-section of an exogenous stem (Elm), of two years growth: 1, the pith; 2, 3, two layers of wood; 4, the bark. *Fig.* 360. Cross-section of an endogenous stem (Corn), showing no distinction of layers.

253. We may call these two provinces severally, the Exogens and the Endogens:—two Greek words denoting outside growers, inside-growers, referring to their modes of growth.

254. Now, taking such an Exogen as the Apple-tree, and such an Endogen as the Indian Corn, we may distinguish them thus: The Exogen has its wood, if any, arranged in concentric rings, or layers, as seen in Fig. 360;—the outer ring

252. Show how the Flowering Plants are divided. 253. Please give the character of an Exogen; an Endogen. Meaning of these two words?

being the youngest; the leaves net-veined; the flowers seldom (or never completely) three-parted; and the seeds two-lobed. On the contrary:

255. The Endogen has its wood, if any, confused, the inner portions being the newest;—its leaves parallel-veined;—its flowers three-parted; and its seeds one-lobed.

LESSON XXXI.

MORE ABOUT THE NATURAL SYSTEM.

256. Thus Exogens and Endogens are so clearly defined that you may know them as far off as you can see them. The next step in the analysis is, to subdivide each of these provinces. First, as to the Exogens: We know that they generally have pistils in their flowers, with the young seeds (ovules) inclosed in their ovaries. But there are exceptions to this rule. The Pines, Yews, &c., have no pistils at all, or, at least, no stigmas, and produce naked seeds, not inclosed in seed-vessels. Hence, we have two classes of Exogens: the naked-seeded and the vessel-seeded. The botanists call the latter the ANGIOSPERMS (Greek, *angios*, a vessel; *sperma*, seed); and the former, the GYMNOSPERMS (Greek, *gymnos*, naked).

257. Secondly, the Endogens: here consider the peculiar forms and flowers of the Grasses. Their flowers are all enveloped in green, alternate scales, called *glumes*, instead of

254. Is the Lily an Exogen or Endogen? The Buttercup? The Maple, &c.?

256. What is the next step in the analysis? State the manner of subdividing the Exogens. What is the meaning and etymology of the word "Angiosperms?" What of Gymnosperms? Give an example of each.

257. Show the subdivision of the Endogens. What of the Petaliferæ? What of the Glumiferæ?

the circles of petals common in other flowers.   Hence we have a class of Glume-plants and of Glumeless-plants, or, as the botanists say, GLUMIFERÆ and PETALIFERÆ.   Thus we divide all the Flowering Plants into four Classes, viz.:

1. *Angiosperms;* Exogens bearing stigmas and seed-vessels.

2. *Gymnosperms;* Exogens with no stigmas, and with naked seeds, as the Pines, Firs, Larches, Cedars, Cypresses, Yews, &c.

3. *Petaliferæ;* Endogens with no glumes and ordinary flowers.

4. *Glumiferæ;* Endogens with glumes instead of petals, as the Grasses, Sedges, Grains.

258. Again, each of these Classes are to be subdivided into Cohorts, as follows: the Angiosperms are divided (not very naturally) into three cohorts, viz.:

1. The *Dialypetalæ*, or Polypetalous Exogens, having flowers with the petals distinct and separate, as in the Buttercup, Rose, Mustard.

2. The *Gamopetalæ*, having flowers with the petals united into one piece, as in the Phlox, Morning-glory, Foxglove.

3. The *Apetalæ*, having flowers without petals, either naked, or with only one circle of floral envelopes (which must then be considered as sepals, whatever be the color); as Ginger-root (*Asarum*), Poke (*Phytolacca*), and Pig-weed (*Chenopodium*).

4. Next, the Gymnosperms are regarded as forming one cohort, called the *Conoids*, having the fruit usually in cones. (Less. XXIII.)

258. After the classes, what is the next step in analysis? How are the Angiosperms subdivided? Please define the Polypetalous Exogens; the Gamopetalous; the Apetalous. What cohort do the Gymnosperms form? Why? What two cohorts do the Petaliferous Endogens form? Define the fifth cohort. Define the sixth cohort. What cohort do the Glumiferous Endogens constitute?

The Endogenous Petaliferæ are divided into two cohorts, *viz.*:

5. The *Spadicifloræ*, having the flowers on a spadix, as in the Egyptian Calla and Jack-in-the-pulpit.

6. The *Floridiæ*, having the flowers separate, not on a spadix, as in Tulip, Gladiolus.

7. The Class Glumiferæ constitutes the seventh cohort, under the name *Graminoids*, i. e., the Grass-like plants.

Six other cohorts are formed from the flowerless plants, but we cannot notice them in this work.

259. Lastly, the cohorts áre themselves divided into, or composed of, the Natural Orders, which we defined in Lesson XXX.

---

260. TABLE I. TABULAR VIEW OF THE NATURAL SYSTEM.

Kingaom. Sub-kingdoms Provinces. Classes. Cohorts.

<pre>
                                             (Dialypetalous.
                                             { Gamopetalous,
                            Angiosperms.. (Apetalous.
             (Exogens.. { Gymnosperms.—Conoids.
             {   (Spadicifloræ,
             {   Petaliferæ.....(Florideæ.
   (Phænogamia. (Endogens. (Glumiferæ....—Graminoids.
Vegetables. (Cryptogamia. (Its divisions here omitted.)
</pre>

261 TABLE II. VIEW OF THE NATURAL SYSTEM.

1. Flowering Plants. (Next pass to No. 2.)   PHÆNOGAMIA.
1. Flowerless Plants. (Pass to No. 9.)   CRYPTOGAMIA
   2. Leaves net-veined. Flowers never quite 8-parted....3. *EXOGENS.*
   2. Leaves parallel-veined. Flowers 8-parted....4. *ENDOGENS.*
      3. Stigmas present. Seeds in seed-vessels....5. Angiosperms.
      3. Stigmas none, seeds naked. Pines. Spruces, &c.. .6. Gymnosperms,
         4. Flowers without glumes, having petals, &c....7. Petaliferæ
         4. Flowers with green, alternate glumes, no petals ..9 Glumiferæ

---

259. Finally, into what are the cohorts themselves divided? Give us examples of each of all these cohorts.
260 Explain the use of Table I 261 Of Table II.

6

5. Petals distinct and separate.   POLYPETALÆ
5. Petals united more or less.   GAMOPETALÆ.
5. Petals none.   APETALÆ.
    6. The cone-bearing plants. Cedars, Larches.   CONOIDS
      7. Inflorescence a spadix.   SPADICIFLORÆ.
      7. Inflorescence not a spadix.   FLORIDEÆ.
        8. Grass-like plants.   GRAMINOIDS.
9. Such as Ferns, Mosses, Lichens, Sea-weeds, Mushrooms all omitted in
this book. (See Class-Book, Chapter XIV

262. TABLE III. ANOTHER VIEW OF THE NATURAL SYSTEM

VEGETABLE KINGDOM, divided into two sub-kingdoms, viz.:
  Sub-kingdom First, PHÆNOGAMIA, the Flowering Plants, including
    Province 1., the *EXOGENS*, or Dicotyledons. including two classes
      Class 1, the Angiosperms, having three Cohorts, viz.:
        Cohort A, POLYPETALOUS Exogens (as Roseworts, &c.);
        Cohort B, GAMOPETALOUS Exogens (Phloxworts, &c.); and
        Cohort C, APETALOUS Exogens (Pokeworts, &c.).
      Class 2, the Gymnosperms, with one Cohort, viz.:
        Cohort D, CONOIDS, or cone-bearing plants (Pineworts, &c.).
    Province II., the *ENDOGENS*, or Monocotyledons, two Classes, viz.
      Class 3, the Petaliferous Endogens, having two Cohorts;
        Cohort E, SPADICIFLORÆ (the Aroids, &c.);
        Cohort F, FLORIDEÆ (Lilyworts, &c.).
      Class 4, the Glumiferous Endogens, one Cohort, viz.:
        Cohort G, GRAMINOIDS (Grasses, Sedges, &c.).
  Sub-kingdom Second, CRYPTOGAMIA the Flowerless Plants
  Province III. &c , &c

LESSON XXXII.

OF THE ANALYSIS OF PLANTS.

263. To study any subject by the separate examination of
the parts of which it is composed, is a process called *analysis.*
For example, in Grammar, we *analyze* a sentence when we
point out and separately consider the subject, predicate,

---

262. Of Table III. 263. What is the general meaning of analysis? Illustrate.

object, &c. In Chemistry, we analyze water when we separate its two elements, oxygen and hydrogen, and examine each by itself.

264. In Botany, however, we use the word analysis in a wider sense. It implies not only the separate study of each particular organ composing the plant, but doing all this in connection with certain tables, in order to determine its name and history.

265 This kind of analysis is the constant and delightful pursuit of the active botanist. Without it, the study of books loses half its pleasure and usefulness. The student can acquire a better knowledge of a species by the study of a living specimen, than by memorizing the longest description found in books.

266. During the flowering months, he will often meet with species in blossom which are yet unknown to him. If he is duly interested in his study, he will not fail to seize and analyze each new specimen, and even extend his walk in search of more. In this manner, he may in a few seasons become acquainted with every species in his locality.

267. But we do not expect that all this will be accomplished by our young friends while using as their only text-book this little work. We only aim *now* to furnish them with the means of making a *fair beginning*, so that they may be able, in future seasons, to advance rapidly with the " Class Book," or other works of that rank.

268. In the following pages we present the pupil with numerous tables, designed to conduct our inquiries in every process of botanical analysis; also accompanied by a plain, miniature Flora, or a partial description of all the flowering plants in the United States.

264. What is its signification in botany?

269. Specimens gathered for analysis should have flowers in full bloom, full-grown leaves, and also, if possible, the mature fruit. If it be an herb, it is well to have the whole of it, as the root and lower leaves often afford characters by which the species is known. Suppose you now have good specimens of some one unknown plant, gathered for analysis, —how will you proceed with them?

270. We first examine the several parts of the plant, beginning with the root and ending with the pistil or ovary, determining the character of each according to the definitions given in the former lessons. After this, we refer to the table commencing on page 121, entitled, "Review of the Natural System," and read, compare, and decide according to the directions contained in Lesson XXXIII., in order to determine the Natural Order to which the specimen belongs. Having determined the Order, we next turn to that Order, and determine the Genus and Species by means of other similar tables.

271. In examining the specimen, previous to the use of the tables, the first inquiries may be somewhat as follows:

As to the plant—whether it be an herb, shrub, or tree.

As to the root—whether tuberous, fibrous, or fibro-tuberous.

As to the stem—whether a scale-stem or leaf-stem; bulbous, rhizome, or erect, &c.

As to the leaves—whether alternate or opposite; parallel-veined or net-veined; whether the figure be ovate, lanceolate, oblong, &c.

---

269. What kinds of specimens are to be preferred for analysis?

270. Please state the first thing to be done with them. After you have found the Order, what then?

· 271. What special care should be taken? As to character, what do we inquire concerning plants? What concerning the root? the stem? the leaves? the stipules? What concerning the flowers? the calyx? the corolla? stamens? What concerning the pistil or fruit?

As to stipules—whether present or absent.

As to the flowers—whether symmetrical or unsymmetrical; regular or irregular; whether the calyx be free or adherent; the petals, whether distinct or united; the stamens, whether hypogynous or perigynous, whether opposite to the petals or alternate with them.

As to the pistil and fruit,—whether the carpels be more than one, and whether distinct or united. (See Lesson XVIII.)

LESSON XXXIII.

HOW TO ANALYZE A PLANT BY THE TABLES.

272. Our readers are already informed that the Flora which accompanies these instructions is not intended to make them acquainted with the plants of the country, but simply to teach the pupil *how to analyze.* Hence they will not expect to find in it any thing like a full account of all our flora, but a few plain exercises by which they may trace every flowering plant in the country to its Natural Order, about one in every two to its Genus, and about one in every five to its Species. In conducting an exercise in this Flora with a class of pupils who have well studied the former part of the work, some method like the following would be interesting and profitable.

273. Suppose the class present, and each furnished with a specimen of some one unknown species, both in flower and fruit.

Teacher. Are you all ready? Turn to the Flora and let us find out together the family relations and the names of this fine plant. We will commence at the "Analysis of the Natural Orders" (page 132), and read the first pair of lines, which we will call a *couplet.*

John (reads). "Plants bearing flowers (Phænogamia).

"Plants not bearing flowers (Cryptogamia)."

Teacher. To which of these sub-kingdoms does your specimen belong?

John. To the flowering plants, for it has both flowers and fruit.

Teacher. Now tell us to which couplet we shall next pass.

John. To the second.

Teacher. Very well. Edward, you may read and determine the second couplet in the same manner.

Edward. "Leaves net-veined. Flowers never completely three-parted.

"Leaves parallel-veined (rarely net-veined). Flowers three-parted." This specimen seems to answer to the first line, having net-veined leaves and five-parted flowers. It is, then, an Exogen. Pass to No. 3.

Teacher. Now let it pass along, and if a wrong decision is made, let the observer signify it by raising his hand.

Sarah. "Stigmas present. Seeds inclosed in seed-vessels.

"Stigmas none. Seeds naked." These flowers have pistils and stigmas. I think it is an Angiosperm. Pass to No. 5.

Eliza. "Corolla with distinct petals.

"Corolla with united petals.

"Corolla none; sepals sometimes none." My specimen has five distinct petals, and five sepals. It is therefore *Polypetalous.* Pass over to **A**.

Jane. "Herbs.

"Shrubs, trees, or undershrubs." This plant is an herb. Pass on to No. 2.

Mary. "Leaves alternate or all radical.

"Leaves opposite, on the stem." The leaves of the stem are alternate, but many are radical. Pass to No. 15.

Louisa. "Flowers regular or nearly so. Fruit never a legume.

"Flowers irregular," &c. I do not remember the *legume.* (Several hands are raised.)

Teacher. Edward will define a legume.

Edward. A legume, sir, is a fruit like a pea-pod.

Teacher. Can Mary improve this definition?

Mary. The legume is a simple, or one-carpeled fruit, with two valves and one cell.

Louisa. But this plant has regular flowers, in any case. See No. 17.

Nancy. "Stamens numerous." &c. I count more than twenty stamens here. Turn to No. 21.

Lucy. "Stamens on the torus," &c. I think they are on the torus, and not on the calyx. Next to No. 22.

Emily. " Pistils few or many, distinct (at least as to the styles).

"Pistils (styles, also, if any), completely united." I see many little green pistils, quite distinct, in the centre of the flower. Go to No. 23.

Caroline. "Petals five or more, deciduous. Leaves not peltate," &c. This flower has five petals, but I do not know whether they are *deciduous* or not.

Teacher. Will some of you relieve Caroline's doubts?

Emily. I think they are deciduous, for they have already fallen off from several of my flowers.

Teacher. True. Then what is Caroline's decision?

Caroline. I suppose, then, that the plant belongs to the "Order of the Crowfoots," which is the first natural order.

Teacher. Well done. This brings us to the order of which our plant seems to be a member. Let us now turn to that order and learn the *genus* of the plant. But before we look into the "Analysis of the Genera," we should carefully compare our plant with the characters of the order, so that we may be sure that we have not erred in our conclusion. John will read aloud these characters, and the class will consider whether their specimens answer to each.

John (reads). "Herbs, rarely shrubs, with a colorless, acrid juice" (&c., to the end).

Teacher. Since we are now confident that we have a plant belonging to the order of the Crowfoots, let us commence the "Analysis of the Genera." Edward, the first couplet.

Edward. " Sepals four, valvate in the bud. Achenia tailed.

"Sepals imbricate in the bud." The sepals are imbricate. No. 2.

Sarah. " Ovaries one-seeded, achenia in fruit.

"Ovaries with two or more seeds." I find one seed in each ovary,— indeed, the ovary is itself like a little seed. Pass to No. 3.

Eliza (after reading the couplet). The greenish sepals are quite different from the yellow petals. Go to the triplet marked *d.*

Jane (after reading the three lines). As this plant has leaves on the stem, and a little scale with honey at the base of each petal, I must pronounce it a Crowfoot, genus No. 4.

Teacher. We now turn to that genus (page 147), and read its character for the sake of confirmation and a better knowledge.

Mary (reads the character of the genus Ranunculus aloud).

Teacher. We are now ready for the analysis of the species. Mary is next.

Mary. " Petals yellow. Seeds (carpels) rough with prickles. Flowers small. South. . . . .

" Petals yellow, seeds smooth and even. . . . .

" Petals white (claws yellow). Seeds wrinkled crosswise." This spe cimen has smooth seeds and yellow petals. Pass to *a*.

Louisa. " Leaves more or less divided," &c. This second line of the triplet describes the plant. Pass on to *b*.

Nancy. " Root leaves neither divided nor cleft, merely crenate.

" Lower leaves three-cleft, but not divided to the base.

" Leaves all ternately divided and much cleft." Pass to *c*.

Lucy. " Sepals reflexed in flower. Plants erect.

" Sepals spreading in flower, shorter than the petals." The sepals are reflexed. Read Nos. 14, 15.

Emily, after reading both descriptions, finally concludes that she holds in her hand a specimen of the Bulbous Crowfoot, or Ranunculus bulbosus, in which conclusion all concur.

LESSON XXXIV

VARIOUS SUGGESTIONS AND CAUTIONS.

274. THE work of analysis is often attended with difficul ties which severely try the skill and perseverance of the young botanist. So it often is in the study of Algebra, or of Logic; indeed, in nearly every valuable branch of learning His decisions may be wrong through a want of a thorough acquaintance with botanical terms, or through his ignorance of the real characters of his specimens. Of course his success will always be in proportion to the accuracy of his knowl edge,—here, as well as in all other pursuits.

---

274. Mention two sources of error in the analysis of plants

275. But the minuteness of the organs or parts to be studied is often discouraging even to the accurate student, much more to the careless one. To overcome this, skill in dissection and a dauntless courage in observation are indispensable. Moreover, there is often much ambiguity in the nature of the subject. For example, whether the Geraniums are herbs or shrubs; whether the flowers of Petunia are regular or irregular; whether the Pear leaf is ovate or oval, &c. Experience will at length diminish this difficulty.

276. The exact limits between the classes, the cohorts, &c., are not always easily defined. For example, is Trillium an Exogen or an Endogen? Its netted leaves indicate the former, but its flowers being three-parted throughout, and its seeds with one cotyledon, prove it to be an Endogen. Again, is Spring Beauty an Exogen or an Endogen? Its leaves seem, at first, parallel-veined, but as its flowers are five-parted it is an Exogen.

277. Angiosperms will be readily distinguished from Gymnosperms, if we remember that almost all the latter are evergreen trees, like the Pines, Cedars, Larches, &c.

278. The industrious student will very soon find himself so well acquainted with the different characters of the cohorts that he will in analysis refer his plant at once to its right cohort, without consulting the previous parts of the table. This is desirable; and a thorough acquaintance with the

---

275. What of the minuteness of the organs of some plants? What of the ambiguity of the plants themselves? What will soon diminish this difficulty? Mention examples of this ambiguity.

276. Are the limits of the classes, cohorts, genera, &c., always clear? How do we know that the Trillium is an Endogen? that Spring Beauty is an Exogen?

277. How may the Gymnosperms be quickly distinguished?

characters of the five great orders following will prove a great saving of time and trouble.

279. The CRUCIFERS are herbs with alternate leaves, cruciform flowers (§ 87), two stamens shorter than the other four, and two-celled pods. Example, Mustard.

280. The PEAWORTS are plants with one-celled pods, mostly papilionaceous flowers and compound leaves. Examples, Pea, Bean.

281. The UMBELLIFERS have alternate leaves, small, regular, five-parted flowers, in umbels, and two-seeded fruit. Caraway

282. The ASTERWORTS are herbs with compound flowers, that is, with heads composed of many little five-parted flowers appearing together like a single flower. Asters, Sunflower.

283. The LABIATES are herbs with square stems, opposite leaves, labiate flowers, and fruit deeply cleft into four parts. Peppermint.

Among Endogens we select two or three orders.

284. The ORCHIDS. Herbs with very irregular and grotesque flowers, and stamens united to the style. Orchis.

285. The SEDGES. Herbs with solid stems; linear, grasslike leaves (if any), on entire sheaths; and with green glumes and flowers.

286. The GRASSES. Herbs with hollow stems, linear leaves on split sheaths, and with green glumes and flowers.

---

279. Define the Crucifers   280. The Peaworts
281. The Umbellifers.   282. The Asterworts.
283. The Labiates.   284. Define also the Orchids.
285. The Sedges.   286 The Grasses.

ABBREVIATIONS AND SIGNS,

Often used in Descriptive Botany.

ach., achenia.	*fil.*, filament.	*pet.*, petals.
æst., æstivation.	*fl.*, flower; *fls.*, flowers.	*r.*, rare, uncommon.
alter., alternate.	*fr.*, fruit.	*recp.*, receptacle.
anth., anther.	*hd.*, head; *hds.*, heads.	*reg.*, regular.
a.rill., axillary.	*hyp.*, hypogynous.	*rhiz.*, rhizoma
c., common.	*imbr.*, imbricate.	*rt.*, root.
cal., calyx.	*inf.*, inferior.	*sds.*, seeds.
caps., capsule.	*invol.*, involucre.	*seg.*, segments.
cor., corolla.	*irreg.*, irregular.	*sep.*, sepals.
decid., deciduous.	*leg.*, legume.	*st.*, stem.
diam., diameter.	*lf.*, leaf; *lvs.*, leaves.	*sta.*, stamens.
emarg., emarginate.	*lfts.*, leaflets.	*stig.*, stigma.
f. or *ft.*, feet.	*ova.*, ovary.	*sty.*, styles.

Apr., April. *Aug.*, August. *Dec.*, December. *Feb.*, February. *Jan.*, January. *Jl.*, July. *Jn.*, June. *Mar.*, March. *Nov.*, November. *Oct.*, October. *Sept.*, September.

N., Northern, that is, the northern portions of the United States.
N.-E., New England, or the Northeastern States.
N.-W., the Northwestern States.
E., the Eastern, or the Atlantic States.
W., the Western, or the States bordering on the Mississippi and Ohio rivers.
M., the Middle States or portions of the United States.
S., the Southern States.
S.-E., the Southeastern States, and *S.-W.*, the Southwestern States.
N. Y., New York. *Mass.*, Massachusetts. *Pa.*, Pennsylvania, &c.

f. (with or without the period), a foot.
′ (a single accent) denotes an *inch* (a twelfth of 1 foot).
″ (a double accent) a second, a *line* (a twelfth of an inch).

① An annual plant.	§ (placed after), a naturalized plant.
② A biennial plant.	† (placed after), cultivated for ornament.
♃ A perennial plant.	‡ (placed after), cultivated for use.
♄ A plant with a woody stem.	∞ Indefinite or numerous.
♀ A pistillate flower or plant.	♂ A staminate flower or plant.

☿ A perfect flower, or a plant bearing perfect flowers.
⚥ Monœcious, or a plant bearing staminate and pistillate flowers.
♀ ♂ Diœcious; pistillate and staminate flowers on separate plants.
♀ ☿ ♂ Polygamous; the same species, with pistillate, perfect, and staminate fls.
0 (a cipher) signifies wanting or none, as, "Petals 0."

ANALYSIS OF THE NATURAL ORDERS,

Being a Key for the ready determination of any plant, native or cultivated, growing within any State east of the Mississippi river, or bordering on its western shore.

NOTE.—A star (*) prefixed to the name of the Order, denotes that that Order, with Its genera and species, is described in its place in the Flora. The Orders net thus marked are not noticed in the Flora beyond this Table. The Orders are here numbered ∞ correspond with the "Class Book of Botany."

CLASSES AND COHORTS

1 Flowering Plants ... 2.   *Sub-kingdom*, PHÆNOGAMIA.
1 Flowerless Plants....Ferns, Mosses, Lichens, Mushrooms,
 Sea-weeds, &c. (not further noticed here). *Sub-kingdom*, CRYPTOGAMIA.
2. Leaves net-veined.   Flowers never completely 3-parted ...3. *EXOGENS.*
2. Leaves parallel-veined (rarely net veined). Flowers 3-parted....4. *ENDOGENS.*
 3. Stigmas present. Seeds inclosed in a seed-vessel....5. Angiosperms.
 3. Stigmas none. Seeds naked (Pines, Spruces, &c.).. .6. Gymnosperms.
4 Flowers without glumes, colored or green....7. Petaliferæ
4. Flowers with green, alternate glumes, no perianth....8. Glumiferæ
 5. Corolla with distinct petals....**A.** *Cohort* 1. POLYPETALOUS.
 5. Corolla with united petals....**B.** *Cohort* 2. GAMOPETALOUS.
 5. Corolla none. Sepals sometimes none   .**C.** *Cohort* 3. APETALOUS.
6. The cone-fruited plants (same as Gymnosperms)..**D.** *Cohort* 4. CONOIDS.
7. Fls. on a spadix, apetalous or incomplete....**E.** *Cohort* 5. SPADICIFLORÆ.
7. Fls. complete, perianth double. No spadix..**F.** *Cohort* 6. FLORIDEÆ.
8. The grass-like plants (same as Glumiferæ).   **G.** *Cohort* 7. GRAMINOIDS

A Orders of the Polypetalous Exogens

1. Herbs....2.
1. Shrubs, trees, or undershrubs. .3.
2. Leaves alternate or all radical....15.
2. Leaves opposite on the stem....11.
 3. Flowers regular or nearly so....4.
 3. Flowers irregular (or fruit a legume, § 180) ..57
4. Stamens 8 times as many as the petals, or more.. 5.
4. Stamens 1 or 2 times as many as the petals, or fewer . 7.
 5. Leaves opposite....60.
 5. Leaves alternate. . 6.

6. Stamens on the torus or on the hypogynous (§ 83) petals.. 63.

6. Stamens and petals on the calyx tube (perigynous, § 83).. .68.

7. Ovaries simple, distinct or one only.   Vines or erect shrubs....69.

7. Ovary compound....8.

    8. Ovary inferior,—wholly adherent to the calyx....70.

    8. Ovary superior,—free from the calyx, or nearly free....9.

9. Stamens opposite to the petals, and of the same number....72.

9. Stamens alternate with the petals, or of a different number....10

    10. Leaves opposite on the stem....73.

    10. Leaves alternate, compound....76.

    10. Leaves alternate, simple ...78.

11. Stamens 3 times as many as the petals, or more....47.

11. Stamens 1 or 2 times as many as the petals, or fewer....12

    12. Pistils distinct and simple, few or one only....48.

    12. Pistils united into a compound ovary....13.

        13. Ovary free from the calyx. ..14.

        13. Ovary adherent to the calyx. ..49.

14. Stamens opposite to the petals, and of the same number....51.

14. Stamens alternate with the petals, or of a greater number....52.

    15. Flowers regular or nearly so.   Fruit never a legume....17.

    15. Flowers irregular (rarely regular), and the fruit a legume.. 16

        16. Stamens 3 or more times as many as the petals....42.

        16. Stamens few and definite, 5–12....43.

17. Stamens numerous, 3 or more times as many as the petals....21.

17. Stamens few and definite....18.

    18. Ovary free from the calyx,—superior... 19.

    18. Ovary adherent to the calyx,—inferior....39.

19. Pistils *one* or indefinite (1–15), distinct, simple.. .80.

19. Pistils definite in number, as follows, viz.....20.

    20. Carpels (or pistils) 2, united, the short styles combined into 1....31

    20. Carpels 3 or 4, united, the styles or stigmas 3, or 4, or 6....32.

    20. Carpels 5, distinct or united, with 5 distinct styles....37.

    20. Carpels 5, united, and the styles combined into 1. .38.

21. Stamens on the torus (hypogynous)....22.

21. Stamens situated on the corolla at base....27.

21. Stamens situated on the calyx at the base....28.

    22. Pistils few or many, distinct (at least as to the styles)... 23.

    22. Pistils (and styles also, if any) completely united....24.

23. Petals 5 or more, deciduous.   Leaves not peltate.   Order of the * *Crowfoots.*   1

23. Petals 3, persistent and withering.   Floating leaves peltate.   *Water-shields.*   7

23. Petals many, deciduous.   Leaves all peltate.   * *Water-beans.*   8

    24. Sepals 2 only....26.

    24. Sepals 4, 5, or 6, mostly 5 ...25

25. Petals numerous, imbricate in the bud.   * *Water Lilies.*   9
25. Petals 5, imbricate in bud.   Leaves tubular.   * *Water-pitchers.*   10
25. Petals 5, convolute in bud.   Flowers of 2 sorts.   *Rock-roses.*   17
    26. Petals 5, imbricate in bud.   * *Purselanes.*   22
    26. Petals 4, usually crumpled in bud.   * *Poppyworts.*   11
     27. Filaments united into a tube.   Anthers 1-celled.   * *Mallows.*   24
28. Sepals 2, persistent.   Fruit a pyxis (§ 178).   * *Purselanes.*   22
28. Sepals 3 to 5....29.
    29. Petals imbricate in bud.   Fruit a long pod.   South.   *Lindenblooms.*   26
    29. Petals imbricate in bud.   Fruit not a pod.   * *Roseworts.*   47
    29. Petals convolute in bud.   Fruit compound.   *Loasads.*   53
30. Stamens opposite to the imbricated petals.   Pistil *one.*   * *Berberids.*   6
30. Stamens alternate with the petals, or more numerous.   * *Crowfoots.*   1
    31. Stamens 6 (tetradynamous, § 108).   Pods 2-celled.   * *Crucifers.*   13
    31. Stamens 4, or 8-12.   Pod 1-celled.   *Capparids.*   14
32. Sepals and petals in 3's.   Stamens 6.   Small herbs.   *Limnanths.*   36
32. Sepals and petals in 4's.   Stamens 8.   Climbing.   * *Indian Soupworts.*   41
32. Sepals and petals in 5's....33.
    33. Stamens definitely 5....34.
    33. Stamens indefinite, 3-30....36
     34. Stamens monadelphous.   Stems climbing.   *Passionworts.*   57
     34. Stamens distinct....35.
      35. Stem climbing.   Flowers greenish.   (*Mexican vine.*)   *Order* 104
      35. Stem erect.   Flowers yellow.   *Turnerworts.*   56
      35. Stem erect.   Flowers cyanic.   * *Sundews.*   19
       36. Flowers perfect, very many and small.   *Rock-roses.*   17
       36. Fls. monœcious. Plants woolly, scurfy, or downy. *Order* 112
    37. Stamens 5, alternate with the 5 petals.   Seeds many.   * *Flaxworts.*   30
    37. Stamens 5, opposite to the 5 petals.   Seed 1.   (*Leadworts.*)   *Order* 80
    37. Stamens 10 (twice as many as the petals), united at base. * *Wood-sorrels.*   82
    37. Stamens 6-24 (twice as many as the petals), distinct.   * *Houseleeks.*   60
     38. Ovary 1-celled.   Leaves radical, spinous.   S.   * *Sundews.*   19
     38. Ovary 3-5-celled.   Leaves mostly radical, dotless.   * *Order* 73
     38. Ovary 3-5-celled.   Leaves cauline, dotted, pinnate.   *Rueworts.*   37
39. Style 1, but the carpels (§124) as many as the petals (2-6).   * *Onagrads.*   52
39. Styles 2, carpels 2, fewer than the (5) petals....40.
39. Styles 3-5....41.
    40. Seeds several.   * *Saxifrages.*   61
    40. Seeds two only.   * *Umbelworts.*   63
     41. Sepals 2, with 5 petals.   * *Purselanes.*   22
     41. Sepals as many as the petals.   *Araliads.*   64
      42. Ovaries many or few, rarely 1, always simple.   * *Crowfoots.*   1
      42. Ovary compound, 3-carpeled, open before ripe.   *Mignonettes.*   15

43. Sepals fewer or more in number than the petals....44.
43. Sepals and petals each of the same number....45.
   44. Sepals 2 (or vanished); petals 4 (2 pairs), with 1 or 2 spurs. *Fumeworts.* 12
   44. Sepals 4, petals 2; the largest sepal spurred behind. * Jewelweeds. 34
   44. Sepals 5, petals 3. No spur. * Milkworts. 45
     45. Flowers 4-parted, not very irregular. No spur. Capparids. 14
     45. Flowers 5-parted....46.
       46. Stamens 8. Spur slender. Trophyworts. 85
       46. Stamens 5. Spur blunt, or none. * Violets. 16
       46. Stamens 10 (or more). Fruit a legume. No spur. * Peaworts. 46
47. Pistils many, entirely distinct, simple. * Orowfoots. 1
47. Pistils 3 to 5, united more or less completely. * St. Johnsworts. 18
47. Pistils 5 to 10, united, with sessile stigmas and many petals. Ice-plants. 28
   48. Pistil only 1, simple. Petals 6-9. Stamens 12-18. * Berberids. 6
   48. Pistils 3 or more, distinct, simple. Flowers all symmetrical. *Houseleeks. 60
   48. Pistils 2, covered up by the stamens. Juice milky. * Order 97
49. Carpels as many as the sepals....49.°
49. Carpels fewer than the sepals. ..50.
   49.° Anthers opening at the top. Melastomes. 50
   49.° Anthers opening along the side. * Onagrads. 52
     50. Seeds numerous. Styles 2. * Saxifrages. 61
     50. Seed 1 in each cell. Styles 2 or 3. Araliads. 64
     50. Seed 1 in each cell. Style 1 (double). * Cornels. 65
       51. Style 3-cleft at the summit. * Purselanes. 22
       51 Style and stigma 1, undivided. * Order 78
52. Leaves pinnate, with stipules between the petioles. Bean-capers. 84
52. Leaves simple, toothed or lobed....53.
52. Leaves simple, entire.. .54.
   53. Flowers cruciform, with 6 stamens. * Crucifers. 13
   53. Flowers 5-parted, with 10 stamens. * Geraniums. 31
     54. Petals and stamens on the throat of the calyx. Loosestrifes. 51
     54. Petals on the torus (hypogynous)....55.
       55. Flowers irregular, unsymmetrical (§ 110). * Milkworts. 45
       55. Flowers regular, 2 (or 3)-parted throughout. Water-peppers. 20
       55. Flowers regular, 5-parted....56.
         56. Leaves dotted with pellucid or black dots. *St. Johnsworts. 18
         56. Leaves not dotted. * Pinkworts. 21
57. Pistil a simple carpel, becoming a legume. Stamens 10-100. * Peaworts. 46
57. Pistil compound, 3-carpeled....58.
57. Pistil compound, 5-carpeled....59.
   58. Flowers perfect. Leaves digitate. * Buckeyes. 41
   58 Flowers monoecious (§ 109). Leaves 1-sided. Cultivated. Begoniads. 59
     59. Stipules present. Plants half-shrubby. Cultivated. * Geraniums. 31
     59. Stipules none. Shrubs or half-shrubs. Native. * Order 73

60. Stamens on the torus, in several sets. Leaves dotted.   * *St. Johnsworts.*   18
60. Stamens on the calyx (perigynous, § 83)....61.
    61. Ovaries many, free, but inclosed.   *Calycanths.*   48
    61. Ovary compound, free in the bell-shaped calyx.   *Loose-strifes.*   51
    61. Ovary compound, adherent to the calyx....62.
       62. Leaves with a marginal vein.   *Myrtleblooms.*   49
       62. Leaves with no marginal vein.   * *Saxifrages.*   61
63. Petals imbricate or valvate in the bud....65.
63. Petals convolute in the bud....64.
    64. Anthers 1-celled, turned inwards.   # *Mallows.*   24
    64. Anthers 2-celled, turned outwards. South.   *Silk-cottons.*   25
       65. Ovaries distinct, many or few....66.
       65. Ovary compound....67.
          66. Petals 6, valvate (§ 129). Erect shrubs.   *Papaws.*   3
          66. Petals 3-9, imbricate. Trees or erect shrubs.   * *Magnoliads.*   2
          66. Petals 6-9, imbricate. Climbing shrubs.   *Moonseeds.*   5
             67. Leaves dotted with pellucid dots.   *Orangeworts.*   28
             67. Leaves dotless. Sepals valvate. Fls. small. *Lindenblooms.*   26
             67. Leaves dotless. Sepals imbricate. Fls. large. *Teaworts.*   27
68. Style 1, with many stigmas. Green, fleshy shrubs (acti). *Indian Figs.*   54
68. Styles several, or 1 with 1 stigma. Woody trees or shrubs.   * *Roseworts.*   47
    69. Pistils many, spicate on the slender torus. Climbers.   *Schizands.*   4
    69. Pistils 2-6, capitate on the short torus. Climbers.   *Moonseeds.*   5
    69. Pistil 1 only. Stamens opposite the petals.   * *Berberids.*   6
70. Flowers 4-parted, with 8 stamens.   * *Onagrads.*   52
70. Flowers 4-parted, with 4 stamens.   * *Cornels.*   65
70. Flowers 5-parted, with 5, 10, or many stamens....71.
    71. Ovary 5-carpeled; 5-styled.   *Araliads.*   04
    71. Ovary 2-carpeled. Leaves palmate-veined.   * *Currants.*   55
    71. Ovary 2-carpeled. Leaves pinnate-veined.   * *Saxifrages.*   61
       72. Leaves opposite. Stem climbing by tendrils.   *Vineworts.*   44
       72. Leaves alternate. Erect, or vine without tendrils. *Buckthorns.*   43
73. Carpels 3-5....74.
73. Carpels 1 or 2....75.
    74. Styles short. Leaves simple.   *Staff-trees.*   42
    74. Styles long and slender. Leaves pinnate, serrate.   * *Soapworts.*   41
       75. Styles 2, slender. Samara double.   * *Mapleworts.*   40
       75. Style 1, short. (Drupe, or single samara.)   * *Order*   99
76. Filaments 10, united into a tube. Leaves bi-pinnate. *Pride-of-India.*   29
76. Filaments 5, distinct....77.
    77. Leaves pellucid-punctate.   *Rueworts.*   87
    77. Leaves not dotted. Ovary 3-carpeled, 1-seeded.   *Sumacs.*   88
    77. Leaves not dotted. Ovary 3-carpeled, 3 seeded.   * *Soapworts.*   41

78. Petals 4, yellow.   *Witch-hazels*   62
78. Petals 4–7, cyanic....79.
   79. Fruit becoming fleshy drupes....80.
   79. Fruit becoming dry capsules....81.
     80. Stigmas 3, but the drupe is 1-seeded.   *Sumacs*.   38
     80. Stigmas 4–6, and the drupe 4–6-seeded. (*Hollyworts*.) *Order*   74
       81. Capsule 3-seeded. Seed with a scarlet aril. *Staff-trees*   42
       81. Capsule 2 or 3-seeded, seed not ariled.   § 3. * *Order*   73
       81. Capsule many-seeded. § 2. * *Ord.* 73, and *Pittospores*.   39

B. GAMOPETALÆ, OR MONOPETALOUS EXOGENS

1. Stamens (6–100) more numerous than the lobes of the corolla....3.
1. Stamens (2–12) fewer than the corolla lobes, or of the same number....2.
  2. Ovary adherent to the calyx tube, that is, inferior....3.
  2. Ovary free from the calyx tube, that is, superior....4.
    3. Stamens cohering by their anthers....11.
    3. Stamens entirely distinct....12.
      4. Flowers regular....5.
      4. Flowers irregular....28.
        5. Stamens as many as the petals....6.
        5. Stamens 2, fewer than the petals....26.
    6. Stamens opposite to the lobes of the corolla (and free).. .14.
    6. Stamens alternate with the lobes of the corolla (rarely connate)....7.
      7. Shrubs, trees, with the stigmas or carpels 3 to 6... 15.
      7. Herbs 1–10-carpeled, or shrubs 2-carpeled....16.
    8. Stamens 6, united below into 2 equal sets. Herbs.   * *Order*   12
    8. Stamens 10, united into a split tube around the 1 style.   * *Order*   46
    8. Stamens many, united into an entire tube around the styles.   * *Order*   24
    8. Stamens many, united only at the base into 1 or 5 sets....9.
    8. Stamens entirely distinct....10.
      9. Calyx of 5 leafy, imbricated sepals. Shrubs, trees. (*Teaworts*.) *Order*   27
      9. Calyx tubular, 5-toothed, or truncate. Shrubs, trees. *Styracaceæ*.   75
        10. Stamens 8 or 10.   Flowers all perfect.   * *Heathworts*.   73
        10. Stamens 8 or 16. Fls. not all perfect (diœcious). *Persimmons*.   76
     11. Flowers in a compact head surrounded by an involucre. * *Asterworts*.   70
     11. Flowers separate, irregular, perfect. Plants erect.   * *Lobeliads*.   71
     11. Flowers separate, regular, imperfect. Weak vines. (*Cucurbits*.) *Order*   58
12. Leaves alternate.   Flowers 5-parted, regular, separate.   * *Bellworts*.   72
12. Leaves opposite, with stipules between, or verticillate.   * *Madderworts*.   67
12. Leaves opposite. Stipules none....13
    13. Stamens 4 or 5.   Ovary 2–5-celled.   * *Honeysuckles*.   66
    13. Stamens 2 or 3.   Ovary 1-celled, 1-seeded.   *Valerians*.   68
    13. Stamens 4.   Ovary 1-celled, 1-seeded.   *Teazelworts*.   69

14. Herbs.   Ovary with 5 styles and but 1 seed.   *Leaduorts.*   80
14. Herbs.   Ovary with 1 style and many seeds.   * *Primworts.*   78
14. Shrubs, trees.   Ovary 1-styled, 5-celled, 1-seeded.   *Soapworts.*   77
 15. Style none.   Drupe 4–6-seeded.   *Hollyworts.*   74
 15. Style one.   Drupe 4-seeded.   *Vervains.*   88
 15. Style one.   Capsule 3–5-celled, many-seeded.   * *Heathworts.*   73
16. Ovary 1, deeply 4-parted or 4-partible, forming 4 achenia. * *Borrageworts.*   90
16. Ovaries 2, distinct, often covered by the stamens....18.
16. Ovary 1, compound....17.
 17. Ovary 1-celled....20.
 17. Ovary 2–6-celled....22.
  18. Stigmas united or connate. ..19.
  18. Stigmas distinct.   Flowers minute, yellow   * *Bindweeds.*   93
   19. Flower-bud with convolute pieces.   * *Dogbanes.*   96
   19. Flower-bud with valvate pieces.   * *Asclepiads.*   97
    20. Seeds several....21.
    20. Seed one.   Corolla limb entire.   * *Order* 101
     21. Leaves cleft and lobed.   * *Hydrophylls.*   91
     21. Lvs. or lfts. entire.   Fls. not spicate. *Gentianworts.*   95
     21. Leaves entire.   Flowers spicate.   *Ribworts.*   79
22. Leaves opposite....23.
22. Leaves alternate....24.
 23. Ovary 2-celled.   * *Loganiads.*   85
 23. Ovary 3-celled.   Plants not twining. ....｝
  24. Ovary 3-celled.   Plants not twining. ｝   * *Phloxworts.*   92
 24. Ovary 2-celled, 2–6-seeded.   Twining   * *Bindweeds.*   93
 24. Ovary 2-celled, 4-seeded.   Stem erect.   * *Borrageworts.*   90
 24. Ovary 2-celled, many-seeded....25.
  25. Styles 2.   * *Hydrophylls.*   91
  25. Style one.   * *Nightshades.*   94
 26. Herbs.   Corolla 4-parted, dry, scarious.   *Ribworts.*   79
 26. Shrubs....27.
  27. Corolla 5-parted, imbricate in bud.   *Jasmineworts.*   98
  27. Corolla 4-parted, valvate or none.   * *Oliveworts.*   99
28. Ovary deeply 4-parted, forming 4 (or fewer) achenia....29.
28. Ovary entire, of one piece....30.
 29. Leaves opposite.   Stems square.   * *Labiates.*   89
 29. Leaves alternate.   Stems round.   * *Borrageworts.*   90
  30. Ovary with 4 or fewer seeds.   Leaves opposite.   *Vervains.*   87
  30. Ovary with many seeds, or more than 4....31.
   31. Trees or climbing shrubs.   Seeds winged. * *Bignoniads.*   88
   31. Trees.   Seeds wingless.   * *Paulownia,* in *Order* 86
   31. Erect shrubs.   Seeds wingless.   * *Heathworts.*   73
   31. Herbs....32.

82. Leafless and without verdure.   *Broomrapes.*   82
82. Leaves only at base. Fls. spurred.   *Butterworts.*   81
82. Leafy....83. Fruit 4 or 5-celled.   *Pedaliads.*   84
     83. Fruit 2-celled....34.
     84. Corolla convolute in bud.   *Acanths.*   87
     84. Corolla imbricate in bud.   * *Figworts.*   86
     84. Corolla plicate in bud.   * *Nightshades.*   94

C. Orders of the Apetalous Exogens.

1. Plants herbaceous, the flowers not in aments (except in the Hop, 115)....2.
1. Plants woody,—shrubs or trees....22.
   2. Flowers with a regular calyx or calyx-like involucre....3.
   2. Flowers naked, having neither calyx nor corolla....20.
     3. Calyx tube adherent to the ovary, limb lobed, toothed, or entire....8.
     3. Calyx free from the ovary, sometimes inclosing it....4.
   4. Ovaries several, entirely distinct, each 1-styled, 1-seeded.   * *Order*   1
   4. Ovary one only, simple or compound....5.
     5. Style or stigma one only....6.
     5. Styles or stigmas 2–12....7.
       6. Ovary 1-ovuled, bearing but one seed....11.
       6. Ovary many-ovuled, bearing many seeds....12.
       7. Ovary 1–3-ovuled, 1–3-seeded....13.
       7. Ovary 4–∞-ovuled, 4–∞-seeded....17.
8. Stamens 1–12, as many or twice as many as the stigmas....9.
8. Stamens 2–10, not symmetrical with the 1 or 2 stigmas....10.
   9. Stigmas and cells of the ovary 1–4.   * *Order*   52
   9. Stigmas and cells of the ovary 6.   * *Birthworts.*   100
     10. Ovary many-seeded. Styles 2.   * *Order*   61
     10. Ovary 1 or 2-seeded. Style 1.   *Sandalworts.*   108
   11. Flowers perfect. Calyx 4-lobed. Stamens 1–4.   * *Order*   47
   11. Flowers perfect. Calyx entire, funnel-shaped, colored.   * *Marvelworts.*   101
   11. Flowers imperfect. Calyx lobed, green.   *Nettleworts.*   115
     12. Stamens 4, opposite the sepals.   (*Loosestrifes.*)   *Order*   51
     12. Stamens 5, alternate with the sepals.   * *Order*   78
   13. Fruit 3–6-seeded, with 3 (often cleft) stigmas.   *Spurgeworts.*   112
   13. Fruit 1-seeded....14.
     14. Stipules sheathing the stems.   * *Knot-grasses.*   102
     14 Stipules none....15.
       15. Calyx with scarious bracts outside.   *Amaranths.*   106
       15. Calyx double. Climbing.   *Mexican Vine.*   104
       15. Calyx naked....16.
         16. Leaves alternate.   *Goosefoots.*   105
         16. Leaves opposite.   § 3. * *Order*   21

17. Leaves opposite....18.
17. Leaves alternate....19.
    18. Fruit a pyxis, opening by a lid.   * *Order* 22
    18. Fruit a capsule, opening by 4 or 5 valves.   * *Order* 21
        19. Fruit a capsule, 5-celled, 5-horned.   * *Order* 60
        19. Fruit a fleshy, 4–∞-seeded berry.   * *Pokeweeds.* 102
        19. Fruit dry, 1-seeded, opening by a lid.   *Amaranths.* 106
20. Flowers on a spadix with a spathe.   * *Order* 131
20. Flowers in a long, naked spike. Stamens 6 or 7.   *Lizard-tails.* 123
20. Flowers solitary, axillary, minute. Water-plants....21.
    21. Stamen 1. Leaves opposite.   *Starworts.* 124
    21. Stamens 2. Leaves alternate, dissected.   *Threadfoots.* 125
    21. Stamens 12–24. Leaves whorled, dissected.   *Hornworts.* 126
22. Flowers, none of them in aments....23.
22. Flowers (imperfect), the sterile only in aments....34.
22. Flowers (imperfect), both the sterile and the fertile in aments...35.
    23. Leaves opposite....24.
    23. Leaves alternate....27.
        24. Stamens 2.   * *Order* 99
        24. Stamens 3. Parasites.   *Mistletoe—Loranths.* 108
        24. Stamens 4–9....25.
            25. Fruit a double, 2-winged samara.   * *Order* 40
            25. Fruit not winged....26.
                26. Seeds 6. Low shrubs. *Box.*   *Spurgeworts.* 112
                26. Seed 1. Shrubs.   *Oleasters.* 111
27. Style or stigma 1. Seed 1....28.
27. Styles or stigmas 2....31.
27. Styles or stigmas 8–9....32.
    28. Calyx free from the ovary....29.
    28. Calyx adherent to the ovary....30
        29. Anthers opening by valves.   * *Laurels.* 107
        29. Anthers opening by slits.   *Daphnads.* 110
            30. Seeds 2–4. Shrubs.   *Sandalworts.* 109
            30. Seed 1. Trees.   * *Order* 65
31. Stamens numerous.   * *Order* 62
31. Stamens as many as the calyx lobes.   *Elmworts.* 113
    32. Leaves pinnate. Pistils 5.   (*Prickly Ash.*) *Order* 37
    32. Leaves simple, linear, evergreen.   *Crowberries.* 116
    32. Leaves simple, not linear....33.
        33. Flowers 8-parted. Fruit dry.   (*Stillingia.*) *Spurgeworts.* 112
        33. Flowers 4 or 5-parted. Fruit fleshy.   (*Buckthorns.*) *Order* 48
34. Nut or nuts in a cup or involucre. Leaves simple.   * *Mastworts.* 119
34. Nut naked, a tryma (§ 172). Leaves pinnate.   *Hickoryworts.* 118

85. Fruit fleshy, compound. Juice (sap) milky. *Artocarps* 114
85. Fruit dry (except in Myrica, 121). Sap watery....86.
    86. Aments globular, racemed. Nutlets 2-celled. (*Liquidambar*.) *Order* 62
    86. Aments globular, solitary. Nutlets 1-celled. *Sycamores* 117
    86. Aments cylindrical or oblong....87.
        87. Ovary 1-celled, 1-seeded. Fruit dry or fleshy. *Galeworts*. 121
        87. Ovary 2-celled, 2-ovuled, 1-seeded. * *Birchworts*. 120
        87. Ovary many-ovuled, many-seeded. * *Willoworts*. 122

D. Orders of the Conoids.

Leaves simple. The fertile flowers in cones. Stems branched. *Pineworts*. 127
Leaves simple. The fertile flowers solitary. Stems branched. *Yews*. 128
Leaves pinnate. Stems not branched, palm-like. *Cycads*. 129

E. Orders of the Spadiciflorae.

1. Trees or shrubs with palmately-cleft leaves all from one terminal bnd, and
   a branching spadix arising from a spathe. *Palms*. 130
1. Herbs with simple (rarely ternate) leaves. Spadix simple....2.
    2. Plants minute, floating loose on the water. *Duckmeats*. 132
    2. Plants with stem and leaves rooting in the soil....3.
        3. Spadix evident, in a spathe or on a scape. * *Aroids*. 131
        3. Spadix obscure or spike-like. Stems leafy....4.
            4. Flowers with no perianth, densely packed. *Cat-tails*. 133
            4. Flowers with a perianth or not. In water. *Naiads*. 134

F Orders of the Florideae.

1. Flowers (not on a spadix) in a small, dense, involucrate head. ..17
1. Flowers (not on a spadix) solitary, racemed, spicate, &c....2.
    2. Perianth tube adherent to the ovary....4.
    2. Perianth free from the ovary....3
        3. Petals and sepals differently colored (except in Medeola, 147)....9
        3. Petals and sepals similarly colored....12.
    4. Flowers imperfect....5.
    4. Flowers perfect .6
        5. Low herbs, in water *Frogbits*. 136
        5. Woody climbers. *Yamroots*. 144
        6. Stamens 1 or 2, growing to the pistil (gynandrous). * *Orchids*. 138
        6. Stamen only 1, with half an anther. *Arrowworts*. 139
        6. Stamens 3 to 6, distinct....7.
            7. Perianth woolly or mealy outside. Ovary half-free. *Bloodworts*. 142
            7. Perianth glabrous outside....8.

8. Stamens 3.   Anthers turned inwards.   *Burmaniads.* 137
8. Stamens 3.   Anthers turned outwards.   * *Irids.* 143
8. Stamens 6.   * *Amaryllids.* 140
9. Pistils many, distinct, achenia in fruit.   * *Water-plantains.* 135
9. Pistils 3, more or less united....10.
  10. Leaves verticillate, in 1 or 2 whorls. Stigmas 3. * *Trilliads.* 147
  10. Leaves alternate....11.
    11. Stigmas 3. Plants growing on other plants. *Bromeliads.* 141
    11. Stigmas united into one.   * *Spiderworts.* 152
12. Leaves net-veined, broad....13.
12. Leaves parallel-veined....14.
  13. Flowers perfect, 4-parted.   *Croomia—Roxburgs.* 146
  13. Flowers diœcious, 6-parted.   *Greenbriers.* 145
14. Styles, and often the stigmas also, united into 1....15.
14. Styles and stigmas 3, distinct....16.
  15. Flowers colored, regular. Stamens 6 (4 in one species). * *Lilyworts.* 148
  15. Flowers colored, irregular, or else 3-stamened.   *Pontederiads.* 150
  15. Flowers greenish, glume-like, or scarious.   * *Rushes.* 151
16. Leaves rush-like. Ovary of three 1-seeded carpels.   * *Arrow-grasses.* 135
16. Leaves linear, lanceolate, &c. Ovary 6-∞-seeded.   * *Melanths.* 149
  17. Petals yellow, small, but showy. Leaves radical.   *Xyrids.* 153
  17. Petals white, minute, fringed. Leaves radical.   *Pipeworts.* 154

G. Orders of the Graminoids.

Flower with a single bract (glume). Stem solid. Sheaths entire.   *Sedges.* 155
Flower with several bracts (glumes and pales). Stem hollow. Sheaths
  split on one side.   *Grasses.* 156

THE FLORA:

OR,

SELECTIONS FROM THE NATIVE AND CULTIVATED PLANTS OF THE UNITED STATES.

Designed as first exercises in Analytical Botany.

EXPLANATIONS.—The Tables in this work are designed to be *complete*, that is, each Ordinal Table includes all the genera belonging to that order known within the limits of the Flora (*i. e.* the States east of the Rocky Mountains); and each Generic Table includes, in like manner, *all* its known species. The numbers annexed to the genera in the Ordinal Tables, refer to the descriptions immediately following. If no number be annexed, the pupil will understand that that genus is not further noticed.

COHORT I.
THE POLYPETALOUS EXOGENS.

Essential Character.—Flowering Plants (PHÆNOGAMIA), with their stems growing by additions to their outside in layers (EXOGENS), their seeds inclosed in a seed-vessel or pericarp (ANGIOSPERMS), their flowers with a double perianth and their petals distinct (POLYPETALÆ). (But to this last condition there are many exceptions.)

ORDER I. RANUNCULACEÆ. The Crowfoots.

Herbs, rarely *shrubs*, with a colorless, acrid juice, with
leaves mostly alternate and much divided, without stipules;
sepals 3–15, deciduous, distinct, and colored when apetalous;
petals 3–15, distinct, often deformed or contracted or wanting;
stamens ∞, distinct, hypogynous;
pistils x (rarely 1 or few), distinct, becoming in
fruit either achenia, follicles, or berries.

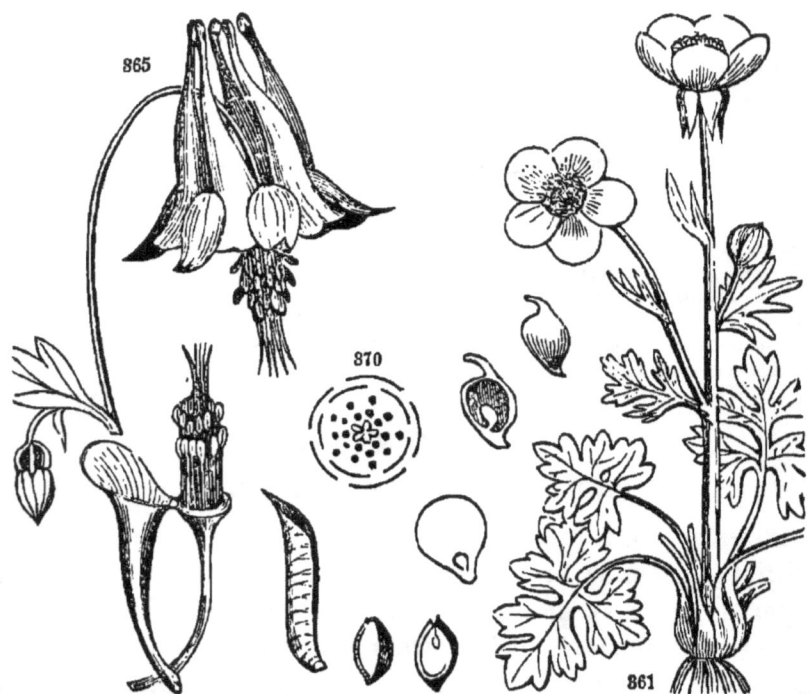

Fig. 361. Bulbous Crowfoot; 2. a petal, showing the honey-scale at base; 3, a single ovary 4, section of it, showing the ovule.

Fig. 365. Wild Columbine; 6, one of the hollow petals attached to the receptacle with the stamens and styles; 7, a ripe follicle; 8, a seed ; 9, section of it, showing the embryo.

Fig. 370. Plan of the flower.

Analysis of the Genera.

§ Sepals 4, valvate in the bud. Achenia tailed....a
§ Sepals imbricate in the bud....2
    2 Ovaries 1-seeded, achenia in fruit....3
    2 Ovaries with 2 or more seeds....4
        3 Corolla 0, or undistinguishable from the colored calyx....5
        3 Corolla and calyx distinct either in color or form....d
    4 Sepals as permanent as the stamens. Fruit dry....6
    4 Sepals falling off sooner than the stamens....k
    4 Sepals persistent with the fruit. Petals very large....m
        5 Sepals persistent with the stamens....b
        5 Sepals caducous (falling) sooner than the stamens....e

6 Flowers regular....7
6 Flowers irregular....h
    7 Petals none....e
    7 Petals smaller than the sepals....f
    7 Petals larger than the sepals....g

a Petals none or stamen-like. Leaves all opposite. *Virgin's Bower.* CLEM'ATIS. 1
 b Stem leaves opposite, remote from the flower. *Anem'one.* ANEMÒNE. 2
 b Leaves all radical. 8 bracts close to the flower. *Liverleaf.* HEPAT'ICA.
 c Flowers mostly imperfect. Leaves compound. *Meadow Rue.* THALIC'TRUM.
 c Flws. perfect. Lvs. simple, palmately lobed. *Prairie Rue.* TRAUTVETTE'RIA.
  d Leaves all radical, linear. Torus linear. Small plant.
                                     Mousetail. MYOSU'RUS.
  d Lvs. cauline. Petals with a honey-scale at base. *Crowfoot.* RANUN'CULUS. 4
  d Leaves cauline. No honey-scale. Petals red or yellow.
                                Pheasant's-eye. ADO'NIS. 5
 e Sepals white, 5 in number. Leaves compound. *False Rue.* ISOPY'RUM.
 e Sepals yellow, 6–9. *Marsh Marigold.* CAL'THA. 6
  f Petals tubular at apex. Roots yellow. *Gold-thread.* COP'TIS. 7
  f Petals tubular at base, 1-lipped. *Globe-flower.* TROL'LIUS. 8
  f Petals tubular, 2-lipped. Sepals persistent. *Hellebore.* HELLEB'ORUS.
  f Petals concave, 2-lobed. Flowers racemed. *Yellow-root.* ZANTHORHI'ZA.
 g Petals larger than the colored sepals, 8-lobed. *Fennel-flower.* NIGEL'LA.
 g Petals larger than the colored sepals, spurred alike. *Columbine.* AQUILE'GIA. 9
  h Upper sepal spurred, inclosing spurred petals. *Larkspur.* DELPHIN'IUM. 10
  h Upper sepal hooded, covering 2 deformed petals. *Monk's-hood.* ACONI'TUM. 11
 k Flowers numerous, in long, slender racemes. *Bugbane.* CIMICIF'UGA.
 k Flowers many, in short racemes. Berries simple. *Baneberry.* ACTÆ'A.
 k Flower one only. Leaves 2. Berry compound. *Turmeric-root.* HYDRAS'TIS.
  m Disk sheathing the ovaries. Very Showy. *Peony.* PÆO'NIA.

1. CLEM'ATIS. Virgin's Bower.

Calyx of 4 colored sepals, valvate-induplicate in the bud. Petals none, or small and stamen-like. Stamens many, shorter than the sepals. Ovaries many, styles becoming long and feathery upon the seed-like achenia. —♃ Mostly climbing vines, with twisted petioles for tendrils, and with opposite, compound leaves.

§ Leaves verticillate. Outer stamens petal-like. Vine....No. 1
§ Leaves opposite. Petals none. Calyx colored....a
  a Erect herbs near 1 foot high. Flowers solitary....10, 11, 12
  a Vines climbing....b

7

b Flowers clustered in panicles....2, 3, 4, 5
b Flowers solitary, large, showy....6, 7, 8, 9

1 Clem'atis verticilla'ris. *Purple Virgin's Bower.* Leaves ternate, 4 at each node. Flowers purple, 2 at each node. Hills. N. W.
2 C. Virginia'na. *Virginian V.* Leaves ternate. Leaflets smooth, lobed, and toothed.
6 C holoseric'ea. *Silky V.* Leaves ternate, leaflets downy or silky, entire. S.
4 C Catesbya'na. *Catesby's V.* Lvs. bi-ternate, lfts. 3-lobed. Clusters axillary. S.
5 C. Flam'mula. *Sweet V.* Lvs. pinnate. Lfts. entire, pointed. Fls. terminal. †
6 C. cylin'drica. *Crisped V.* Lfts. acute, thin, 3-15. Sepals wavy at edge, b–p. S.
7 C. reticula'ta. *Veiny V.* Lfts. obtuse at each end, thickish. Sep. wavy. b–p. S.
8 C. Vior'na. *Leather-flower.* Lfts. ovate, acute, pinnate. Sep. not wavy. P.
9 C. Viticella. *Vine-Bower.* Lfts. oval, 3-15. Sepals not wavy, obovate. P. †
10 C. ochrolen'ca. *Ground V.* Lvs. undivided, ovate, entire, silky beneath. p–y.
11 C. ova'ta. *Egg-leaved V.* Lvs. undivided, broad-ovate, smooth, glaucous. p. S.
12 C. Baldwin'ii. *Baldwin's V.* Lvs. 3-cleft, the upper lance-ovate, entire. p. S.

2. ANEMO'NE. Anem'one, or Wind-Flower.

Calyx regular, of 5–15 colored sepals resembling petals. Petals properly none. Stamens many, much shorter than the sepals. Pistils many, collected into an oblong or roundish head. Achenia generally without tails. Leaves mostly radical, palmately lobed, those of the stem opposite, forming a sort of involucre.

§ Carpels with long, feathery tails in fruit. Flower large....1
§ Carpels without tails....a
a Stem leaves (involucre) sessile....2, 3
a Stem leaves petiolate....b
b Flower-stalk 1 or several, all leafless....4, 5, 6
b Flower-stalks 2-5, all but the first 2-leaved in the middle....7, 8

1 A. Nuttal'lii. *Pasque-flower.* Lvs. cleft into linear lobes, very hairy. *Apr.* N.-W.
2 A. Carolinia'na. *Carolina A.* Flower only one, with 15 sepals. S. W.
3 A. Pennsylvan'ica. *Pennsylvanian A.* Flowers 1-5, with 5 obovate, pure white sepals. Height 12-20′. N. W.
4 A. nemoro'sa. *Wood A.* Flower 1, stalk 2 or 3′ long. Sepals rose-white. *Apr.*
5 A. cylin'drica. *Gray's A.* Flowers mostly several, whitish, stalks 6-12′ long. Fruit heads oblong or cylindrical. *May.* N. W.
6 A. thalictroi'des. *Rue A.* Flowers several or many, rosy or white, on short (1-2′) stalks. Leaves of the invol. 2, twice ternate. *Apr.* Common.
7 A. Virginia'na. *Virginia A.* Leaf-lobes, lance-ovate. Flowers greenish-white. Height 2-3 feet. Common.
8 A. multif'ida. *Red A.* Leaf-lobes linear. Flowers red. Height 6-12′. r. N.

3. HEPAT'ICA. Noble Liverwort.

Calyx (generally called an involucre) of 3
entire, ovate, green sepals (or bracts), situated
a very little below the corolla. Corolla of 5–9
petals, arranged in 2 or three rows. Achenia
without tails.—♃ Pretty little plants blossoming
in early spring. Leaves all radical, thick,
3-lobed, green through the winter. Flowers
numerous, one on each scape, blue, roseate, or
white.

Fig. 371. Flower and leaf of H. triloba.

1 H. tril'oba. Leaf-lobes and sepals obtuse. Scapes hairy, several inches high.
2 H. acutil'oba. Leaf-lobes and sepals acute. Fls. and scapes like the other.

4. RANUN'CULUS. Crowfoot. Buttercups.

Calyx of 5 ovate sepals. Corolla of 5 roundish, shining petals, each
with a honeyed scale or pore at the base inside. Stamens ∞. Achenia
numerous, flattened, crowded in a roundish or oblong head.—A large
genus of herbs mostly perennial (♃) and with yellow flowers. Leaves divided or entire. Juice very acrid.

§ Petals yellow. Seeds (carpels) rough with prickles. Fls. small. S....1, 2
§ Petals yellow. Seeds smooth and even ....⎫
§ Petals white (claws yellow). Seeds wrinkled crosswise ....⎬ 2
⎭
a Leaves all undivided. Plants growing in wet places....5–8
a Leaves more or less divided, not growing under water....b
a Leaves in fine, thread-like divisions, growing under water....3, 4
  b *Root* leaves neither divided nor cleft, merely crenate....9, 10
  b Lower leaves 3-cleft but not divided to the base. Height 1–2f....11–13
  b Leaves all ternately divided and much cleft....c
    c Sepals reflexed in flower. Plants erect....14, 15
    c Sepals spreading in flower, shorter than the petals....16–18
1 R. murica'tus. *Rough-fruited C.* Plant smooth. Seeds with large, stout, hooked
    beaks. Sepals spreading. Leaves 3-lobed and cleft. *South.*
2 R. parviflo'rus. *Small-flowered C.* Plant hairy. Seeds with a very short beak.
    Sepals finally reflexed. Leaves 3-lobed and cleft. *South.*
3 R. aquat'ilis. *Water Crowfoot.* In ponds and rivers. The *white* petals with a
    cavity at base. Only the flowers above water. *Summer.*
4 R. Pursh'ii. *Pursh's Crowfoot.* In stagnant water. The yellow petals with
    a scale at base. Floating leaves, 3–5-parted. *Spring.*

5 **R. Flam'mula.** *Spear-leaved C.* Stems ascending (1–2f). Leaves all lanceolate, narrow, entire, on sheathing stalks. *Sum.*

6 **R. pusil'lus.** *Tiny C.* Stems nearly erect. Leaves ovate and lanceolate. Petals mostly but 3, with about 10 stamens. *May.*

7 **R. rep'tans.** *Creeping C.* Stems creeping and rooting (4–8′). Leaves lance-linear, and linear. Flower 1 at a node. *July.*

8 **R. Cymbala'ria.** *Boat C.* Stems creeping and rooting (1 foot). Leaves all round-cordate, crenate. *Salt-marshes. June.*

9 **R. aborti'vus.** *Abortive C.* Plant glabrous, 1–2f. high. Root leaves, round-cordate. Petals smaller than the sepals. *Spring. v.*

10 **R. rhomboi'deus.** *Rhombic C.* Plant hairy, bushy, 4–10′ high. Root leaves rhombic-ovate. Sepals spreading. N. W.

11 **R. palma'tus.** *Palm C.* Stem hairy. Seeds with a straight beak in a round head. Leaves palmately 3–5-cleft, with sinus closed. *South.*

12 **R. recurva'tus.** *Hook-seed C.* Stem hairy. Seeds with a recurved beak in a round head. Leaves all similarly 3-parted. Flowers small. *Spring.*

13 **R. scelera'tus.** *Villainous C.* Stem glabrous. Seeds not beaked, in an oblong head. Flowers small. Leaves palmately 3–5-parted. *June, July.*

14 **R. bulbo'sus.** *Bulbous C.* Erect (6–12′) from a solid bulb. Petals large. Head of fruit round. Root leaves ternate. *Spring.*

15 **R. Pennsylvan'icus.** *Bristly C.* Erect (1–3f.), very hairy. Head of fruit oblong. Leaves ternate. *Summer.*

16 **R. repens.** *Large creeping C.* Stems first ascending, then creeping. Flower-stalks furrowed. Petals obovate, large. Wet places. *June.*

17 **R. fascicula'ris.** *Early C.* Stem erect. Root fibres thickened. Flower-stalks terete. Petals narrow. Leaves appear pinnate. *May.*

18 **R. acris.** *Tall Buttercup.* Stem erect (2–3f.). Leaves palmately divided, and cleft. Petals roundish. Flower-stalk terete, calyx spreading. *Summer. v.* In the gardens, it becomes double-flowered.

5. ADO'NIS. Pheasant's-eye.

Sepals 5, colored. Petals 5–15, with no scale on the claws. Achenia in a spike, egg-shaped, and pointed with the hardened, persistent style. Leaves numerously cleft into linear and very narrow segments. Flowers terminal, solitary, red or yellow.

1 **A. autumna'lis.** *Late Ph.* A fine, hard annual, from Europe, cultivated in gardens, and naturalized in some places. Stem rather thick for its height, branched. Leaves pinnately parted, with very numerous segments. Petals 5–8, of a bright crimson color, 1½′ across.

2 **A. verna'lis.** *Early Ph.* Petals 10–12, oblong, yellow, dentate. Upper leaves sessile, all much divided. Flowers large. *Spring.* †

6. CAL'THA. Marsh Marigold.

Calyx colored, of 5 roundish sepals resembling petals. Corolla 0. Stamens ∞. Follicles 5–10, oblong, compressed, erect, many-seeded.— ♃ Smooth marsh plants.

C. palus'tris. *Cowslips. Marsh Marigold.* In wet meadows. Root large, thick. Stem about 1f. high, hollow, round, branched. Leaves large (4–6' wide), roundish, cordate, crenate—lower on long, half-round petioles, upper sessile— all of a dark, shining green, and very smooth. Flowers of a golden yellow in all their parts, 1½' broad. Outer row of stamens club-shaped, long. *Spring.*

7. COP'TIS. Gold-thread.

Calyx of 5 or 6 oblong, colored sepals. Corolla of 5 or 6 small club-shaped sepals, hollow and 1-lipped at top. Stamens 20–25. Follicles 5–10, stalked, beaked, spreading, 4–6-seeded.— ♃ Herbs with radical leaves, and long, creeping root-stocks.

C. trifo'lia. *Gold-thread.* Leaves 3-foliate, all radical, the divisions broad, 4–8'' long, crenate, smooth shining, sessile. Petiole 1–2' long. Stems underground, creeping extensively, bright yellow, and very bitter. Peduncles 3–4' high, each 1-flowered. Calyx white. Petals yellow, much smaller than the sepals, barely distinguishable among the stamens by their color. *May.*

8. TROL'LIUS. Globe-flower.

Calyx of 5, 10, or 15 concave sepals colored like petals. Corolla of 5–25 small, linear petals, which are tubular at base. Stamens many, much shorter than the sepals. Pods many, each many-seeded.— ♃ smooth, with palmately-parted leaves.

1 T. laxus. *American G.* Sepals 5. Petals 15–25, shorter than the stamens. Grows in swamps, M. r. Calyx yellow, greenish outside. *June.*

　T. Europæ'us. *European G.* Sepals 10–15. Petals 5–10, as long as the stamens. Flowers globular, bright yellow. †

8 T. Asiat'icus. *Asiatic G.* Sepals 10, orange-colored. Petals 10, longer than stamens. †

9. AQUILE'GIA. Columbine.

Sepals 5, ovate, colored, spreading. Petals 5, tubular with a wide mouth, the outer margin erect, the inner attached to the receptacle, and behind extended into a long, spurred nectary. Stamens 30–40, the inner ones longer and sterile. Styles 5. Follicles 5, many-seeded.—♃ Leaves twice and thrice ternate. Flowers nodding.

1 A. Canaden'sis. *American C.* Spurs straight, longer than limb. Stamens exserted. Flowers scarlet.
2 A. vulga'ris. *European C.* Spurs incurred, shorter than limb. Stamens included. Flowers purple. †

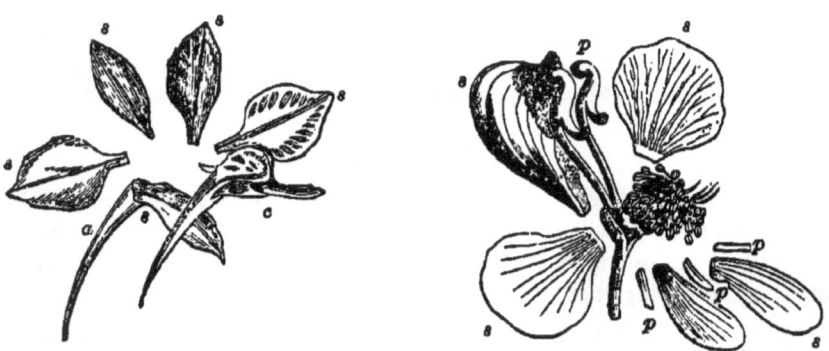

Fig. 872. Flower of Larkspur, displayed: *s, s, s, s, s,* the five petals; *a,* the spurred sepal; *c,* the two petals, spurred, which spur was sheathed in the spurred sepal.
Fig. 873. Flower of Garden Aconite, displayed: *s, s, s, s, s,* the five sepals; *p, p, p, p, p,* the five petals.

10. DELPHIN'IUM. Larkspur.

Sepals 5, colored, the upper one spurred. Petals very irregular, the two upper ones extended behind into a tubular, honeyed spur, sheathed in the spur of the calyx. Styles 1-5. Follicles 1-5.—Showy herbs with the leaves much divided. Flowers blue, red, or purple, never yellow.

§ Petals united into 1 piece. Pistil and pod 1....1, 2
§ Petals 4, distinct. Pistils and pods 2-5....(a)
ᵈ Leaves many-parted into linear segments....3
a Leaves divided into 3-7 wedge-shaped lobes....(b)
    b Tall (2-5f.), with slender, many-flowered racemes....4, 5
    b Low (6-18'), with few (6-12) flowered racemes....6-8
1 D. consol'ida. *Field L.* Fls. loosely scattered. Ovary smooth. Lvs. finely cut. ① †
2 D. Aja'cis. *Rocket L.* Flowers covering the branches. Ovary pubescent. Leaves finely cut. ① †
    3 D. azu'reum. *Azure L.* Fls. in strict, slender racemes. Ovaries 3-5. ♃ W. †
4 D. exalta'tum. *Tall Wild L.* Leaf-lobes 3-5, curvate. Spur straight. *M. Summer.* ‡
5 D. elatum. *Bee L.* Leaf-lobes 3-7, curvate. Spur curved downwards. †

6 **D. tric'orne.** *Low Wild L.* Leaf-lobes linear. Pods recurved. Height 6–12′. M.W.

7 **D. vires'cens.** *Green-flowered L.* Leaf-lobes lanceolate. Fls. greenish-white. S. W.

8 **D. grandiflorum.** *Great-flowered L.* Leaf-lobes 5–7, linear. Fls. large, b–p. †

11. ACONITUM. Monk's-hood. A'conite.

Sepals 5, irregular, colored, upper one vaulted or hooded. Petals 5 or 6, the two upper on long claws, concealed beneath the upper sepal, recurved and honeyed at top; the other 3 or 4 very small. Styles 3–5. Follicles 3–5.—♃ Leaves palmately cleft or divided. Flowers odd and showy, in terminal spikes.

1 **A. unctua'tum.** *Wild M.* Stem reclining, widely branched. Helmet conical. M. S.

2 **A. Napel'lus.** *Garden A.* Stem erect, nearly simple. Helmet semicircular. †

Order II. MAGNOLIA'CEÆ. The Magnoliads.

Trees and *shrubs* with membranous stipules sheathing the buds, with *leaves* alternate, leathery, simple entire, or lobed, never serrate; *flowers* solitary, large and showy, mostly odorous and perfect; *sepals* 3–6, colored like the 6–12 hypogynous imbricated petals; *stamens* numerous, hypogynous, distinct, and many ovaries; *fruit* compound, composed of the united carpels.

Analysis of the Genera.

§ Pistils arranged in a cone....2
§ Pistils whorled in a single row. Shrub. *South.*   *Star Anise.* ILLIO'IUM.
    2 Anthers opening inwards.   MAGNO'LIA. 1
    2 Anthers opening outwards.   LIRIODEN'DRON. 2

1. MAGNOLIA.

Sepals 3. Petals 6–9. Anthers longer than the filaments, opening inwards. Carpels 2-valved, 1–2-seeded, imbricated into a hard, cone-like fruit. Seeds berry-like, suspended when ripe by a long seed-stalk.—A noble genus of trees or shrubs, with large, fragrant flowers.

§ Native Magnolias, flowering with the leaves....a
§ Exotic Magnolias, flowering before the leaves expand....8

a Leaves acute at the base (not cordate)....b
a Leaves cordate or auriculate at the base. Trees 8^–40f. high....5-7
b Leaves shining above, white or rust-colored beneath. Petals 9–12....1, 2
b Leaves dull green both sides, thin, deciduous. Petals 6–9....3, 4

1 M. grandiflo'ra. *Big Laurel.* Tree evergreen, 60–70f. high. Leaves thick, rust downy beneath. Flowers 8 or 9' broad, white. S.

2 M. glauca. *White Bay.* Shrub deciduous, 6–25f. high. Leaves very smooth glaucous beneath. Flowers 2–8' broad, cream-color. Swamps. †

3 M. acumina'ta. *Cucumber-tree.* Tree large. Leaves oval, acuminate, scat tered. Flowers small (3–4' broad), petals obovate. M. S.

4 M. umbrel'la. *Umbrella-tree.* Tree small (20–30f.). Leaves wedge-lanceo late, whorled, very large, as well as the flowers. M. S.

5 M. corda'ta. *Yellow Cucumber-tree.* Petals 6–9, yellow, with reddish lines. Lvs. broad-ovate, slightly cordate. Flowers 4' broad. S.

6 M. Fra'seri. *Ear-leaved M.* Petals 6, pure white. Leaves ear-shaped at base, obovate-spatulate, near 1f. long. *Spring.* S.

7 M. macrophyl'la. *Great-leaved M.* Petals 6, white, each 6–8' in length. Leaves 2–3f. long, obovate-spatulate, cordate. Tree 30–50f. high. S. W. †

8 M. conspic'ua. *Yulan.* Flowers in Spring, large, rose-colored or white, with 6–9 petals or sepals, nearly erect. Japan.

2. LIRIODEN'DRON. Tulip-tree.

Sepals 3. Petals 6, in two rows. Anthers opening outwards. Carpels 1–2-seeded, imbricated into a cone, indehiscent, separating from each other in fruit.—Trees with large and fragrant flowers.

Fig. 374. Young branch of Tulip-tree, unfolding from the bud: *s, s,* stipules.

L. tulipif'era. *Tulip-tree. Whitewood. Poplar.* This is one of the finest and largest trees of our forests. The trunk is generally straight and cylindric, dividing at the top rather abruptly in a few coarse and crooked branches. Leaves dark green, smooth, square at the end, with 2 lobes each side, 8–5' in length and breadth. Flowers large and elegant, greenish-yellow, orange within, 4–6' broad. *May, June.*

ORDER VI.—BERBERIDACEÆ. The Berberids.

Herbs and *shrubs*, with alternate leaves and perfect flowers, with *sepals* imbricated in the bud in 2 or more rows; *petals* opposite the sepals, also imbricated in two or more rows; *stamens* opposite to the petals, the anthers usually opening by two lids; *ovary* 1-celled, solitary and simple, forming a capsule or berry.

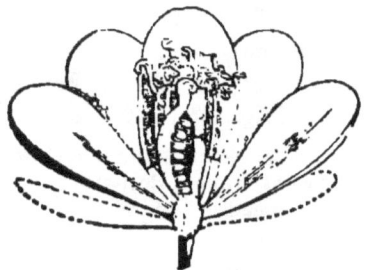 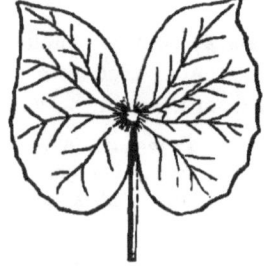

Fig. 875. Section of the flower of Jeffersonia. *Fig.* 876. A leaf of the same.

Analysis of the Genera.

§ Herbs, anthers opening by two valves hinged at top....a
§ Herbs, anthers opening by 2 slits lengthwise....b
§ Shrubs, with yellow flowers and acid berries. BER'BERIS. 1
a Stamens 6. Fruit 2, drupe-like, soon-naked seeds. *Cohosh.* LEON'TICE.
a Stamens 6. Fruit a 2-4-seeded berry. *Umbrella-leaf.* DIPHYLLEI'A.
a Stamens 8. Fruit a capsule opening by a lid. *Twin-leaf.* JEFFERSE'NIA. 2
   b Stamens 9-18. Flower 1, with 2 leaves. *May Apple.* PODOPHYL'LUM. 3

1. BER'BERIS. Barberry.

Sepals 6, obovate, colored, the 3 outer ones smaller. Petals 6, roundish, with two glands at the base of each, inside. Stamens 6. Stigma sessile, disk-like, on the top of the ovary. Berry oblong, sour, 1-celled, 2-3-seeded.—Fine, hardy shrubs, with the wood, inner bark, and flowers yellow.

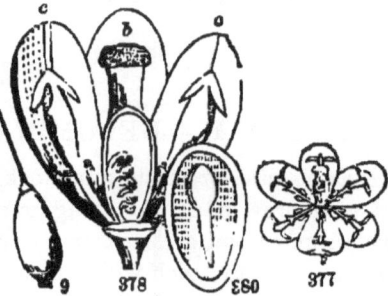

Fig. 877. *a*, Flower of Berberis vulgaris; *b*, the pistil (magnified), with the ovary cut open c, c. petals with stamens opposite; 9, a berry; 880, a seed cut open, showing the embryo.

1 **B. vulga'ris.** *Common Barberry-bush.* A well-known, bushy, handsome shrub, in hard soils. Grows 3–8f. high. Leaves oval, near 2′ long, rounded-obtuse at apex, tapering to a petiole, with bristly serratures on the margin. Flowers yellow, a dozen or more in each hanging raceme, with entire petals. Stamens irritable, springing against the stigma when touched. Berries red, very sour. *June.*

2 **B. Aquifo'lium.** *Holly-leaved B.* Leaves pinnate, of 7–13 thick, spiny-toothed leaflets. Shrub 3–5f. high. Cal. †

2. JEFFERSO'NIA. Twin-leaf.

Sepals 4, colored, caducous. Petals 8, spreading. Stamens 8, with linear anthers. Pod on a short stipe, opening by a lid.— ♃ Flowers and leaves from the root. Scape 1-flowered. (Figs. 375, 376.)

J. diphyl'la. *Twin-leaf.* A very curious plant, acaulescent. Leaves each with two blades, about 1f. high. Flowers same height, white. Root-stock black, with a thick mass of fibres, supposed good in rheumatism. M. W.

3. PODOPHYL'LUM. Mandrake.

Sepals caducous. Petals 6–9, obovate, concave. Stamens 12–18, with linear anthers, the lids scarcely opening. Berry large, egg-shaped, 1-celled, crowned with the solitary stigma.—Low, somewhat poisonous herbs, with one or two leaves and one flower.

P. pelta'tum. *May Apple. Wild Mandrake.* A singular and interesting plant, in woods and fields. Height about 1f. The barren plants bear but a single leaf, which is 5–8′ broad, 5–7-lobed, and centrally peltate. The flowering plants have a pair of leaves, with the flower at the fork of the two petioles—the leaves not peltate, but with a deeply-hollowed base, about 7-lobed. Flower drooping, white, about 2′ across. Fruit yellowish, with the flavor of the Strawberry. *May.*

ORDER VIII.—NELUMBIACEÆ. The Water-beans.

Herbs aquatic, prostrate root-stock, and radical, peltate leaves, with *flowers* large, solitary, on long, upright scapes, 4 or 5-sepaled; *petals* numerous, arranged in many rows, as are also the many stamens; *ovaries* separate, each with a simple style and stigma, becoming in *fruit* 1-seeded nuts, half sunk in the hollows of the very large torus, the *seeds* with a very large embryo and no albumen.

NELUM'BIUM. Nelumbo.

The character of the genus the same as that of the order.

N. luteum. *Yellow Nelumbo.* A magnificent flowering plant, frequent in the stagnant waters of the South and West, rare in N. Y. and Conn. The leaves are 1–2f. broad, round, entire, peltate in the centre, which is concave, and elevated above the water more or less on the long petioles. Flowers several times larger than the White Water Lily, but without fragrance. Petals concave, of a brilliant white at edge, becoming yellow towards the base. Nuts (called Water-beans) about as large as acorns, eatable. *June, July.*

Order IX.—NYMPHÆACEÆ. The Water Lilies.

Herbs aquatic, with roundish leaves from a prostrate rhizoma;
flowers large and showy, the sepals, petals, and stamens gradually passing
  into each other, imbricated and arranged in many rows;
sepals few, colored inside, persistent; *stigmas* radiating and crowning the
ovary, which in fruit becomes a capsule compound and 5-celled;
seeds minute, numerous, with the embryo at the end of the albumen.

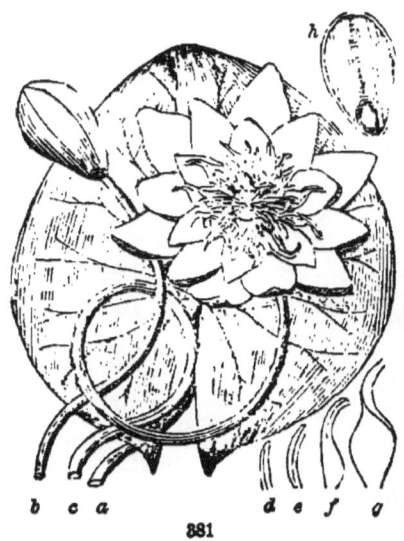

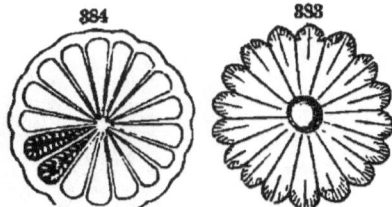

Analysis of the Genera.

Petals large as the sepals, white, red, or blue. NYMPHÆ'A. 1
Petals smaller than the sepals, stamen-like, yellow. *Frog Lily.* NUPHAR.

Fig. 381. Nymphæa odorata: *a,* the leaf; *c,* the flower; *b,* the bud; *d, e. f, g,* stamens gradually changing into petals; *h,* a seed cut open, showing the embryo in a little sac. *Fig.* 383, the many-rayed stigma; 384, cross-section of the many-celled ovary.

NYMPHÆ'A. Water Lily.

Sepals 4 or 5, green outside. Petals in many rows inserted on the receptacle beneath the ovary. Stamens inserted above the petals. Anthers slender, opening inwards, the outer filaments gradually widening and passing into petals. Capsule ripening under water.

N. odora'ta. *White Water Lily.* One of the loveliest of flowers, frequent in ponds and sluggish streams. The root-stock is long and thick, running in the mud where the water is from 3-10f. deep, sending up leaves and flowers to the surface. Leaves 5-6' long, roundish, cleft at the base to the centre, where the long petiole is inserted, margin entire. Petals lanceolate, 1½-2' long, of the most delicate texture and whiteness, often tinged with purple. Filaments yellow. *July.*

ORDER X. SARRACENIACEÆ The Water-pitchers.

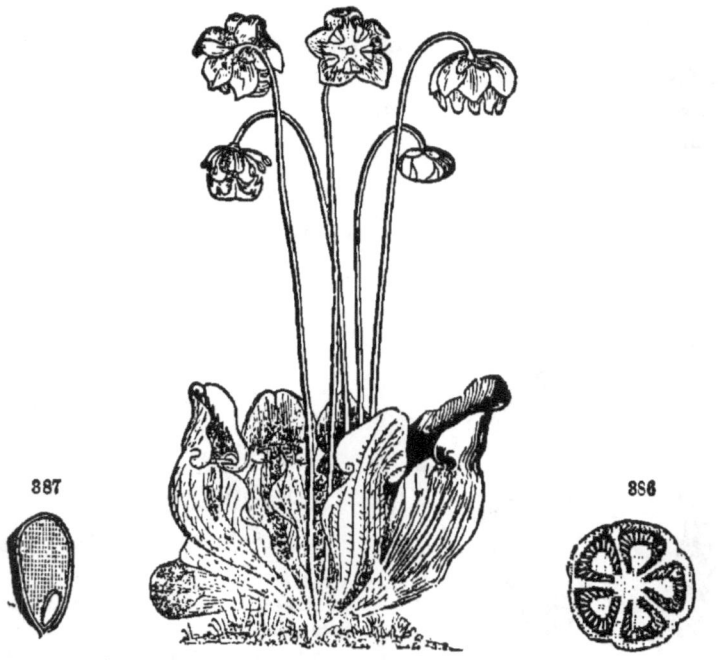

387   386

Fig. 885. Sarracenia purpurea, with bud, flower, and fruit.
Fig. 386. Section of the 5-celled ovary.
Fig. 887. A seed (magnified), with small embryo and large albumen.

Herbs aquatic, in bogs, with fibrous roots, and with the
leaves all radical, urn-shaped, hollow, and large flowers on scapes;
sepals 5, with 3 little bracts at base; *petals* 5, clawed, incurved;
stamens hypogynous; *ovary* 5-celled, with a single style, the
stigma very broad, peltate, and 5-angled, crowning, in fruit, the
capsule, which is 5-celled and full of minute, albuminous seeds.

SARRACE'NIA. Pitcher-plant Trumpet-leaf.

Character essentially as expressed in the order. Nos. 2, 3, 5, 6, are
probably *varieties*, not species.

¶ Leaf-blade inflected over the throat of the tube....7, 8
¶ Leaf-blade erect, or nearly erect; throat of the tube open....a
  a Leaf-tube pitcher-shaped, with a broad wing....1–3
  a Leaf-tube trumpet-shaped, with a narrow wing....4–6

1 S. purpu'rea. *Purple Pitcher-plant.* Flowers purple. Leaves all inflated alike,
  dark green with purple veins, 6–9′ long. Scapes 1-flowered, 1–2f. high. *June.*
2 S. heterophyl'la. *Yellow Pitcher-plant.* Fla. yellow. Lvs. pale, the outer slender. *Jn. r.*
3 S. ala'ta. *Narrow-winged P.* Fls. yellow. Lvs. all more slender than in No. 1. S.-W.
  4 S. (Grono'vii) flava. *Yellow Trumpet-leaf.* Flowers yellow. Leaves 18–36′
    high, all yellowish green,-veins not purple.
5 S. rubra. *Red Trumpet-leaf.* Fls. reddish purple. Lvs. (1–2f.) purple-veined. S.
6 S. Drummondii. *Drummond's T.* Fls. purple. Lvs.(18–36′) mottled with colors. S.
7 S. psittacina. *Parrot's Pitcher-plant.* Fls. purple. Lvs. (3–5′) spotless, hooded. S.
8 S. variolaris. *Mottled P.* Flowers yellow. Lvs. (12–18′) spotted with white. S.

ORDER XI. PAPAVERACEÆ. **The Poppyworts.**

Herbs, generally with a colored juice, with alternate leaves;
flowers on long peduncles, solitary, never blue;
sepals 2 or 3, falling off when the flower expands;
petals generally 4, sometimes 8; *stamens* 4, 8, 12, 16, or 20, &c.;
stigmas 2, or if more, star-like on the flat apex of the compound ovary;
fruit a pod-shaped or roundish capsule; *seeds* numerous and minute.

Analysis of the Genera.

¶ Plants with a yellow juice. Petals yellow, crumpled in bud....a
  a Stigmas and placentæ 3, 4, or 6. Capsule ovoid....c
  a Stigmas and placentæ 2 only. Capsule long, pod-shaped....b

¶ Plants with an orange-red juice.   *Bloodroot*. Sanguina'ria. 1
¶ Plants with a milk-white juice.   *Poppy*. Papa'ver. 2
¶ Plants with a colorless juice.   Calyx a cap, falling off whole.
     Petals 4, orange-yellow. Lvs. dissected. *California Poppy*. Eschsoholt'zia.
ʙ Pod 1-celled, smooth. Leaves pinnately divided. Fls. yellow.
     Stamens 24–32, shorter than the 4 petals.   *Celandine*. Chelido'nium.
ʙ Pod 2-celled, rough. Leaves palmate. Stig. 2 horned. *Horn Poppy*. Glau'cium.
   ᴄ Style present, stigmas 3 or 4. Stem lvs. 2, opposite. *Yellow Poppy*. Meconop'sis
   ᴄ Style none, stigmas 4 or 6. Stem lvs. alternate. *Prickly Poppy*. Argemo'ne

1. SANGUINA'RIA. Bloodroot.

Sepals 2, caducous. Petals 8–12, the outer longer. Stamens about 24. Stigma sessile, 1 or 2-lobed. Capsule pod-like, oblong, 1-celled, 2-valved, acute at each end, and many-seeded.—♃ A low, acaulescent plant, with white flowers, and full of a red or orange-colored juice.

S. Canadensis. *Bloodroot.* An interesting plant, in shady, rich soils, flowering in early spring. Rhizoma thick, fleshy, and when broken or wounded exudes a blood-colored juice, as does every other part. From each joint of the root-stock springs a single large glaucous leaf, and a scape about 6′ high, bearing a single flower. Leaf kidney-shaped, with lobes separated by rounded sinuses between. Flower of a square outline, white, scentless, and of short duration.

Fig. 388. Sanguinaria Canadensis: *b*, the pod; *c*, cross-section of it; *d*, seed cut open, showing the embryo.

2. PAPA'VER. Poppy.

Sepals 2, caducous. Petals 4. Stamens ∞. Stigmas many, united into a star-like crown, sessile upon the thick ovary. Capsule 1-celled, opening by pores beneath the edges of the stigma. Exotic herbs, mostly ⊙, with a milk-white juice abounding in opium.

388

   * Bristly or hairy. Leaves pinnatifid. Flowers scarlet....2, 3
   * Smooth, glaucous. Leaves cut-trothed, clasping. Flowers white ...1

1 **P. somnif'erum.** *Opium P.* Fls. large, often double. † *Summer.*
 2 **P. dubium.** *Small Red P.* Pod club-shaped, smooth. Leaves coarsely
  divided. Flowers light red, smaller than in No. 1. M. S. *Summer.*
 2 **P. Rheas.** *Corn P.* Pod globular, smooth. Leaves more finely divided.
  Flowers large and brilliant, deep scarlet, often double. † *Sum.*

ORDER XII. FUMARIACEÆ. The Fumeworts.

Herbs smooth and delicate, with brittle stems and watery juice;
leaves usually alternate, many cleft or compound;
flowers irregular, purple, white or yellow; *sepals* 2, very small;
petals 4, more or less cohering, one or both of the outer saccate, the two
  inner inclosing the anthers in their coherent tips;
stamens 6, in 2 sets of 3 in each; *pistil* 1; *pod* 1-celled.

Analysis of the Genera.

 * Corolla equally 2-spurred or 2-saccate at base....a
 * Corolla unequal, only 1 of the petals spurred....b
 a Petals slightly united or distinct, deciduous. Not climbing. **DICEN'TRA.** 1
 a Petals firmly united, persistent. Plants climbing. *Mountain Fringe.* **ADLU'MIA.**
  b Ovary with several seeds, forming a slender pod. *Corydal.* **CORYD'ALIS.** 2
  b Ovary with 1 seed, becoming a globular nut. *Fumitory.* **FUMA'RIA.**

1. DICEN'TRA. Ear-drop.

Sepals 2, very small, sometimes disappearing. The 2 outer petals alike,
saccate at base, with spreading tips; the 2 inner alike, spoon-shaped,
meeting face to face over the stamens and pistils. Filaments flat, separate
or not. Middle anther of each set 2-celled, the outer 1-celled. Pod
many-seeded.—♃

 § Low herbs (6'), with white flowers in simple racemes....1, 2
 § Taller (1-2f.), with purple flowers racemed or panicled....3, 4
 1 D. cuculla'ria. *White Ear-drop.* Root bulb-like. Spurs of the flowers divergent
  acute, straight. Flower nearly as broad as long. *Spring.*
 2 D. Canaden'sis. *Squirrel-corn.* Root bearing yellow tubers as large as peas. Fls
  much longer than broad, spurs rounded, incurved. *May, Jn.*
 3 D. exim'ia. *Wild Purple Ear-drop.* Racemes panicled. Flowers oblong
  with very short blunt spurs. Sepals manifest. M. S. †
 3 D. spectab'ilis. *Chinese E.* Raceme simple. Flowers nearly as broad as long
  (1'), very fine and showy; sepals obsolete. †

2. CORYD'ALIS. Corydal.

Sepals 2, very small. Petals 4, one of which is spurred at base. Filaments with broad bases united into 2 sets, sheathing the ovary. Pod 2-valved, slender, many-seeded. Leaves twice ternate, on the stem. In rocky places. *Spring.*

O. glau'ca. *Pink C.* Erect. Fls. pink-yellow, panicled. Leaf-lobes obtuse. ②
O. au'rea. *Golden C.* Diffuse. Fls. yellow, racemed. Leaf-lobes acute. ①

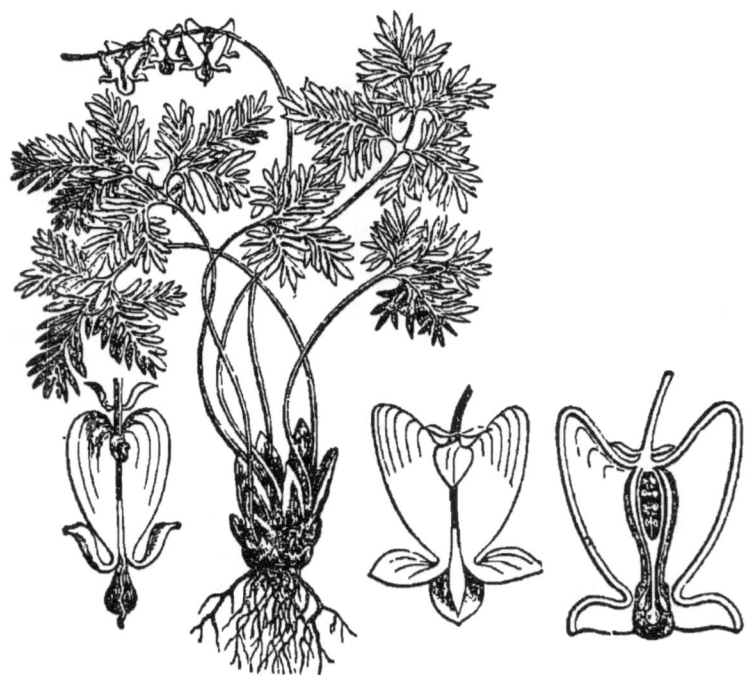

Fig. 389. Dicentra cucullaria, entire plant. *Fig.* 390. Enlarged view of a flower. *Fig.* 391. A section of the same. *Fig.* 392. A flower (enlarged) of D. Canadensis.

ORDER XIII. CRUCIFERÆ. The Crucifers, or Mustardworts.

Herbs with alternate *leaves* and no stipules, and regular flowers, with *sepals* 4, and *petals* 4, spreading in the form of a cross;

stamens 6, 2 of them on opposite sides shorter than the rest ; an
ovary of 2 united carpels, forming in fruit a
silique or *silicle*, with 2 cells and few or many seeds ;
seeds without albumen, the large *embryo* variously bent and folded.

NOTE.—Under this large Order, as under others, we present to our young readers a complete analysis, by which they may trace to its genus *any Mustardwort* growing in the United States But as the genera are so nearly alike, great care and close observation will be needful in avoiding mistakes. The plants for examination *must be in fruit* as well as in flower.

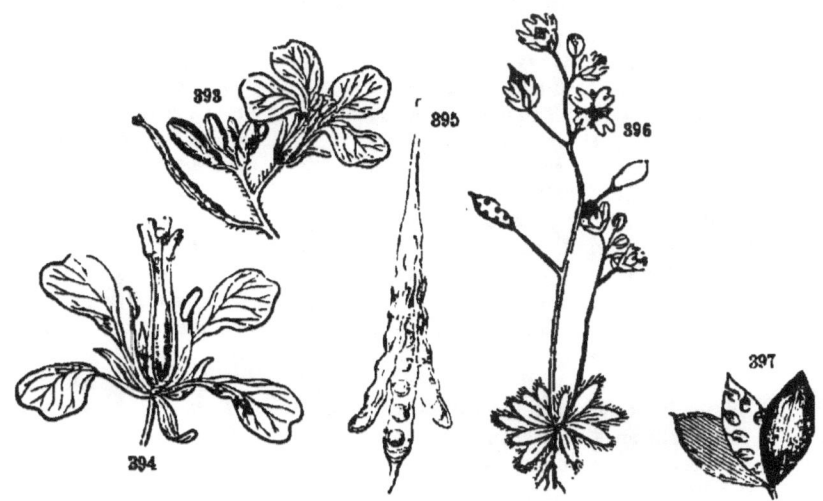

Fig. 393. Flower of White Mustard. *Fig.* 394. Same, with its parts separated. *Fig.* 395. A silique, ripe and open. *Fig.* 396. Draba verna. *Fig.* 397. A pod open.

Analysis of the Genera.

* Garden plants cultivated for ornament or art.

1 Fruit a silique or long pod (§ 363)....5

1 Fruit a silicle or short pod (§ 364)....2

2 Silicle 2-celled, with 2 or more seeds....3

2 Silicle 1-celled, with one seed only.   *Woad.* ISA'TIS.

3 Petals all equal....4

3 Petals unequal, the 2 outside ones larger.   *Candy-tuft.* IBE'RIS. 1

4 Some of the stamens toothed. Gardens.   *Madwort.* ALYS'SUM.

4 Stamens all toothless. Silicles very large and thin.   *Satin-flower.* LUNA'RIA.

5 Seeds flat. Stigma rounded or head-shaped.   *Wall-flower.* CHEIRAN'THUA.

5 Seeds flat. Stigmas 2-horned, spreading.   *Stock.* MATTHI'OLA.

5 Seeds egg-shaped, Stigma with 2 converging lobes.   *Rocket.* HES'PERIS.

** *Plants growing wild, or cultivated for food.*

1 Fruit a silique, 2-celled lengthwise, { flowers yellow....8
flowers white, purple, &c.....6

1 Fruit a silicle, 2-celled lengthwise, { flowers yellow....5
flowers white, purple, &c.....2

1 Fruit a jointed pod, with the partitions crosswise....11

2 Silicle flattened or turgid, with a broad partition....4

2 Silicle flattened contrary to the narrow partition....3

    3 Silicle triang., seeds several in each cell.   *Shepherd's-purse.* CAPSEL'LA. 3

    3 Silicle roundish, with one seed in each cell.   *Pepper-grass.* LEPID'IUM. 4

    3 Silicle double, with one seed in each lobe. S.   *Swine Cress.* SENEBIE'RA. 5

4 Silicle flattened. Leaves cauline or radical.   *Whitlow-grass.* DRABA. 6

4 Silicle turgid. Leaves cauline.   *Horse Radish.* ARMORA'CIA

4 Silicle turgid. Leaves all radical.   *r. Awlwort.* SUBULA'RIA.

    5 Silicle obovoid, *i. e.*, inversely egg-shaped, turgid. *False Flax.* CAMELI'NA.

    5 Sil. globose, turgid, membranous. Style long. *Bladder-pod.* VESICA'RIA.

    5 Silicle oblong, turgid, and somewhat terete.   *Cress.* NASTUR'TIUM.

6 Seeds arranged in two rows in each cell, not winged.   *Cress.* NASTUR'TIUM.

6 Seeds in two rows in each cell, wing margin.   *Tower-mustard.* TUR'RITIS.

6 Seeds arranged in a single row in each cell....7

    7 Sil. linear, flattish, each valve with 1 central vein. *Rock Cress.* AR'ABIS. *i*

    7 Silique lanceolate, flat, the valves veinless.   *Tooth-root.* DENTA'RIA. 8

    7 Siliques linear, veinless, terete. Flws. purple. *False Rocket.* IODAN'THUS.

    7 Siliques linear, veinless, flat. Fls. whitish. *Cuckoo-flower.* CARDAMI'NE. 9

8 Seeds ovate or oblong....9

8 Seeds globose....10

8 Seeds flat, with a broad, winged margin. S.   LEAVENWORTH'IA.

    9 Calyx ½-open. Lvs. runcinate, or finely dissect. *Hedge-mustard.* SISYM'BRIUM.

    9 Calyx closed. Leaves lyrate-pinnatifid.   *Winter Cress.* BARBA'REA. 10

    9 Calyx closed. Leaves lanceolate.   *False Wall-flower.* ERYS'IMUM. 11

10 Calyx spreading. Valves of the pod 1-3-veined.   *Mustard.* SINA'PIS. 12

10 Calyx mostly erect. Valves of the pod 1-veined. *Cabbage, &c.* BRAS'SICA.

    11 Pods short, 2-jointed, with 1 seed in each joint. *Sea-rocket.* CAKI'LE.

    11 Pods with several transverse joints and cells. *Radish.* RAPH'ANUS.

1. IBE'RIS. Candy-tuft.

The two outside petals larger than the two inside ones. Pods flattened, truncate, emarginate, the cells one-seeded.—Foreign, ornamental plants.

1 Flowers white. Plant about 1f. high....2-4

  1 I. umbella'ta. *Purple C.* Fls. purple, in umbels. Lvs. serrate, upper entire.

2 I. ama'ra. *Bitter C.* Corymbs lengthening into racemes. Lvs. slightly toothed.

3 I. pinna'ta. *Wing-leaved C.* Corymbs scarcely lengthening. Leaves pinnatifid,

4 I. saxat'ilis. *Rock C.* Corymbs not lengthening. Shrubby. Lvs. linear, entire,

2. LUNA'RIA. Satin-flower.

Sepals somewhat 2-lobed at base of the flower. Petals nearly entire.
Stamens without teeth. Silicle oval or lanceolate, flat. usually very large,
with a stalk. Seed-stalk adhering to the partition.—Foreign, ornamental
plants.

> **L. redivi'va.** *Perennial S.* Pods lanceolate, narrowed to each end. Lvs. sharp-
> toothed. ♃.
> **L. bien'nis.** *Biennial S.* Pods broad-oval, rounded at each end. Lvs. blunt-
> toothed. ⊚

3. CAPSEL'LA. Shepherd's-purse.

Calyx equal at base. Silicles triangular, wedge-shaped at base, notched
at top, compressed laterally, that is, contrary to the narrow partition.
Valves boat-shaped. Style short. Seeds 00, oblong, small.—A common
weed, with white flowers.

> **C. Bursa-Pasto'ris.** *Shepherd's-purse.* Found everywhere, in fields, pastures, and
> road-sides. Stem growing to a foot in height, hairy below, branching. Root
> leaves many (when the plant has room), half a foot long, deeply-lobed and
> toothed. Stem leaves much shorter, with two ear-shaped stem-clasping lobes
> at base. Flowers very small, in racemes which become very long, and are suc-
> ceeded by the little purse-shaped pods. *Apr.–Sept.* (See Fig. 331.)

4. LEPID'IUM. Pepper-grass.

Sepals ovate. Petals ovate, entire. Silicles roundish or oval, notched
at the end, flattened contrary to the very narrow partition. Cells 1-seed-
ed. Valves boat-shaped, dehiscent. Flowers white, racemed, numerous.

> 1 Stem leaves undivided. Flowers from *June* to *Sept.*....2
> 1 L sativum. Leaves all divided and lobed. Pods round. Gardens. *July.*
> 2 L Virginicum. *Tongue-grass.* Pods round, wingless. Stem leaves toothed.
> 3 L rudera'le. Pods roundish-oval, wingless. Petals 0. Stem leaves entire. W.
> 4 L campes'tre. Pods ovate, winged, rough-scaly. Leaves arrow-shaped. W.

5. SENEBIE'RA. Swine Cress.

Silicle 2-lobed, appearing double. Valve somewhat turgid and inde
hiscent. Cells each with 1 roundish and 3-cornered seed. Flowers
white, in short racemes which stand opposite to the leaves.

S. pinnatif'ida. A prostrate, weed-like plant, common at the South, in fields and on river-banks. Leaves divided in a pinnate manner, into oblong, toothed lobes. Flowers obscure, with scarcely any petals. Silicles flattened, notched at apex, wrinkled on the surface. *Feb.–July.*

6. DRA'BA. Whitlow-grass.

Calyx equal at base. Petals equal. Filaments without teeth. Silicle oval-oblong, entire, flattened parallel to the broad partition. Cells 2, many-seeded. Seeds not margined.—Low herbs, with small white or yellow flowers in racemes. (See Fig. 396.)

§ Petals 2-cleft, white. Leaves all radical....1
§ Petals entire or merely notched. Stems more or less leafy....a
   a Style present. Plants perennial....2, 3
   a Style none. Plants annual or biennial....b
      b Pedicels as long or longer than the pod....4, 5
      b Pedicels shorter than their pods....

1 **D. verna.** *Spring W.* Leaves oblong, hairy. Scape 1–5' high. ⊙ (See Fig. 396.)
2 **D. arabi'sans.** *Rock W.* Leaves minutely toothed. Silicle twisted, longer than the pedicel, oblong-lanceolate, 4–6'' long. Lake shores. (Figs. 155–159.)
3 **D. ramosis'sima.** *Bushy W.* Leaves with remote and slender teeth. Silicle as long as its pedicel, style half as long. Flowers white. Rocks.
4 **D. nemora'lis.** *Wood W.* Petals notched at end. Pod half as long as its pedicel. Seeds near 30. Flowers yellowish-white. N-W.
4 **D. brachycar'pa.** *Short-fruited W.* Petals entire. Pod as long as pedicel, 10–12-seeded. Leaves round-ovate. S. W. *March, April.*
5 **D. Carolinia'na.** Leaves round-ovate, entire. Pods linear, in a sort of corymb. Flowers white. Plant hispid, 1–3' high. E. S. *April–June.*
6 **D. cuneifo'lia.** *Wedge-leaved W.* Leaves wedge-oblong. Pods lance-oblong, 20–30-seeded, racemed. Plant 3–8' high. S-W.

7. AR'ABIS. Rock Cress. Sickle-pod.

Sepals erect. Petals clawed, entire. Silique linear, flattened, valves one-veined in the middle. Seeds in a single row in each cell. Flowers white. *April–June.*

§ Leaves all (or at least the radical) pinnatifid....1, 2
§ Leaves all undivided; toothed or entire, often clasping....a
a Siliques short (6–12'') and straight. Seeds not winged....3, 4
a Siliques longer (1–2'), straight or curved. Seeds not winged....5, 6
a Siliques long (3'), curved, hanging. Seeds winged....7, 8

1 **A.** Ludovicia'na. *Louisiana R.* All the leaves feather-cleft. Seeds oor-
dered. Plant slender, 6–10′ high. S. *Mar.*, *Apr.*

2 **A.** lyra'ta. *Lyre-leaved R.* Only the root leaves feather-cleft. Seeds not
bordered. Plant 6–12′ high. Pods 1¼–2′ long.

8 **A.** Thalia'na. *Mouse-ear R.* Stems erect. Leaves nearly entire. Petals twice
longer than the sepals. Pods erect. Plant downy. *May.*

4 **A.** denta'ta. *Toothed R.* Stems diffuse. Leaves sharply toothed. Petals scarce
longer than sepals. Pods spreading. Rough. M. W.

5 **A.** patens. *Patent R.* Downy. Pods spreading and curved upwards, beaked
with a distinct style. *w.* W. S.

6 **A.** hirsu'ta. *Hairy R.* Plant hairy. Siliques straight, erect. Style none.
Leaves arrow-shaped. Fls. g.

7 **A.** læviga'ta. *Smooth Sickle-pod.* Stem leaves arrow-shaped, clasping, narrow.
Pod spreading. Plant glabrous, 2f. high.

8 **A.** Canaden'sis. *True Sickle-pod.* Stem leaves pointed at both ends, sessile. Pod
curved, pendulous. Tall, downy.

8. DENTA'RIA. Tooth-root. Pepper-root.

Sepals converging or closed. Silique lanceolate, with flat, veinless
valves opening elastically. Seeds in a single row in each cell, ovate, not
bordered.—Plants ♃. Rhizoma prostrate, jointed. Stem leaves but 2 or
3. Flowers white or purplish.

¶ Stem leaves almost opposite or whorled... 1, 2, 8
¶ Stem leaves alternate or scattered. Root-stock moniliform....4, 5

1 D. diphyl'la. *Two-leaved P.* Stem leaves 2 only, leaflets 8, ovate, toothed.

2 D. lacinia'ta. *Cut-leaved P.* Stem leaves 3, leaflets 3–5, linear, cleft.

8 D. multif'ida. Stem leaves 2–3, numerously divided into linear leaflets. S.

4 D. max'ima. Leaflets 3, ovate, cut and cleft. Lvs. 3–7. Fls. purple. M.

5 D. heterophyl'la. Lflts. 3, nearly entire; of the rt. lvs. round-ovate, toothed.

9. CARDAMINE. Bitter Cress.

Calyx a little spreading. Silique linear, with flat, veinless valves which
are narrower than the partition. Stigma entire. Seeds not margined,
with a slender seed-stalk. Flowers white or purple.

* Leaves pinnate with many leaflets. *April–June*....1, 2
* Leaves simple or partly ternate. Mostly perennials....a
   a Style slender. In low, wet grounds....
   a Style none. In high mountains....

1 C. hirsu'ta. *Pennsylvanian C.* Smooth, about 1f. erect. Leaves 5–11-foliate, the
terminal lobe largest, 3-lobed. Stigma sessile. Wet. ⊚ *c.*

2 C. praten'sis. *Cuckoo-flower.* Stem simple, ascending. 1f. Leaves 7–15-foliate,
with stalked, roundish leaflets. Style present. Flowers large. Wet. ♃

8 C. rhomboid'ia. *Rhombic C.* Stems upright, bearing tubers at base. Pods linear-lanceolate. Leaves roundish and rhomboidal. *w.* or *p. c.*

4 C. rotundifo'lia. *Round-leaved C.* Stems decumbent, branched. Pods linear-subulate. Leaves roundish, lower 8-lobed. *w.* By streams.

5 C. billidifo'lia. *Daisy-leaved C.* Leaves smooth, roundish. Pods erect. Height 1-3'. N. H.

6 C. spatula'ta. *Spath-leaved C.* Leaves hairy, spatulate. Pods spreading. 6'. S.

10. BARBA'REA. Winter Cress.

Sepals erect, nearly equal at base. Silique columnar, 2–4-cornered. Valves concave or keel-shaped by means of a strong central vein. Seeds in a single row. Leaves lyrate-pinnatifid. Flowers yellow.

B. vulga'ris. *Winter Cress.* Common in old fields, also brook-sides. Whole plant glabrous. Stem 1-2f. high, branching above. Leaves lyrate with the terminal lobe roundish, upper leaves obovate, pinnatifid at base, crenate, or repand-dentate—all dark green, shining. Flowers showy, in racemes. Pods obscurely 4-cornered, slender, ¾' long, curved upwards. *May, June.* ♃

11. ERYS'IMUM. False Wall-flower.

Calyx closed. Silique linear, 4-sided. Stigma capitate. Seeds in a single row in each cell. Mostly ⊚. Flowers yellow.

1 E. cheiranthoi'des. Stem ascending. Fls. small. Pods spreading, 1' in length. M.
2 E. Arkansa'num. *Yellow Phlox.* Stem strictly erect. Flowers large (¾' broad). Pods 2-3' long. River bluffs. A fine plant. W.

12. SINA'PIS. Mustard.

Sepals spreading. Petals ovate, with straight claws. Silique nearly terete, valves 3-veined. Style short. Seeds in a single row, globular.— ① or ② with yellow flowers. (Figs. 893, 894.)

1 S. nigra. *Black M.* Upper leaves lance-linear, entire. Pod 4-cornered, smooth.
2 S. arven'sis. *Field M.* Leaves all repand-toothed. Pods torose, smooth.
8 S. alba. *White M.* Leaves all lyrate-pinnatifid. Pods bristly, shorter than beak.

ORDER XVI. VIOLA'CEÆ. Violets.

Herbs with simple (often cleft), alternate leaves with stipules; *flowers* irregular, spurred, with the sepals, petals, and stamens in 5's; *corolla* spurred at base; *anthers* united: 2 of the filaments appendaged;

style 1, with a one-sided stigma; *capsule* 1-celled, 3-valved; *seeds* many, with the embryo nearly as long as the albumen.

Analysis of the Genera.

Sepals unequal, with ear-shaped lobes at base.   VIOLA. 1
Sepals nearly equal, not appendaged at base.   *Green Violet.* So'LEA.

1. VI'OLA. Violet.

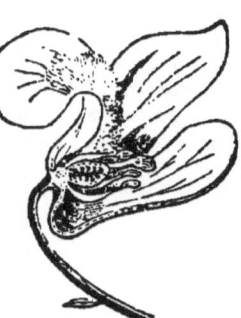

Sepals 5, prolonged at base into two auriculate lobes. Petals more or less unequal, the largest one spurred at base, the 2 opposite ones at the sides equal, the 2 upper ones all equal. Stamens cohering by their anthers, 2 of them spurred at base. Seeds attached to the valves of the capsule. —♃ Low herbs, caulescent or acaulescent. Peduncles angular, solitary, 1-flowered, nodding at the top.

Fig. 898. Violet No. 1: section.

* Acaulescent: leaves and flowers all radical....a
* Caulescent: stems leafy....d
    • Flowers blue....b
    a Flowers white....Nos. 2–4.
    a Flowers yellow....No. 1.
      b Petals beardless....5–7
      b Petals bearded....c
        : Leaves divided....8, 9   [otic 22.
        c Leaves undivided....10–12, and the Ex-
    d Pet. yellow. Stems leafy at the top only....13–15
    d Petals not yellow, or but partly yellow....e
      e Stipules entire. *Summer*....16
      e Stipules fringe-toothed. *May, June*....17–19
      e Stipules lyrate-pinnatifid, very large....20–21

Fig. 899. Ripe, open capsule of Violet.

1 V. rotundifo'lia. *Early Yellow Violet.* Lvs. round-ovate, cordate, smooth. Sepals blunt. *April.*
2 V. lanceola'ta. *Lance-leaved V.* Lvs. lanceolate, tapering to the base. Some bearded.
3 V. primulæfo'lia. *Primrose V.* Lvs. lance-ovate, abrupt at base. Fls. beardless.
4 V. blanda. *Sweet Wild V.* Leaves round-cordate. Fls. beardless, fragrant. *May.*
5 V. palus'tris. *Bog V.* Lvs. reniform-cordate. Spur very short. Stips. ovate. White Mts.
6 V. Selkir'kii. *Selkirk's V.* Lvs. round-cor. Spur near as long as petals, blunt. *May.*
7 V. peda'ta. *Foot-leaved V.* Lvs. pedate, 5–9-part., segm. narrow, entire. Root premorse.
8 V. delphinifo'lia. *Larkspur V.* Leaves in 7–9 linear, 3-cleft segments. W. *April.*
9 V. palma'ta. *Palm-leaved V.* Leaves hastate-lobed. cordate. *Variety* of No. 10.

10 V. cuculla'ta. *Hood-leaved V.* Leaves reniform-cordate, base lobes involute. *com.*
11 V. villo'sa. *Woolly V.* Leaves round-ovate, cordate, obtuse, flat, downy. **M. S.**
12 V. sagitta'ta. *Arrow-lvd. V.* Lvs. lance-oblong, some sagittate or cut-toothed at base.
13 V. hasta'ta. *Halberd-leaved V.* Smooth. Lvs. hastate. Stip. ovate, minute. **S.**
14 V. tripar'tita. *Three-cleft V.* Hairy. Lvs. deeply 3-parted. Stip. lanceolate. **S.**
15 V. pubes'cens. *Downy V.* Downy. Lvs. broad-cordate. Stip. ovate, large. *c.*
6 V. Canaden'sis. *Canada V.* Plant 1f. high. Leaves cordate, pointed, smooth.
17 V. stria'ta. *Cream-colored V.* Spur ¼ the length of the corolla. Stip. large, oblong.
18 V. Muhlenber'gii. *Muhlenberg's V.* Spur ¼ the length of corolla. Stip. lanceolate.
19 V. rostra'ta. *Long-spurred V.* Spur longer than corolla. Stipules lanceolate.
20 V. tric'olor. *Pansy. Heartsease.* Stipules as large as the leaves. Fls. three-colored.
21 V. grandiflo'ra. *Great-flowered V.* Stip. much smaller than the leaves. Purple. †
22 V. odora'ta. *Sweet English V.* Stolons creeping. Lvs. cordate. Fragrant. †

ORDER XVIII. HYPERICACEÆ.   **St. Johnsworts.**

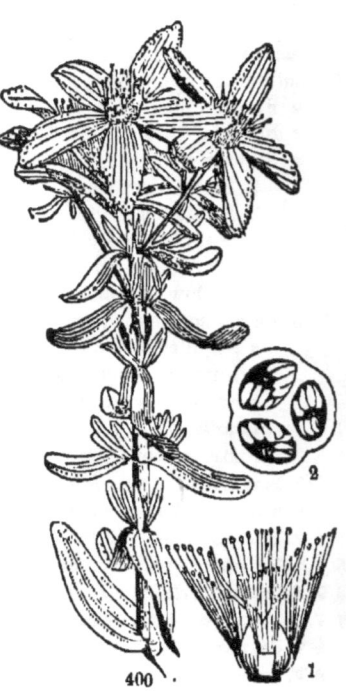

Herbs or *shrubs* with opposite, entire,
    dotted leaves, and no stipules;
flowers mostly yellow, in cymes;
sepals unequal, 4–5, dotted;
petals 4–5, twisted in the bud, dotted,
    and with the veins oblique;
stamens hypogynous, in 3 or more par-
    cels;
ovary superior; *style* 1;
fruit a capsule or berry, many-seeded.

Analysis of the Genera.

Petals and sepals 5....2
Petals and sepals 4. Flowers yellow.
          St. Peterswort. As'OYRUM.
  2 Fls. yellow. *St. Johnswort.* HYPER'ICUM. 1
  2 Flowers purplish.   ELODE'A.

Fig. 400. Hypericum perforatum (Common St.
Johnswort): stem, leaves, and flowers. *Fig.* 401.
The stamens in 3 sets surrounding the ovary with 3
styles. *Fig.* 402. Cross-section of the ovary.

HYPER'ICUM. St. Johnswort.

Sepals 5, connected at base, nearly equal, leaf-like. Petals 5, oblique. Stamens many (sometimes few and distinct), united into 3–5 parcels with no glands between them. Styles 3–5, either distinct or united at base. Capsule 1-celled, or 3–5-celled.—Herbs or shrubs, with branching stems, opposite, entire leaves, and yellow flowers. (Figs. 210, 211, 400–402.)

§ Stamens 25 to 100, more or less united into sets....a
§ Stamens 5 to 15, not at all united....g

  a Carpels (pistils) and styles 5 or more. Capsule 5-celled....Nos. 1, 2
  a Carpels 3. Capsule 3-celled (the partitions meeting)....b
  a Carpels 3. Capsule 1-celled (the partitions not quite meeting)....c
    b Shrubby. Petals not dotted. Leaves lanceolate or oblanceolate....3–5
    b Shrubby. Petals not dotted. Leaves linear....6, 7
    b Herbaceous. Petals sprinkled with black dots....8–10
  c Shrubs. Styles united into 1....d
  c Half-shrubby. Styles united into 1....e
  c Herbaceous. Styles distinct, at least at the top....f
    d Flowers solitary or in 3's, axillary. Stems 2-edged....11, 12
    d Flowers clustered in a compound, terminal cyme....13, 14
    e Flowers in a leafless, stalked cyme. Leaves obtuse....15, 16
    e Flowers in a leafy (few-leaved) cyme. Leaves acute....17, 18
    f Stem or branches 4-cornered or square....19, 20
    f Stem and branches terete, not angular....21, 22
  g Flowers in corymbous cymes....23, 24
  g Flowers racemed on the slender branches....25, 26

1 H. pyramida'tum. *Giant S.* Herb 3–4f., flowers 2' broad. Leaves lance-oblong.
2 H. Kalmia'num. *Kalm's S.* Shrub 1–2f., flowers 1' broad. Leaves lance-linear.
  3 H. Buckle'yi. *Buckley's S.* Leaves obovate. Flowers terminal, solitary. S.
  4 H. prolif'icum. *Prolific S.* Lvs. lance-oblong. Cymes compound. W.
  5 H. galeoi'des. *Bedstraw S.* Lvs. lance-linear. Clusters axillary. S.
6 H. rosmarinifo'lium. *Rosemary S.* Lvs. petioled, shorter than internodes. S.
7 H. fascicula'tum. *Clustered S.* Lvs. sessile, longer than the internodes. S.
  8 H. perfora'tum. *Punctured S.* Stem 2-edged. Lvs. small, light-dotted. c.
  9 H. corymbo'sum. *Corymbed S.* Stem terete. Lvs. large, black-dotted. c.
10 H. macula'tum. *Spotted S.* St. terete. All over black-dotted. Sty. long.
11 H. au'reum. *Golden S.* Lvs. thick, obtuse, sessile. Fls. large (1½'). Stam. 500! S.
12 H. ambig'uum. *Dubious S.* Lvs. thin, acute, sessile. Fls. 8'' broad. Pet. toothed. S.
  13 H. myrtifo'lium. *Myrtle S.* Branches terete. Lvs. clasping. Cyme leafy. S.
  14 H. cistifo'lium. *Rockrose S.* Branches 2-edged. Lvs. sessile. Cyme leafless. S.
15 H. nudiflo'rum. *Naked-flowered S.* Lvs. lance-ovate. Pod ovoid-conic. M. S.
16 H. sphærocar'pon. *Round-fruited S.* Lvs. linear-oblong. Pod globular. W.

8

17 H. adpres'sum. *Closed S.* Lvs. half-erect. Petals obovate, longer than sep.
18 H. dolabrifor'me. *Hatchet S.* Lvs. spreading. Pet. *dolabriform*, long as sep.
19 H. angulo'sum. *Angled S.* Lvs. ovate, acute. Style thrice longer than ovary.
20 H ellip'ticum. *Elliptic S.* Lvs. elliptic, obtuse. Style as long as ovary. N. M.
21 H. grave'olens. *Strong-scented S.* Smooth. Lvs. oblong-ovate, clasping. S.
22 H. pilo'sum. *Hairy S.* Hairy. Lvs. lance-ovate, appressed. S.
23 H. mu'ticum. *Dwarf S.* Lvs. ovate, clasping, 5-veined. Cymes leafy. *c.*
24 H. Canaden'se. *Canada S.* Lvs. linear, black-dotted. Cymes leafless. *c.*
25 H. Saro'thra. *Pine-weed S.* Lvs. awl-shaped, minute. Fls. sessile.
26 H. Drummon'dii. *Drummond's S.* Lvs. linear. Fls. stalked. W.

ORDER XIX. DROSERACEÆ. The Sundews.

Herbs growing in bogs, often covered with glands, with
leaves alternate, circinate (rolled from top to base) in the bud;
flowers regular, of 5 persistent *sepals* and 5 withering *petals;*
stamens 5, distinct, and a single, compound *ovary;*
styles 1–5, and *fruit* a 1–3-celled many-seeded capsule, and with
seeds having a small embryo at the base of the albumen.

Analysis of the Genera.

	Stamens 5.	DROS'ERA. 1
coiled (circinate) in the bud.	Stamens 10–15.	DIONÆ'A. 2
Leaves { not coiled in the bud.	Sterile stamens many.	PARNAS'SIA. 3

1. DROS'ERA. Sundew.

Sepals 5, united at base, persistent. Petals 5. Stamens 5. Styles 3–5,
each deeply 2-parted, so that there seems to be 6–10. Capsule 3–5-valved,
1-celled, many-seeded.—♃ Small aquatic herbs. Leaves (all radical in
the American species) clothed with long, reddish, gland-bearing hairs,
exuding a clear, sticky fluid. Flowers in a raceme on a slender scape,
which is at first coiled downward, but uncoils as the flowers open.

* Scape 4–6 times longer than the spreading leaves....1–3
* Scape 1–2 times longer than the ascending leaves....4–6
1 D. rotundifo'lia. *Round-leaved S.* Leaves round, on long hairy stalks. Fls white,
small (about 3'' broad). Scapes 5–8' high. *c.*
2 D. minor. *Lesser S.* Lvs. wedge-obovate, on smooth stalks. Scape 3–6'. *p. S.*
3 D. brevifo'lia. *Tiny S.* Lvs. spatulate, on short, hairy stalks. Scape 2–3'. *p. S.*

4 D longifolia. *Long-leaved S.* Lvs. spatulate, on long, smooth stalks. 4–7′. White. (Fig. 20, 21.)

5 D. linearis. *Linear-leaved S.* Lvs. linear, obtuse; stalks smooth. 3–6′. White.

6 D. filiformis. *Thread-leaved S.* Lvs. filiform, long. Scape 1f. Purple.

2. DIONÆ'A. Venus' Fly-trap.

Sepals 5, spreading. Petals 5, obovate, with pellucid veins. Stamens 10–15. Style 1. Stigmas 5, many-cleft. Capsule breaking irregularly in opening, 1-celled, many-seeded.—24 Glabrous herbs. Leaves all radical, sensitive, closing convulsively when touched. Scape umbelled.

D Muscip'ula. A very remarkable plant, in sandy bogs, at the South, sometimes cultivated. Leaves spreading, the petiole broadly winged, ending in a roundish blade which is fringed with spines, instantly closing upon insects which alight upon it. Scape 6–12′ high, bearing an umbel of 8–10 white, handsome flowers. *Apr., May.* †

Fig. 403. *Venus' Fly-trap. Fig.* 404. *Ovary and style.*
Fig. 405. *Section of ovary.*

3. PARNAS'SIA. Grass-of-Parnassus.

Sepals 5. Petals 5, inserted on the calyx (perigynous). Stamens also perigynous, in 2 rows, the outer row of numerous sterile filaments, united in 5 sets, the inner row of 5 perfect stamens. Stigmas 4, sessile. Capsule 4-celled. Seeds very numerous.—24 Elegant herbs, with radical leaves and 1-flowered scapes.

405 403 404

1 P. Carolinia'na. *Meadow G.* Sterile filaments, 3 in each set. Leaves about 7-veined, broadly oval or ovate, radical ones on long stalks, cauline few, near the ground, sessile, clasping. Scape about 1f. high, bearing one flower at top, which is about 1′ across. Petals marked with green veins. *July, Aug.*

2 P asarifolia. *Broad-leaved G.* Sterile filaments, 3 in each set. Lvs. reniform. S.

3 P. palus'tris. *Swamp G.* Sterile filaments, 9–15 in each set. Lvs. cordate. N. W.

Order XXI. CARYOPHYLLACEÆ. **Pinkworts.**

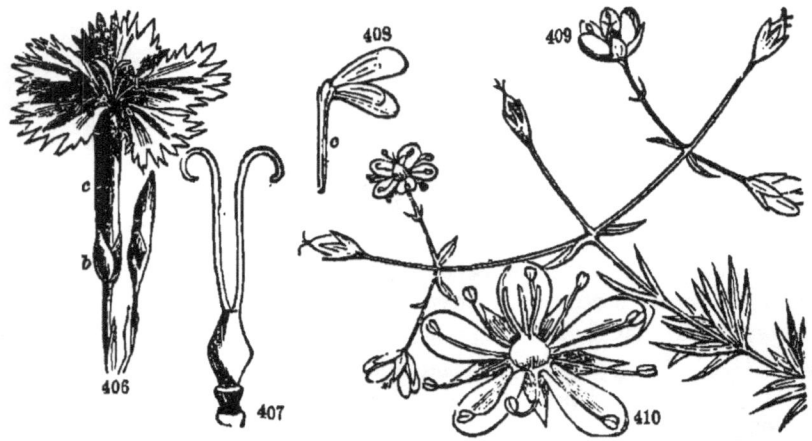

Fig. 406. Pink (Pheasant's-eye): *b*, the bracts; *c*, the tubular calyx. *Fig.* 407. The ovary with its 2 styles, *Fig.* 408. A petal of the Diurnal Lychnis, 2-cleft: *c*, the claw. *Fig.* 409. Arenaria stricta, showing the spreading cyme. *Fig.* 410. A flower enlarged, calyx not tubular.

Herbs with the stems swelling at the nodes; opposite, entire *leaves ;*
sepals 4 or 5, sometimes distinct and sometimes united into a tube;
petals 4 or 5 (sometimes 0), with or without claws, hypogynous;
stamens generally twice as many as the petals; *styles* 2–5;
fruit a 1-celled (rarely 2–5-celled) capsule with numerous seeds, and an
embryo coiled around fleshy albumen.

Analysis of the Genera.

§ Stipules dry, scale-like, between the leaves at base....6
§ Stipules none....2
    2 Sepals united into a tube.   Petals with long claws....3
    2 Sepals distinct or nearly so.   Petals sessile or none....4
        3 Calyx with 2 or more bractlets at base....a
        3 Calyx naked, *i. e.*, with no bractlets....b
    4 Pod 1-celled and with several seeds.   Petals generally present....5
    4 Pod 1-celled, with 1 seed.   Petals none, calyx green....g
    4 Pod completely 3-celled.   Petals none, calyx white....h
        5 Petals 2-parted or 2-lobed....c
        5 Petals undivided and entire....d

6 Styles or stigmas 3 or 5. Pod 1-celled, many-seeded .... e
6 Styles or stigmas 2 or united into 1. Pod 1-seeded .... f
a Styles 2. Petals variously notched or fringed.   *Pink.* DIAN'THUS. 1
  b Styles 2. Capsule 4-toothed when open.   *Soapwort.* SAPONA'RIA.
  b Styles 3. Capsule 6-toothed when open.   *Silene.* SILENE. 2
  b Styles 5. Calyx 5-toothed, teeth short or long. *Rose Campion.* LYCHNIS. 3
c Styles 5. Pod opening at top by 10 teeth.   *Mouse-ear.* CERASTIUM. 4
Styles 3. Pod splitting into 6 valves.   *Chickweed. Starwort.* STELLA'RIA.
  d Styles 3. Valves of the ripe pod 3, each 2-toothed. *Sandwort.* ARENA'RIA.
  d Styles 3. Valves of the pod 3, entire.   *Grove Sandwort.* ALSI'NE.
  d Styles 4 or 5, always as many as the sepals.   *Pearlwort.* SAGI'NA.
  d Styles 3 and 5. Plant fleshy. Disk 10-lobed. *Sea Sandwort.* HONKEN'YA.
e Styles 5. Leaves linear, whorled. Flowers white.   *Spurry.* SPER'GULA.
e Styles 3 and 5. Lvs. linear, opposite. Fls. red. *Sand Spurry.* SPERGULA'RIA.
e Styles 3 in all the fls. Leaves in 4's. Stipules ovate. *All-seed.* POLYCAR'PON.
e Styles 3 in all the fls. Leaves opposite. Stipules many-cleft.   STIPULIO'IDA.
  f Sepals green, distinct or nearly so ....   *Nailwort.* PARONYCH'IA.
  f Sepals white above, united into a tube below.   SYPHONYCH'IA.
g Styles 2. Utricle inclosed in the hardened calyx tube. *Knawell.* SCLERAN'THUS.
h Styles 3. Stamens 3 or 5. Herb flat on the ground. *Carpet-weed.* MOLLU'GO. 5

1. DIAN'THUS. Pink. Carnation.

Calyx tubular, cylindrical, striate, with 2 or more pairs of opposite, imbricated scales at base. Petals 5, with long claws, limb unequally notched. Stamens 10. Styles 2, with revolute stigmas. Capsule cylindrical, one-celled.

¶ Bracts as long as the calyx tube 2, 3, 4
¶ Bracts much shorter than the calyx 5, 6, 7

1 D. Arme'ria. *Wild Pink.* Bracts erect. Leaves linear. Flowers small, pink-red
  in cymes of about 3. Stem 18–24' high. In sandy fields. *July.* E.
2 D. barbatus. *Sweet William,* or *Bunch Pink.* Bracts erect. Leaves lanceolate,
  cymes large, many-flowered. Red or variegated with white. *May–July.* †
D. Chinen'sis. *China Pink.* Bracts spreading. Leaves lance-linear. Flowers
  solitary, red, large. Plant evergreen, not glaucous. †
4 D. caryophyl'lus. *Carnation Pink.* Bracts rounded. Petals crenate, beardless. Whole plant glaucous. Many beautiful varieties. †
5 D. pluma'rius. *Pheasant's-eye.* Bracts ovate. Petals fringe-toothed, bearded.
  Plant glaucous. Flowers solitary, white and purple. †
6. D. super'bus. *Superb P.* Bracts mucronate, ovate. Petals pinnatifid-fringed,
  bearded, cymes level-topped. White. †

2. SILE'NE. Catch-fly. Campion.

Calyx tubular, swelling, without scales at base, 5-toothed. Petals. 5, 2-cleft, the claws often crowned with a stiff scale. Stamens 10. Styles 3. Capsule partly 3-celled, opening by 6 teeth at top. (Fig. 116.)

* Petals many-cleft and fringed. Fls. white or roseate, large. Perennial....1-3
* Petals bifid or entire, not fringed....a
    a Calyx inflated and netted with veins. Perennial....4, 5
    a Calyx close upon the pod, not inflated....b
      b Flowers spicate, alternate. Annual....6, 7
      b Flowers not spicate....c
        c Petals white, closed in sunshine....8, 9
        c Petals red, purple, &c.,—(d) bifid....10, 11
                —(d) entire....12-15

1 S. stella'ta. *Whorled* C. Lvs. in 4's. Calyx inflated. Fls. white, many. *July.*
2 S. ova'ta. *Ovate* C. Leaves opposite. Calyx not inflated. Flowers white. S.
8 S. Baldwin'ii. *Baldwin's* C. Lvs. opposite, obovate. Fls. very large, roseate. S.
  4 S. infla'ta. *Bladder* C. Petals not crowned. Flowers few, white.
  5 S. nivea. *Snowy* C. Petals with a little crown. Flowers many, white.
6 S. quinquevul'nera. *Variegated* C. Woolly. Petals entire, red, white-edged. S.
7 S. noctur'na. *Spiked* C. Downy. Petals narrow, 2-parted, greenish-white.
  8 S. Antirrhi'na. *Snapdragon* C. Sticky in spots. Calyx egg-shaped.
  9 S. noctiflo'ra. *Night* C. Viscid-downy. Calyx cylindric. Petals 2-parted.
10 S. Virgin'ica. *Virginian* C. Leaves spatulate. Fls. large (2'), crimson. M. S.
11 S. rotundifo'lia. *Round-leaved* C. Leaves round, large. Fls. large, scarlet. W.
  12 S. Pennsylvan'ica. Perennial. Petals rose-purple, toothed at end.
  18 S. re'gia. *Royal* C. Perennial. Petals scarlet, entire, oblanceolate.
14 S. Arme'ria. *Garden* C. Annual. Stem sticky in spots. Flowers rose-p. †
15 S. acau'lis. *Stemless* C. Annual. Scape 2' high, 1-flowered. Mountains.

3. LYCH'NIS. Cockle. Rose Campion.

Calyx tubular, 5-toothed, without scales at base. Petals 5, clawed. Stamens 10. Styles 5. Capsule 1-celled, or 5-celled at the base, opening at the top by 5 or 10 teeth. Petals sometimes crowned.

  * Petals broad, entire. Plants very hairy....1, 2
  * Petals 2-cleft, crowned with 2 scales at top of claw....3, 4
  * Petals gashed or 4-cleft. Plants nearly smooth....5, 6
1 L. Githa'go. *Cockle.* Sepals longer than the crownless, purple petals.
2 L. Corona'ria. *Mullein Pink.* Sepals shorter than the stiff-crowned petals. †
  8 L. Chalcedon'ica. *Sweet William.* Fls. scarlet, in a crowded, compound cyme. †
  4 L. diu'rna. *Diurnal L.* Flowers light purple, in an open, loose cyme. †
  (See Fig. 406.)

5 L. corona'ta. *Chinese L.* Petals very broad, fringed with numerous teeth. †
6 L. Floscu'culi. *Ragged Robin.* Petals divided into 4 long teeth, crowned. †

4. CERAS'TIUM. Mouse-ear. Chickweed.

Sepals 5, ovate, acute. Petals 5, bifid or 2-cleft. Stamens 10, sometimes 5 or 4. Styles 5. Capsule cylindrical or roundish, opening at top by 10 tooth-like valves. Seeds numerous. Fls. white, in cymes. (Fig. 114.

Petals about as long as the calyx. Plants hairy. ..1, 2
Petals much longer than the calyx. Plants hairy or downy....3, 4, 5
1 C. vulga'tum. *Common M.* Lva. obovate. Sepals acute. Fls. at first crowded.
2 O. visco'sum. *Sticky M.* Hairs sticky. Leaves lance-ovate. Sepals obtuse.
3 C. arven'se. *Field M.* Lvs. linear. Ripe pods as long as the calyx. N. E.
4 O. oblongifolium. Leaves lance-obl. Pods longer than calyx. M.
3 O. nu'tans. *Nodding M.* Ripe pods curved, thrice longer than calyx. N W.

5. MOLLU'GO. Carpet-weed.

Sepals 5. Petals 0. Stamens 3–5, opposite to the sepals. Styles 3. Capsule 3-celled, 3-valved, many-seeded.—① Low or prostrate herbs, with the leaves appearing whorled.

M. verticilla'ta. Stems slender, jointed, much branched, lying flat on the ground. At each joint stands a whorl of wedge-shaped or spatulate leaves of unequal size, usually about 5 in number, and a few flowers, each solitary on its stalk, which is shorter than the petioles. Flowers small, sepals white inside. In dry places. *July–Sept.*

Order XXII. PORTULACACEÆ. The Purselanes.

Herbs with thick, entire leaves, no stipules, and regular flowers;
flowers with 2 sepals, 5 petals, open only in the sunshine;
stamens opposite to the petals when of the same number, often more;
pistils several, with their ovaries united, free, or half-free, forming in
fruit a pyxis (§ 178) or a capsule.

Analysis of the Genera.

¶ Sepals five. Petals none. Fruit a pyxis. Stamens ∞. *Sea Purselane.* SESU'VIUM.
¶ Sepals 2. Petals 5....a
  a Stamens 5, opposite the petals.   *Spring Beauty.* CLAYTO'NIA. 1
  a Stamens 8–30, on the torus. Pod 3-valved.   TALI'NUM.
  a Stamens 8–30, on the calyx. Pyxis opening by a lid.   PORTULA'CA. 2

1. CLAYTO'NIA. Spring Beauty.

Sepals 2, ovate. Petals 5, emarginate or obtuse. Stamens 5, inserted on the claws of the petals. Stigmas 3, on 1 long style. Capsule 3-valved, 2–5-seeded.—They are small, fleshy, ♃, early-flowering herbs, arising from a small tuber.

1 C. Carolinia'na. Leaves ovate-lanceolate. Sepals and petals obtuse.
2 C. Virgin'ica. Leaves linear or lance-linear. Sepals acute, petals obovate.

2. PORTULA'CA. Purselanes.

Sepals 2. Petals 5, equal. Stamens 8–20. Styles 3–6. Pyxis lid opening off near the middle.—Low and fleshy herbs.

1 P. olera'cea. *Common P.* Leaves thick, wedge-shaped. Stem fleshy, reddish, prostrate. Flowers sessile, small, yellow. A common weed. *Summer.*
2 P. grandiflo'ra. *Great P.* Leaves cylindric and fleshy. Stems ascending. Fls. large, red or scarlet. Cultivated. *June.*

ORDER XXIV. MALVACEÆ. The Mallows.

Herbs, shrubs, or *trees,* with alternate, stipulate, divided leaves, with the *flowers* showy, axillary, regular, often with an involucel at the base; 5 *sepals* valvate and the 5 *petals* convolute in the bud, hypogynous; *stamens* indefinite and monadelphous, the anthers splitting across; *carpels* several, united into a ring or forming a several-celled capsule; *seeds* with a curved embryo in a little albumen.

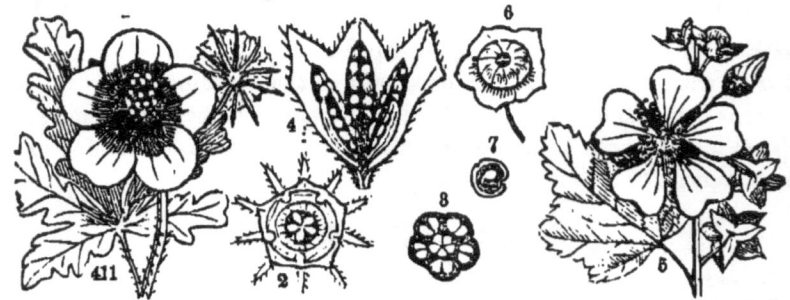

Fig. 411. Hibiscus Trionum (Flower-of-an-hour); 2, cross-section of the flower, showing the arrangement of its parts; 8, cross-section of the 5-celled capsule; 4, capsule open by its five valves; 5, Malva sylvestris; 6, its fruit, consisting of 10 carpels arranged in a circle; 7, section of one of the carpels, showing the curved embryo.

Analysis of the Genera.

§ Calyx naked, *i. e.*, having no involucel....b
§ Calyx furnished with an involucel as if a second calyx....2
  2 Pistils and carpels more than 5....a
  2 Pistils and carpels 5 only, each 1-seeded....c
  2 Pistils and carpels 5 or 3, each 3–∞-seeded....d
a Involucel of 6-9 bractlets. Carpels 1-seeded.   *Marsh M.* ALTHÆ'A. 1
a Involucel of 3 united bractlets. Carpels 1-seeded.   *Tree M.* LAVATE'RA.
a Involucel of 3 distinct bractlets. Carpels 1-seeded.   *Mallow.* MALVA. 2
a Involucel of 3 distinct bractlets. Carpels 2-seeded.   *Basket M.* MODI'OLA.
  b Flowers diœcious. Stigmas 10, linear.   *Napœa.* NAPÆ'A.
  b Flowers perfect. Carpels 5 or more, 1-seeded.   *Sida.* SIDA.
  b Flowers perfect. Carpels 5 or many, 3-9-seeded.   *Indian M.* ABU'TILON.
  c Stigmas 10. Carpels 5, baccate, united.   *Glue M.* MALVAVIS'CUS.
  c Stigmas 10. Carpels 5, dry, distinct.   *Peacock M.* PAVONIA.
  c Stigmas 5. Carpels 5, dry, united into a pod.   *Marsh M.* KOSTELETS'KYA.
  d Involucel of many bractlets. Calyx regular.   *Hibiscus.* HIBIS'CUS. 3
  d Involucel of many bractlets. Calyx split on one side.   *Okra.* ABELMOS'CHUS.
  d Involucel of 3 incisely-toothed bractlets.   *Cotton.* GOSSYP'IUM.

1. ALTHÆ'A. Hollyhock, &c.

Calyx surrounded at base by a 6-9-cleft involucel. Carpels ∞, 1-seeded, not opening, arranged circularly around the axis.

1 A. officina'lis. *Marsh M.* Lvs. downy, entire or 3-lobed. Fls. rose-col., stalked.
2 A. rosea. *Hollyhock.* Leaves rough-hairy, roundish, 5-7-lobed. Flowers sessile.
3 A. ficifo'lia. *Fig-leaved Hol.* Lvs. hairy, deeply 7-parted. Fls. orange-colored.

2. MAL'VA. Mallows.

Calyx 5-cleft, with a 3-leaved involucel at its base. Carpels and styles numerous. Fruit cheese-form, separating when ripe into many 1-seeded pieces, arranged circularly.

  * Flowers white or rose-colored....1, 2, 3
  * Flowers deep red or purple....4, 5, 6
1 M. rotundifo'lia. *Cheese M.* Stem prostrate. Lvs. round-cordate. Fls. small.
2 M. crispa. *Crisp M.* Stem erect, tall. Lvs. abundantly crisped and curled. †
3 M. moscha'ta. *Musk M.* Sts. ascend. Lvs. deeply 5-part. Fls. large, showy. †
  4 M. sylves'tris. *Wood M.* Lvs. roundish, lobed. Petals obcordate.
  5 M. triangula'ta. Lvs. triangular-ovate. Petals wedge-obovate. N.-W.
  6 M. papaver. *Poppy M.* Lvs. palmately parted. Petals erose. Stalks very long. S.-W.

3. HIBIS'CUS. Hibiscus.

Calyx 5-cleft, surrounded by a many-leaved involucel. Styles united, stigmas 5, distinct. Fruit a 5-celled, 5–many-seeded capsule. Flowers large, often nearly a foot broad.

§ Calyx, &c., hispid. Leaves palmately divided....1, 2
§ Calyx, &c., velvet-downy. Leaves undivided, angularly lobed....3, 4
§ Calyx, &c., glabrous, *i. e.*, smooth....a
    a Leaves deeply lobed or parted....5, 6
    a Leaves undivided or slightly lobed....7, 8
1 **H. aculea'tus.** *Prickly H.* Bractlets of involucel forked. Fls. sulph-yellow. S.
2 **H. Trio'num.** *Flower-of-an-hour.* Bractlets entire. Fls. chlorine-yellow. *c.* †
   3 **H. Moscheu'tos.** *Marsh H.* Lvs. ovate, toothed. Sepals abruptly pointed. *Rose-red. c.*
   4 **H. grandiflo'rus.** *Giant H.* Leaves cordate, lower 3-lobed. Sepals gradually pointed. *p-r.* S.
5 **H. milita'ris.** *Sword H.* Lvs. hastately 3-lobed. Flowers tubular-bell-shaped, flesh-color. W.
6 **H. cocci'nus.** *Scarlet H.* Lvs. palmately 5-parted. Cor. expanding, carmine-red. S
   7 **H. Carolinia'nus.** *Lost H.* Herb. Lvs. cordate. Fls. purple. *Very rare.* S.
   8 **H. Syri'acus.** *Tree H.* Tree 8-15f. high. Lvs. wedge-ovate. *w. p.* †

Order XXX. LINACEÆ. The Flaxworts.

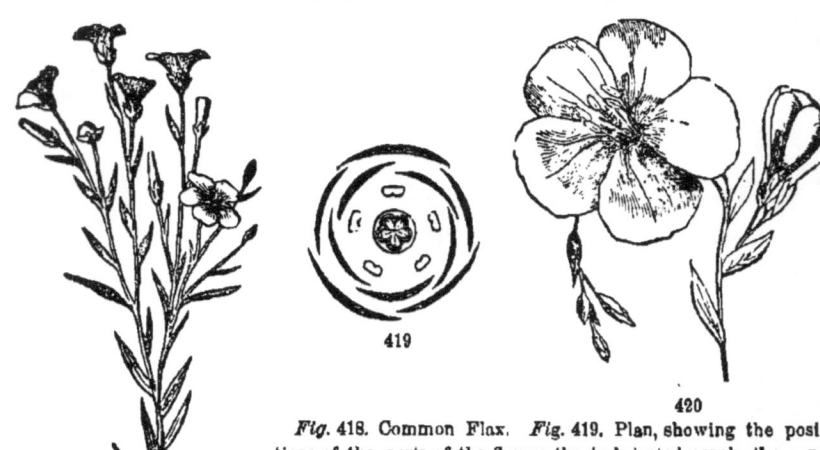

419

420

Fig. 418. Common Flax. *Fig.* 419. Plan, showing the positions of the parts of the flower, the imbricated sepals, the contorted sepals, the 5 stamens, and the 5 carpels. *Fig.* 420. Crimson Flax.

418

Herbs with entire, simple leaves and no stipules; with
flowers regular, symmetrical, perfect, and 5-parted;
calyx imbricate, and *corolla* convolute in the bud;
stamens and *styles* each 5; *capsule* with 5 double-cells, 10-seeded.
Our only genus is

LI'NUM. Flax.

The character is sufficiently indicated in the Order. The long, tough
fibres of the bark constitute the *linen* of commerce.

§ Flowers blue or red, large (1' broad),....Nos. 1-3
§ Flowers yellow. Leaves linear. Sepals ciliate....4, 5
§ Flowers yellow, Leaves lanceolate. Sepals entire....6-8

1 L. usitatis'simum. *Common F.* Flowers blue, in a sort of corymb. Leaves lance-linear, acute. The seed yields *linseed oil.* Fields.

2 L. peren'ne. *Perennial F.* Flowers blue, axillary and terminal. Leaves linear, acute, scattered. Gardens.

3 L. grandiflo'rum. *Crimson F.* Flowers crimson, axillary. Leaves lance-elliptic, acute, sessile. Gardens.

4 L. rig'idum. *Rigid F.* Sepals longer than the globular pod. Styles united at base.

5 L. simplex. *Simple F.* Sepals shorter than ovate pod. Styles distinct. S.-W.

6 L. virginia'num. Stems and branches erect. Flowers 6'' broad. *c.*

7 L. diffu'sum. Stems, branches, leaves diffuse. Flowers 2'' broad. W.

8 L. trig'ynum. *Three-styled F.* Flowers large (1') with 3 styles. †

ORDER XXXI. GERANIA'CEÆ. Gerania.

Herbs or shrubby plants with the lower leaves opposite; with the
flowers regular or irregular, terminal or opposite the leaves; with the
sepals 5, persistent, and *petals* 5, clawed, twisted in the bud; the
stamens 10, monadelphous, and *pistils* 5, united; the carpels in
fruit separating and bending upwards on the elastic style, each with one
seed. Albumen 0.

Analysis of the Genera.

Corolla { regular. { Stamens 10, all of them perfect...............GERA'NIUM. 1
{ Stamens 5 perfect, 5 imperfect..................ERO'DIUM.
{ irregular. Stamens 7 perfect, 3 imperfect............PELARGO'NIUM.

GERA'NIUM.

Sepals and petals 5, regular. Stamens 10, all perfect. Fruit beaked, at last separating into 5, long-styled, 1-seeded carpels. Styles smooth inside, finally curling from the base upward, but still adhering at top to the axis.—Herbs with forked stems, much divided leaves. Flowers mostly purple.

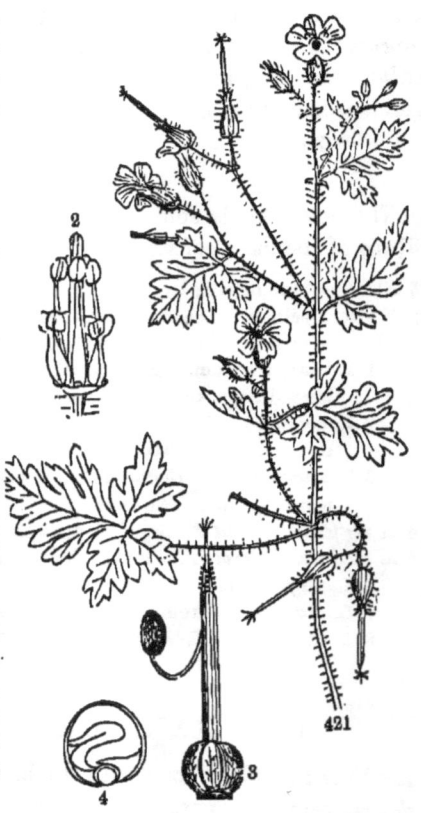

Petals entire, twice as long as the
awned sepals....1, 2
Petals notched or 2-lobed, short.
Leaves palmately 5-7-lobed. Pods
hairy ①....3, 4
1 G. macula'tum. *Spotted G.* Erect. Lvs. palmately 3-5-parted. Flowers large (1′ broad), showy. Sepals mucronate. *Spring. c.*
2 G. Robertia'num. *Herb Robert.* Diffuse, weak. Lvs. primately 3-parted to the base. Flowers small (7″ broad). Sepals mucronate. *June.*
3 G. pusil'lum. *Dwarf G.* Diffuse. Sepals veinless. Leaves parted into 5-7 linear lobes, lobes 3-cleft. Fields and hills. *July.*
4 G. Carolinia'num. Stems diffuse. Sepals with an awn. Lvs. parted into 5 wedge-oblong, many-cleft lobes. Fields. *July.*

Fig. 421. Herb Robert, leaves, flowers, and fruit; 3, fruit enlarged, showing one carpel on its elastic style; 4, cross-section of a seed, showing the large embryo filling the whole space; 2, the 10 stamens.

Observation.—The pupil will perceive by the table above, that the parlor "geraniums" belong to the genus *Pelargo'nium.*

Order XXXII. OXALIDACEÆ. Wood Sorrels.

Low *herbs* with a sour juice, and alternate, compound leaves; with *flowers* regular and symmetrical, 5-sepaled and 5-petaled;

stamens 10, monadelphous, hypogynous, the alternate ones longest; *carpels* 5, united and forming in fruit a 5-celled pod; *seeds* albuminous.

OX'ALIS. **Wood Sorrel.**

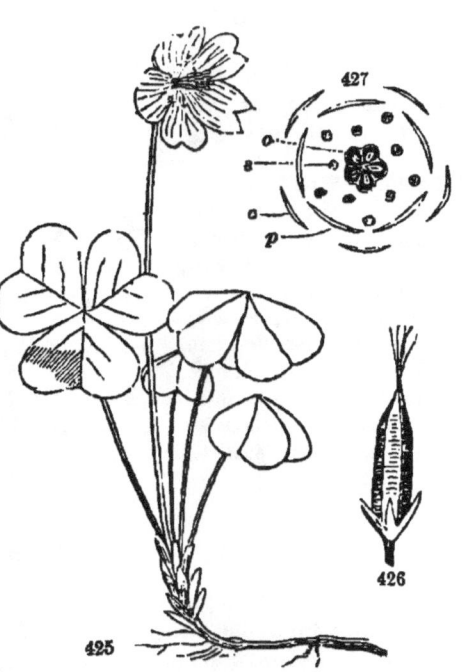

Sepals 5, distinct or united at base, persistent. Petals much longer than the sepals. Stamens united at the base. Styles 5. Capsule roundish or pod-shaped, cells several-seed-ed. Herbs mostly ♃, with trifoliate leaves.

1 O. Acetosel'la. *Wood Sorrel.* Fls. white, with purple veins. Plant acaulescent, arising from a creeping root-stock. *c.* N. *Ju.*

2 O. viola'cea. *Violet W.* Flowers violet-purple. Plant acaules-cent, arising from a scaly bulb. Scape with an umbel. *May.*

3 O. stricta. *Yellow W.* Flowers yellow. Plant with leafy stems, weak, branched. Flowers um-belled. Grows everywhere.

Fig. 425. Oxalis Acetosella. In the plan of the flower, *o*, the 5 carpels in the centre: *s*, the 10 stamens in two rows; *p*, the 5 petals; *c*, the 5 sepals. *Fig.* 426. The ripe pod.

ORDER XXXIV. BALSAMINACEÆ. **The Jewel-weeds.**

Herbs annual, with a fleshy stem, watery juice, and simple leaves; *flowers* very irregular and unsymmetrical; *calyx* spurred; *stamens* 5, on the torus; *pod* bursting by 5 elastic valves.

IMPA'TIENS. **Touch-me-not.**

Sepals colored, apparently but 4 (the 2 upper being united), the lowest (*y*) enlarged into a sac tipped with a bent spur. Petals 4, united into 2 double ones (*p, p*). Stamens 5 short, the anthers united over the pistil.

Fruit a pod of 5 strong elastic valves which break and coil at the slightest touch when ripe, scattering the seeds. Stem tender, thickened at the nodes. Leaves alternate.

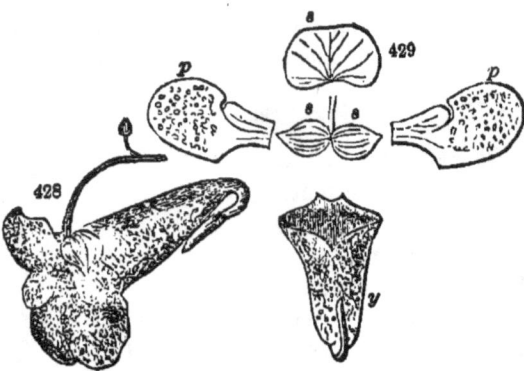

1 I. pal′lida. *Pale Jewel-weed.* Lvs. oblong-ovate. Fls. pale yellow, sparingly dotted, with a very short, recurved spur.

2 I. fulva. *Tawny Jewel-weed.* Leaves rhombic-ovate. Flowers deep orange, thickly spotted, with a long close-reflexed spur.

Fig. 428. Flower of the Pale Jewel-weed. *Fig.* 429. Its parts displayed: *s, s, s, y,* the four sepals, the latter spurred; *p, p,* the 2 petals, each double.

3 I. Balsami′na. *Balsamine.* Leaves lanceolate. Flowers very large and showy, white, crimson, scarlet, flesh-colored, &c. †

Order XL. ACERACEÆ. The Maples.

Trees or *shrubs* with opposite, usually simple palmate-veined leaves; the *flowers* often imperfect, with the 5 *sepals* imbricated in the bud, and the *petals* 5, hypogynous, sometimes 0 ; the *stamens* mostly 8, and the *fruit* a *double samara,* with two opposite wings, 2-seeded.

Analysis of the Genera.

Leaves simple, palmate-veined. Very common.   *Maple.* Acer. 1
Leaves compound, odd-pinnate. Leaflets 3-5, toothed.   *Box-Elder.* Negundo.

1. ACER. Maple.

Calyx of 5 united sepals, 5-lobed. Petals 5 or 0. Styles 2. Stamens 6-8. Leaves simple, palmate-lobed. Flowers mostly polygamous.

     § Pedicels short, in side clusters, flowering *before* the leaves. Trees....1, 2
     § Pedicels long, slender, drooping, flowering *with* the lvs. Large trees....3, 4
     § Pedicels in racemes, flowering *after* the leaves....5-7
1 A. dasycar′pum. *White M.* Leaves deeply lobed, square at base, silver-white beneath. Ovaries downy. Fruit very large, Petals 0. Tree 50f.

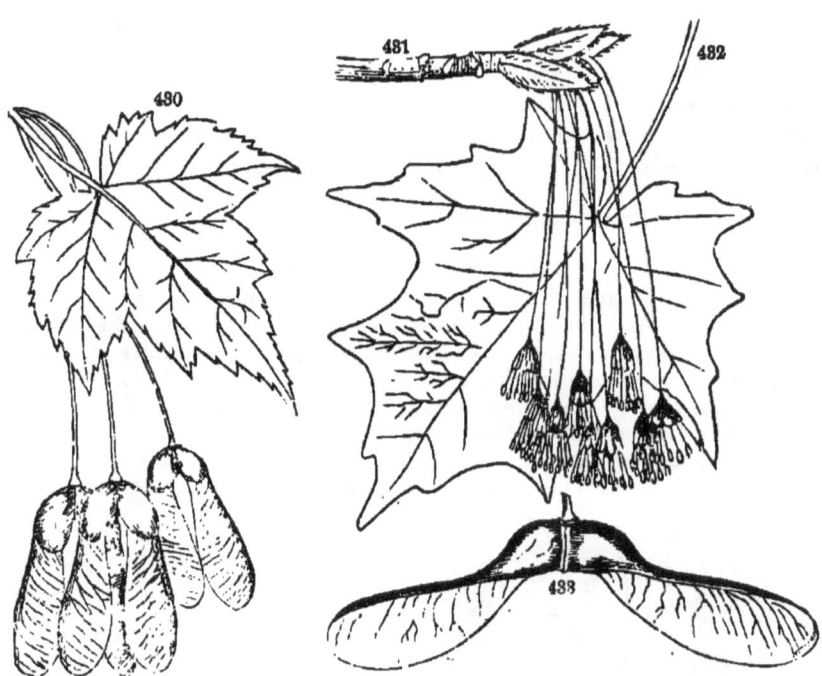

Fig. 430. Red Maple (*Acer rubrum*), a leaf and several samara. *Fig.* 431. Sugar Maple (*Acer saccharinum*), leaf, flowers, and fruit.

2 **A. rubrum.** *Red M. Swamp M.* Leaves lobed, cordate at base, paler beneath. Petals linear-oblong. Ovaries and fruit smooth. 40 to 100f. Flowers red.

3 **A. acchari'num.** *Rock M. Sugar M.* Leaves cordate, 5-lobed, with deep, rounded openings between. Bark light gray. *g-y.*

4 **A. ni'grum.** *Black M. Sugar-tree.* Leaves cordate, with the sinus closed, roundish, with 3 broad, shallow lobes. Bark dark gray. *y*

5 **A. spicatum.** *Mountain-Bush M.* Racemes erect, thyrse-like. Shrub 10–15f. high, in clumps. Bark gray. Leaves 3–5-lobed. *g.*

6 **A. Pennsylvan'icum.** *Striped M. Whistle-wood.* Racemes drooping. Tree small, with striped bark (green and black). Leaves 3-lobed. *g.*

7 **A. Pseudo-Plat'anus.** *Sycamore M.* Racemes long, drooping. A large tree, in parks. Leaves 5-lobed, broad, rounded. Flowers green.

Order XLI. SAPINDACEÆ. Indian Soapworts.

Plants of various habit, mostly with unsymmetrical flowers;
sepals and *petals* both imbricated in the bud;
stamens 5 to 10, inserted on a thick disk under the ovary;
fruit usually colored and showy, lobed, 1 or few-seeded.

The Order includes the following three Tribes.

Analysis of the Genera.

§ 1. The Buckeye Tribe. Leaves opposite, carpels 2-ovaled....a
   a Petals unequal. Stamens 7. Leaves digitate.   *Buckeye.* Æs'oulus. 1
§ 2. The Soapberry Tribe. Leaves alternate. Carpels 1-ovuled....b
   b Trees, with pinnate-leaves and fruit with soapy pulp, covering a
      large seed. Stamens 8-10. South.   *Soapwort.* Sapin'dus.
   b Herbs climbing with tendrils. Leaves biternate. Fruit a large,
      inflated, 3-carpeled pod.   *Balloon-vine.* Cardiosper'mum.
§ 3. The Bladder-nut Tribe. Leaves opposite, pinnate.   Staphyle'a. 3

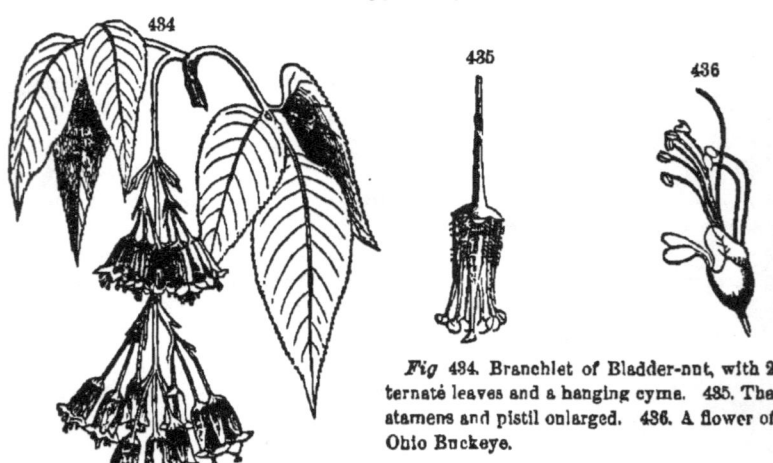

Fig 434. Branchlet of Bladder-nut, with 2 ternaté leaves and a hanging cyma. 485. The atamens and pistil onlarged. 436. A flower of Ohio Buckeye.

1. ÆS'CULUS. Buckeye.

Calyx 5-toothed. Corolla of 4 or 5 unequal petals. Stamens 7, distinct, unequal. Style filiform. Ovary 3-celled, with 2 ovules in each cell, bu only 1 of the 6 ovules grows, becoming a large seed. Flowers in terminal panicles.

§ Fruit covered with prickles. Petals 4 or 5, spreading....1, 2
§ Fruit smooth. Petals 4, erect, 2 of them clawed....3-5

   1 Æ. Hippocasta'neum. *Horse Chestnut.* Leaves of 7, obovate leaflets. Petals 5.
     Fruit prickly. Panicles large, handsome. †

   2 Æ. glabra. *Ohio Buckeye.* Leaflets 5, oval or oblong. Petals 4. Tree ill-
     scented. Flowers yellowish. Seed mahogany-color. W.

3 Æ. fla'va. *Big Buckeye.* A large tree, with pale-yellow flowers. Leaflets 5-7
     Petals very unequal, longer than stamens. W.

4 Æ. Pa'via. *Red-flowered B.* Shrub 3-10f. Fls. large, red, in thyrse-like racemes.
     Very handsome. S. †

5 Æ. parviflo'ra. *White B.* Shrub 2-3f. Petals 4, somewhat alike, spreading,
     thrice shorter than the stamens. S.

2. STAPHYLE'A. Bladder-nut.

Flowers perfect. Sepals 5, colored like the 5 petals. Stamens 5.
Styles 3. Capsules 2 or 3, with thin, inflated walls.—Shrubs.

1 S. trifo'lia. *Ternate B.* A handsome shrub, 6-8f. high. Leaves ternate, leaflets
   ovate. Racemes pendulous. Petals
   ciliate below. Fruit very large, 3-
   celled, inflated like a bladder.

ORDER XLV. POLYGALA-CEÆ. The Milkworts.

Plants without stipules, bearing very
   irregular *flowers;*
stamens 4-8, diadelphous;
anthers opening at the top, 1-celled;
fruit a flattened, 2-celled, 2-seeded
   capsule, free from the calyx.

Fig. 437. Polygala polygama: *a,* the radical
flowers; 8, P. paucifolia; *f,* the crest on the
lower petal; 9, the stamens in 2 sets, and the
style seen beneath the hooded lower petal.

Fig. 440. The ovary and the style: 1, seed of
P. sanguinea, with its 2-lobed caruncle; 2, seed
of P. Nuttallii.

POLYG'ALA. Milkwort.

Sepals 5, persistent, 2 of them
(wings) wing-shaped and colored.

Petals 3, the lower one boat-shaped, and often tipped with a crest. Stamens united by the filaments into a split sheath, or into 2 sets, cohering more or less with the claws of the petals. Fruit a small 2-celled, 2-seeded capsule, flattened on the sides and notched on the top. Seeds with an appendage at one end.—Low, bitter herbs (sometimes shrubs), with simple entire leaves, sometimes bearing underground flowers. (*Fig.* 437, *a.*)

```
* Leaves all alternate and scattered....a
* Leaves whorled, at least the lower ones....e
    a Flowers purple, or reddish, or white....b
    a Flowers yellow or yellowish green....d
        b Flowers solitary or in racemes, purple....Nos. 1-3
        b Flowers in spikes which are oblong or slender....c
            c Leaves lanceolate, large, pointed at each end....4
            = Leaves linear, 1 to 2'' wide....5-7
            c Leaves awl-shaped or bristle-shaped....8-10
        d Spikes solitary, large, thick. Biennial....11, 12
        d Spikes numerous, corymbous, small. Biennial....13, 14
            e Spikes acute, slender....15, 16
            e Spikes obtuse, thick....17, 18
```

1 P. paucifo'lia *Showy M.* Fls. 2 or 3, large (root fls. small). Lvs. ovate. (*Fig.* 438.)
2 P. grandiflo'ra. Fls. racemed, crestless. Lvs. lance-ovate. S.
8 P. polyg'ama. Flowers racemed, crested. Lvs. linear-oblong. (*Fig.* 437.)
   4 P. Sen'ega. *Seneca Snake-root.* Fls. white, in slender spikes. Stem 1f. high.
5 P. sanguin'ea. *Bloody M.* Spikes oblong, obtuse, dense. Wings sessile.
6 P. fastigia'ta. *Roofed M.* Spikes roundish, loose-flowered. Wings clawed.
7 P. Nuttal'lii. *Nuttall's M.* Spikes roundish, acute, dense. Wings elliptic.
   8 P. incarna'ta. *Flesh-colored M.* Lvs. few, subulate. Pet. much longer than calyx.
   9 P. seta'cea. *Naked M.* Leaves very minute. Petals longer than calyx. S.
   10 P. Chapman'ii. *Chapman's M.* Lvs. subulate. Calyx long as petals. S.
11 P. lu'tea. *Yellow M.* Tall (8-12'), with orange-yellow flowers. M. S.
12 P. na'na. *Dwarf M.* Low (3-5'), with greenish-yellow flowers. S.
   13 P. cymo'sa. *Cyme-flowered M.* Lvs. mostly cauline. Seed not bracted. S.
   14 P. ramo'sa. *Branching M.* Lvs. mostly radical. Seed bracted. S.
15 P. verticilla'ta. *Whorled M.* Lvs. linear. Wings roundish. Fls. greenish. W.
16 P. Boykin'ii. *Boykin's M.* Lvs. lance-obovate. Wings round-obovate. S.
   17 P. crucia'ta. *Cross M.* Spikes obtuse, thick, sessile. Wings pointed.
   18 P. brevifo'lia. *Short leaved M.* Spikes obtuse, loose, stalked. Wings acute.

ORDER XLVI. LEGUMINOSÆ. Leguminous Plants.

Plants with alternate, mostly compound stipulate leaves, with 4-5 *sepals;* 5 *petals,* more or less papilionaceous, sometimes regular;

about 10 *stamens*, monadelphous, diadelphous, or distinct; a single, simple *pistil*, producing a legume in fruit, and with no *albumen* in the seeds.

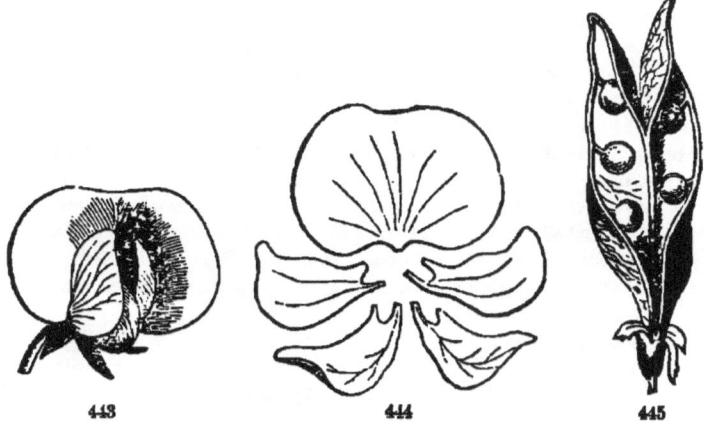

Fig. 443. Flower of the Pea. *Fig.* 444. Its petals displayed; *v*, the banner; *a, a*, the wings; *c, c*, the 2 keel petals. *Fig.* 445. A legume (pea-pod).

Analysis of the Genera.

§ Flowers papilionaceous (§ 89). Upper petal (banner) covering the rest in bud...2
§ Flowers nearly regular, or upper petal covered by the rest in bud....**t**
§ Flowers regular, in dense heads. Petals valvate in bud. Leaves bipennate....**u**
  2 Stamens 10, all distinct....**s**
  2 Stamens 10, all or 9 united....3
    3 Leaves cirrhous (*Fig.* 96), the rachis ending with a tendril. ...**r**
    3 Leaves not cirrhous....4
      4 Pod a loment (§ 180), *i. e.*, jointed between the seeds ...6
      4 Pod a legume, 1, 2, or ∞ seeded, not in joints....5
      5 Erect (or if prostrate, with palmately 3-foliate leaves)....7
      5 Trailing or twining vines, leaves pinnately compound....9
        6 Flowers yellow....**q**
        6 Flowers cyanic (not yellow)....**p**
       7 Leaves simple, with yellow flowers....**o**
       7 Leaves palmately 5-15-foliate (rarely simple) . .**n**
       7 Leaves palmately 3-foliate....**m**
       7 Leaves pinnately 3-foliate....**k**
       7 Leaves pinnate with no odd leaflet, 15-25 pairs. ..**h**
       7 Leaves pinnate with an odd leaflet....8

8 Leaflets dotted with dark glands....g
8 Leaflets not dotted. Herbs....f
8 Leaflets not dotted. Shrubs or trees....ᵤ
  9 Leaves pinnately 5-15-foliate....d
  9 Leaves pinnately 3-(rarely 1-) foliate. Flowers yellow....o
  9 Leaves pinnately 3-foliate. Flowers cyanic....10
    10 Calyx 4-toothed or entire....b
    10 Calyx 5-toothed or 5-cleft....a

ᵃ Keel with the stamens and style spirally coiled.   *Bean.* PHASE'OLUS. 1
a Keel obtuse, on short claws. Fls. very large, blue. S. *Blue Banner.* CENTROSE'MA.
a Keel acute, on long claws. Fls. very large, roseate. *Butterfly Pea.* CLITO'RIA.
  b Calyx 4-cleft, supported by 2 bractlets. Fls. purple. *Milk-vine.* GALAC'TIA.
  b Calyx 4-toothed, with 2 bractlets. Fls. purple. Sds. flattened. DOL'ICHOS.
  b Calyx 4-toothed, without bractlets. Fls. pale p. *Hog-Peanut.* AMPHICARPÆ'A.
  b Calyx entire. Flowers and seeds scarlet. S.   *Red Bean.* ERYTHRI'NA.
    c Legumes 5-seeded. S.   VIG'NA.
    c Legumes 1-2-seeded. S.   RHYNCO'SIA.
d Herbs. Keel (straight in Galactia, 2) spirally twisted.   *Pea-vine.* APIOS. 2
d Shrubs. Keel curved. Fls. blue, in hanging racemes. †   WISTA'RIA.
  e Flowers white or red, in racemes.   *Locust.* ROBIN'IA. 3
  e Flowers yellow, few in a cluster. Pods inflated. *Bladder Senna.* COLU'TEA.
f Pod 2-celled lengthwise, turgid.   *Milk Vetch.* ASTRAG'ALUS.
f Pod half-2-celled lengthwise.   *Bastard Vetch.* PHACA.
f Pod 1-celled. Style hairy outer side.   *Goat's Rue.* TEPHRO'SIA.
f Pod 1-celled. Style not hairy at all. S.   *Indigo.* INDIGO'FERA.
  g Shrubs. Fls. spicate, only 1 petal (the banner). W.S. *Lead Plant.* AMOR'PHA.
  g Herbs. Flowers with 10 stamens, bluish, spicate. W.   DA'LEA.
  g Herbs. Flowers with 5 stamens, white or red, capitate. W. PETALOSTE'MON.
h Pod 1-2-seeded, valves double. Tall, with yellow flowers. S. GLOTID'IUM.
h Pod many-seeded, very long. Tall, with yellowish flowers. S. SESBA'NIA.
  k Pod few-seeded. Flowers scarlet in   ERYTHRI'NA.
  k Pod few-seeded. Flowers white or yellow. *Melilot.* MELILO'TUS. 4
  k Pod 1-seeded. Flowers yellow. Leaves resinous-dotted in RHYNCOSIA.
  k Pod 1-seeded. Flowers cyanic. Leaves dark-dotted. PSORA'LEA.
  k Pod 1-seeded. Flowers cyanic. Leaves not dotted. *Melilot.* MELILO'TUS. 4
m Herbs with curved or spiral pods.   *Medic.* MEDICA'GO.
m Herbs with small 1-4-seeded pods not coiled. *Clover.* TRIFO'LIUM. 5
m Tree with yellow flowers in hanging racemes. † *Golden Chain.* LABUR'NUM.
  n Stamens all united. Calyx 2-lipped.   *Lupine.* LUPI'NUS. 6
  n Stamens all but 1 united. Calyx bill-shaped. PSORA'LEA.
o Shrubby. Keel oblong, straight.   *Scotch Broom.* GENIS'TA.
o Herbs. Keel curved, acuminate.   *Rattle Pod.* CROTALA'RIA.
  p Leaves pinnate, 5-21-foliate. Umbels stalked. CORONIL'LA.
  p Leaves pinnate, 5-21-foliate. Racemes stalked. Vt. HEDYS'ARUM.

p Lvs. pinn'ly 3-fol., stipellate. Pod 3-7-jointed. *Tick Trefoil.* DESMO'DIUM.
p Lvs. pinn'ly 3-fol. Stipels none. Pod 1-jointed. *Bush Trefoil.* LESPEDE'ZA. 7
q Leaves palmately 4-foliate. Stamens *all* united. ZOR'NIA.
q Leaves pinnate, 7–49-foliate. Stamens 9 united. ÆSCHYNOM'ENE.
Leaves pinnately 8-foliate. Pod slender at base. STYLOSAN'THES.
Leaves pinnately 4-foliate. Pod gibbous at base. *Peanut.* AR'ACHIS.
r Leaflets serrate. Pods 2-seeded. *Chick Pea.* CICER.
r Leaflets entire. Style grooved outside, hairy inside. *Pea.* PISUM.
r Leaflets entire. Style flattened, hairy most inside. *Sweet Pea.* LATH'YRUS.
r Leaflets entire. Style filiform, hairy most outside. *Vetch.* VIO'IA.
s Pod legume flat and thin, short-stiped. Lvs. pinnate. Tree. S.W. CLADAS'TRIS.
e Pod inflated, stipitate (stalked at base). Lvs. 1–3-foliate. BAPTIS'IA. 9
t Fls. perfect, purple, papilionaceous. Tree. Lvs. simple. *Judas-tree.* CEROIS.
t Fls. perfect, yellow. Lvs. equally pinnate. *Senna.* CASSIA. 10
t Fls. imperfect, green. Sta. 5. Trees thorny. *Honey Locust.* GLEDITS'CHIA.
t Fls. imp., greenish. St. 10. Trees unarmed. Ky. *Coffee-tree.* GYMNOC'LADUS.
u Pods flat, jointed between the seeds. Shrubby. *Sensitive Plant.* MIMO'SA.
u Pods prickly, 4-sided, 4-valved. *Sensitive Brier.* SCHRAN'KIA.
u Pods smooth, turgid, filled with pulp. Tree. S. *Sponge-tree.* VACHEL'LIA.
u Pods smooth, flat, dry. Petals distinct. Stam. 5–10. Herbs. DESMAN'THUS.
u Pods smooth, flat, dry. Petals united. Stam. 8–200. S. *Julibrassin.* ACA'OIA.

1. PHASE'OLUS. Bean, &c.

Calyx 5-toothed or cleft, the 2 upper teeth half united. Keel including the stamens and style, and with them spirally coiled or twisted. Legume straight or curved, many-seeded. Seeds oblong, kidney-shaped.—Herbs twining or trailing. Leaves pinnately trifoliate, stipellate. *June–Oct.*

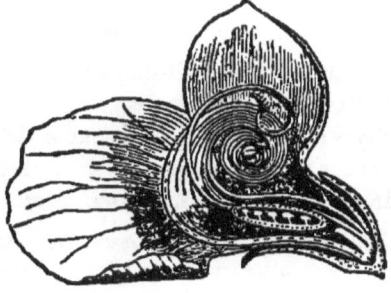

Fig. 446. Section of flower of the Bean, showing the spirally coiled stamens and style, the simple ovary, &c.

* Native species, growing in fields and woods....a
* Exotic species, growing only by cultivation....b
  a Flowers racemed. Pods curved....1
  a Flowers 1 or few in a head. Pods straight....2–4
    b Stems climbing....5–7
    b Stems erect, bushy....8

1 P. peren'nis. *Perennial Wild-bean.* Leaflets ovate, pointed. Racemes in pairs. 4–7f. *p.*

2 P. diversifo'lius. *Trailing W.* Leaflets angular, 2–3-lobed. Peduncle longer than leaf. *c.*

3 P. hel'volus. *Long-stalked W.* Leaflets lance-ovate, not lobed. Peduncle 3–4 times longer than the leaf. M. S.

4 P. pauciflo'rus. *Few-flowered W.* Leaflets linear-oblong, hairy. Peduncle longer than the leaf. W.

5 P. vulga'ris. *Common Garden-bean.* Leaflets ovate, pointed. Racemes solitary, shorter than leaves.

6 P. multiflo'rus. *Scarlet Pole-bean.* Fls. scarlet, showy. Root tuberous. Pedicels opp. †

7 P. luna'tus. *Lima B.* Flowers white. Lfts. ovate-deltoid, acute. Pods broad, large. †

8 P. na'nus. *Bush-bean.* Erect, busby. Leaves broad-ovate, acute. †

2. A'PIOS. Ground-nut.

Calyx bell-shaped, somewhat 2-lipped, the 2 side teeth nearly obsolete, the lower tooth longest. Keel incurved and at length coiled against the very broad, reflexed banner. Ovary sheathed at base.—Twining, smooth herbs. ♃ Root bearing eatable tubers. Leaves pinnately 5–7-foliate.

A. tubero'sa. Stem round, twining about other plants, 2–4f. in length. Leaflets mostly 7, narrow-ovate, more or less acuminate, on short stalks. Racemes axillary, solitary, dense-flowered, shorter than the leaves. Flowers dark purple. The tubers on the root are oval, thick, and very nutritious. In thickets and shady woods. *July, Aug.*

3. ROBIN'IA. Locust.

Calyx short, bell-shaped, 5-cleft, the 2 upper divisions more or less united. Banner large, wings obtuse. Stamens diadelphous (9 & 1). Style bearded inside. Legume flattened, long, many-seeded.—Trees and shrubs with stipular spines. Leaves unequally pinnate. Flowers showy, in axillary racemes. *April, May.*

R. visco'sa. *Clammy Locust-tree.* Racemes rather compact, rose-white, erect. Branchlets and stalks sticky. Leaflets ovate. In parks. Native South.

R Pseudaca'cia. *Common Locust-tree.* Racemes rather loose, drooping, white, fragrant. Leaflets oblong-ovate, smooth, as well as the branchlets.

R. his'pida. *Rose Acacia.* Shrub 4–9f. high, hispid, with clusters of large, purple flowers. Leaflets 5 or 6 pairs, broadly oval.

4. MELILO'TUS. Melilot. Sweet Clover.

Calyx tubular, 5-toothed. Keel petals completely united, shorter than the others. Of the 10 stamens 9 are united, one separate. Pod 1 or few-seeded, longer than the permanent calyx. Leaves pinnately trifoliate. Flowers in racemes.

1 **M. officina'lis.** *Yellow M.* Leaflets obovate-oblong, obtuse, dentate. Calyx half as long as the yellow corolla. Pod 2-seeded. Stem 3f.

2 **M. alba.** *White M.* Leaflets ovate-oblong, square at end. Calyx not half as long as the white corolla. Pod 2-seeded. Height 4–6f. Very fragrant.

5. TRIFO'LIUM. Clover. Trefoil.

Calyx 5-cleft, with bristly teeth, persistent. Petals more or less united at the base, persistent and withering. Banner longer than the wings, which are also longer than the keel. Stamens 10, diadelphous (9 & 1). Legume short, membranous, often included in the calyx, 1–6-seeded, mostly indehiscent. — Herbs with palmately trifoliate leaves. Leaflets straight-veined. Flowers in heads or spikes. *Apr.–Sept.*

* Flowers yellow, in small, dense, oval heads. Pod 1-seeded....1, 2

* Flowers cyanic (not yellow)....a

   a Flowers on little stalks (pedicels) and finally deflexed....b

   a Flowers nearly or quite sessile, never deflexed....c

Fig. 447. Red Clover,—a head of flowers. *Fig.* 448. A single flower. *Fig.* 449. A pod, with a part of the calyx. *Fig.* 450. A seed, cut open. See also *Fig.* 87.

    b Heads small, on stalks some ten times longer....3, 4

    b Heads large, on stalks two or three times longer....5, 6

      c Calyx teeth feathery, longer than the whitish corolla....7

      c Calyx teeth shorter than the purple or roseate corolla....8–10.

1 **T. procum'bens.** *Yellow C.* Stipules much shorter than the petioles. Style 3 or 4 times shorter than the pod. Heads ovate, ¼ in. thick. Stems prostrate. *May.*

2 **T. agra'rium.** *Larger Yellow C.* Stipules longer than the petiole. Style about as long as the pod. Heads oblong, ½ in. thick. Stems ascending. *June, July.*

3 **T. Carolinia'num.** *Southern C.* Stipules leaf-like. Calyx teeth thrice longer than its tube. Legume 4-seeded. Scarcely forms a turf. W. S.

4 **T. repens.** *White C. Shamrock.* Stipules narrow, scale-like. Calyx teeth shorter than its tube. Pod 4-seeded. Forms a dense turf. Fls. white. c.

T. reflexum. *Buffalo C.* Lflts. obovate. Calyx nearly as long as the red corolla.

6 **T. stoloni'ferum.** *Prairie C.* Leaflets obcordate. Calyx not half as long as the white corolla. W.

7 **T. arven'se.** *Rabbit-foot C.* Heads cylindrical, very hairy. Lfts. narrow obovate.
8 **T. praten'se.** *Red C.* Leaflets spotted, oval. Heads roundish, sessile. Flowers rose-red, or white. *c.* † (Figs. 447–456.)
9 **T. me'dium.** *Zigzag C.* Lfts. oblong. Heads roundish, stalked. Fls. deep purple. *r.*
10 **T. incarna'tum.** *Rose Trefoil.* Lfts. round-ovate. Heads oblong. Fls. rose-red. †

6. LUPI'NUS. Lupine.

Calyx deeply 2-lipped, upper lip 2-cleft, lower entire or 3-toothed. Wings united towards the top, keel acuminate. Stamens monadelphous, the filaments forming an entire sheath. Anthers alternately oblong and globose. Pod leathery and knotted.—Herbs, with leaves palmately 5–15-foliate, rarely simple.

1 **L. peren'nis.** *Common L.* Root creeping, perennial. Stem erect, 1–2f. high, hairy. Leaflets soft-downy, 7–11, oblanceolate, 1½–2′ long, broadest above the middle. Flowers alternate, in an erect, terminal raceme, blue, varying to white. It is often called *Sun-dial*, from the fact of its leaves turning to face the sun from morning till night.—Several other species are cultivated in gardens. *May, June.* (Fig. 66.)
2 **L. villo'sus.** *Mullein L.* Stem erect, 1–2f., terminating in a showy raceme. Leaves simple, clothed in a dense coat of silky wool as well as the stem. S.

7. LESPEDE'ZA. Bush Clover.

Calyx 5-parted, with 2 bractlets at base, the sepals nearly equal. Keel very obtuse, on slender claws. Stamens diadelphous (9 & 1). Legume lens-shaped, small, flattened, unarmed, one-seeded, not opening.—♃ Leaves pinnately trifoliate. Flowering in *Aug., Sept.*

¶ Flowers in dense spikes, whitish, with a purple spot on the banner....1, 2
¶ Fls. racemed, &c., violet or purple. Some of the fls. with no corolla....a
 a Stem prostrate, trailing, diffuse. Leaflets oval....3
 a Stem erect and mostly branched, 1–3f. high....4, 5
1 **L. capita'ta.** *Head B.* Leaflets elliptical, silky. Spikes shorter than leaves. Stem nearly simple, 2–4f.
2 **L. hirta.** *Hairy B.* Leaflets roundish-oval. Spikes longer than leaves. Stem branching, very hairy.
3 **L. repens.** *Creeping B.* Downy more or less, except the upper side of the leaves, which is always smooth. Stems slender, many.
4 **L. viola'cea.** *Violet B.* Smoothish. Leaflets oval, varying to oblong and linear obtuse, mucronate. Corolla 3–4″ long. Varies greatly.
5 **L. Steu'vi.** Plant velvety or downy. Lfts. roundish-obovate. Variable.

8. PI'SUM. Pea.

Calyx divisions leaf-like, 2 upper shortest. Banner large, reflexed. Stamens 10, diadelphous (9 & 1). Style flattened, keel-shaped, bearded on the upper side. Legume oblong, tumid. Seeds globose.—Climbing herbs. Leaves pinnate, ending with a branching tendril.

P. sati'vum. *Common Garden Pea.* Leaflets usually 4, ovate, entire. Stipules rather larger than the leaflets (2–3′ long), ovate, half-cordate at base. Flowers 2 or more on axillary peduncles, large, white. Pods 2 or 3′ long, 5–9-seeded. A very valuable leguminous plant, all over smooth and glaucous. There are many varieties. *June.* (Also, Fig. 443.)

Fig. 451. Common Pea: *s*, the large stipules; *p*, the pod; *f*, the flower; *t*, the tendrils on the end of the leaf.

9. BAPTIS'IA. Wild Indigo.

Calyx 4–5-cleft half way. Petals of about equal length, somewhat united. Banner roundish, notched at the end. Stamens 10, distinct, deciduous. Pod inflated, many-seeded, raised on a stalk in the persistent calyx.— ♃ Large herbs with leaves palmately 3-foliate or simple. Flowers in racemes. Leaflets mostly oblong, broadest above. *Apr.–Sept.*

§ Leaves simple. Flowers yellow. (3 species far South, omitted.)
§ Leaves 3-foliate....α Flowers blue, in a few long racemes....1
　　　　　　　　　　　a Flowers white, in a few long racemes....b
　　　　　　　　　　　a Flowers yellow, solitary, or in short racemes....c
　　b Stipules leaf-like, longer than the petioles....2, 3
　　b Stipules much shorter, or not longer than the petioles....4, 5
　　　　c Flower-stalks not longer than the calyx....6, 7
　　　　c Flower-stalks much longer than the calyx. S. Omitted.
1 B. austra'lis. *Austral W.* Smooth. Lfts. obovate or oblong. Fls. large. W. S. †
2 B. leucophæ'a. *Whitish W.* Stipules large, ovate. Racemes nodding. W.
3 B. villo'sa. *Woolly W.* Stipules small, lance-linear. Racemes erect. S.
4 B. leucantha. Stipules lance-linear, about as long as petioles. W. S. †
5 B. alba. Stipules and bracts minute, early falling off. S.
6 B. lanceola'ta. Leaflets narrow-elliptic. Flowers axillary. S.
7 B. tincto'ria. Leaflets small, round-obovate. Racemes terminal. Common.

9

10. CAS'SIA. Senna.

Sepals 5, scarcely united at base, nearly equal. Petals 5, unequal, but not papilionaceous. Stamens 10, distinct, 3 upper anthers often sterile, 3 lower ones beaked. Legume long, many-seeded.—Leaves simply and abruptly pinnate, mostly with a gland on the petiole. Flowers yellow. *July, Aug.*

    ¶ Racemes axillary. 3 of the anthers imperfect, 7 of them perfect....1, 3
    ¶ Racemes above the axils. Anthers all perfect. Stem 1-2f. high....4, 5
1 C. obtusifo'lia. *Blunt S.* Leaflets 4-6, obtuse. Stem 1-3f. high. S.
2 C. occidenta'lis. *Western S.* Leaflets 6-12, acute. Stem 4-6f. high. S.
3 C. Marilan'dica. *American S.* Leaflets 12-18, mucronate. Stems 5f. high.
    4 C. Chamæcris'ta. *Sensitive Pea.* Anthers 10, unlike. Fls. large. Lfts. 16-24.
    5 C. nic'titans. *Sensitive S.* Anthers 5, alike. Fls. small. Leaflets 12-30.

ORDER XLVII. ROSACEÆ. Roseworts.

Trees, shrubs, or *herbs* with stipules mostly, and alternate *leaves;* with *flowers* regular, commonly showy, perfect, and polyandrous; with 5 *sepals* united at base, often supported by as many *bractlets* outside; 5 *petals* (rarely 0), which are perigynous as well as the *stamens;* 1-∞ *pistils,* which are distinct, or sometimes united and adhering to the calyx tube; *fruit* various; *seeds* with no albumen.

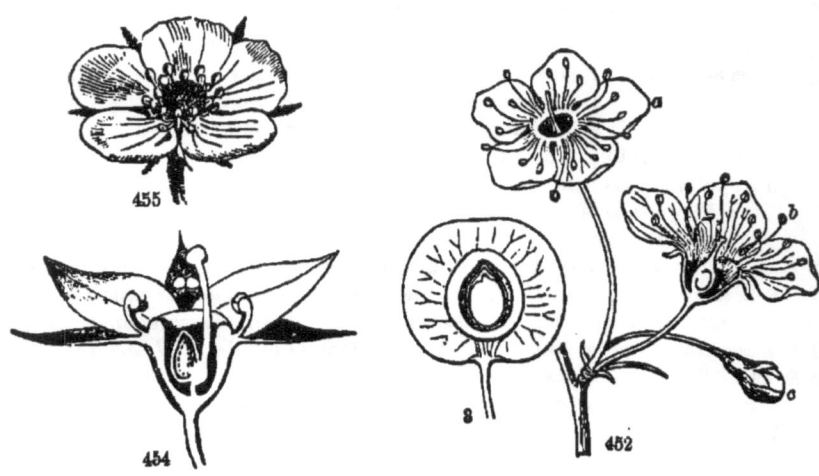

Fig. 452. Flowers of the Great Red Cherry: *b*, section,
showing the perigynous stamens, the single ovary, &c.
Fig. 453. Section of the cherry, showing the seed lying
in the stone and pulp. *Fig.* 454. Section of the flower of
Lady's-mantle (Class Book, p. 325), with the simple
ovary, lateral style, &c. *Fig.* 455. A flower of Strawberry. *Fig.* 456. A section of the same,
showing the perigynous stamens, the many simple pistils on the large torus. *Fig.* 457 Section
of a Rose, showing the many simple pistils sunk in the hollow torus, &c.

Analysis of the Genera.

§ Flowers with 1 pistil and no petals. Herbs....a
§ Flowers with 1 pistil and 5 petals. Shrubs or trees....2
§ Flowers with 2—∞ pistils....3
  2 Style lateral, *i. e.*, arising from the side of the ovary....o
  2 Style terminal, *i. e.*, arising from the top of the ovary....c
3 Pistils (carpels) 2–5, all consolidated with the calyx. Fruit a pome....d
3 Pistils (carpels) 2–50, free, in an open or closed calyx....4
  4 Carpels 1-seeded, achenia inclosed in the calyx tube....e
  4 Carpels 1-seeded, achenia dry or pulpy in an open calyx....5`
  4 Carpels several-seeded, pods in an open calyx....k
    5 Styles persistent on the dry achenia....f
    5 Styles falling off with the rest of the flower....6
      6 Calyx entirely bractless. Flowers never yellow....g
      6 Calyx with bractlets beneath it as if double....h

a Stamens 1–4. Style lateral. Fls. scattered.   *Lady's-mantle.* ALCHEMIL'LA.
a Stamens 4. Style terminal. Fls. in dense spikes.   *Burnet.* SANGUISOR'BA.
a Stamens ∞. Style terminal. Flowers in spikes.   *Burnet.* POTE'RIUM.
  b Stamens about 20. Drupe 1-seeded. S.   *Cocoa Plum.* CHRYSOBALA'NUS.
  c Stone globular, smooth. Fruit not glaucous.   *Cherry.* CER'ASUS. 1
  c Stone flattened, smooth. Fruit glaucous or downy.   *Plum.* PRU'NUS.
  c Stone roughened with pits and furrows. Fruit pulpy.   *Peach.* PER'SICA.
  c Stone roughened with pits and furrows. Fruit dry.   *Almond.* AMYG'DALUS.
    d Petals spat.-oblong. Pome with 5 dble.-cells. *Shad-bush.* AMELAN'CHIER. 2
    d Petals roundish. Pome with bony, 1-seeded cells.   *Thorn.* CRATÆ'GUS.
    d Petals roundish. Pome with thin, 2-seeded cells.   *Apple.* PYRUS. 3
    d Petals roundish. Pome with 5, many-seeded cells.   *Quince.* CYDO'NIA. 4

e Carpels many, in the fleshy calyx. Flowers often double.   *Rose.* ROSA. 5
e Carpels 2 only, in the dry, fluted, prickly calyx.   *Agrimony.* AGRIMO'NIA.
   f Petals and sepals 8 or 9. A small, rare plant on mountains   DRYAS.
   f Petals and sepals 5. Achenia numerous.   *Avens.* GEUM. 6
g Sepals equal. Fruit a heap of pulpy achenia. Fls. cyanic.   *Bramble.* RUBUS. 7
g Sepals unequal. Stems creeping. Flowers white. *False Violet.* DALIBAR'DA. 8
   h Torus small, dry. Flowers yellow. Bractlets minute or 0. WALDSTEI'NIA. 9
   h Torus small, dry. Fls. mostly yellow. Bractlets large.
                                   Cinquefoil. POTENTIL'LA. 10
   h Torus becoming very large and juicy in fruit. *Strawberry.* FRAGA'RIA. 11
   h Torus becoming large and spongy. Fls. purple. Lvs. pinnate. COM'ARUM.
k Petals obovate, not yellow. Stamens very long.   *Steeple-bush.* SPIRÆ'A. 12
k Petals lance-linear, not yellow. Stamens very short. *Indian Physic.* GILLE'NIA.
k Petals multiplied, orange-yellow: Pods 1-seeded. Shrubs.
                               Guelder Rose. KER'RIA.

1. CER'ASUS. Cherry.

Calyx 5-cleft, regular, deciduous. Petals 5, much spreading. Stamens 15–30. Ovary with 2 ovules. Drupe globular, very smooth, destitute of a glaucous bloom. Stone also globular and smooth.—Trees or shrubs. Leaves folded in the bud. Flowers early, white. *May.* (Fig. 452.)

§ Leaves evergreen, leathery, entire....1
§ Leaves deciduous, thin....*a*
  a Flowers in umbel-like clusters from side buds. Drupes red....*b*
  a Flowers in racemes leafy at base. Cherries black or blackish....2, 3
    b Shrubs or trees growing wild, native....4, 5
    b Trees cultivated, not native....6, 7
1 C. Carolinia'na. *Cherry Laurel.* Flowers in dense, short racemes. Fruit black, poisonous. Splendid in cultivation.
2 C. sero'tina. *Wild Black C.* Trees with lance-oblong, blunt-toothed leaves.
3 C. Virginia'na. *Choke C.* Shrubs with oval-obovate, slender-toothed leaves.
4 C. pum'ila. *Sand C.* Shrubs trailing, with lance-obovate, acute lvs. Fr. egg-shaped.
5 C. Pennsylvan'ica. *Wild Red C.* Trees. Lvs. oblong-ovate, acuminate. Fr. roundish.
6 C. A'vium. *Oxheart C.* Leaves oblong-ovate, acuminate, hairy beneath.
7 C. vulga'ris. *Great Red C.* Leaves lance-ovate, acute, narrowed to base.

2. AMELAN'CHIER. June-berry.

Calyx 5-cleft. Petals 5, oblong-ovate and oblanceolate. Stamens short. Styles 5, somewhat united at base. Pome 5-celled, cells cartilaginous, each nearly divided into two 1-seeded divisions.—Small trees or shrubs with simple, serrate leaves, and white early flowers in racemes.

A. Canaden'sis. *Shad-berry. June-berry.* A small tree or shrub found in woods, with a dark-grayish bark. Flowers large white, in racemes at the ends of the branches, appearing in April and May, while the forests are yet naked. Fruit round, purplish, well-flavored, ripe in June. The plant is very variable in size, and in the leaves, &c.

3. PY'RUS. Pear. Apple.

Calyx urn-shaped, limb 5-cleft. Petals 5, roundish. Stamens 00. Styles 2–5. Pome fleshy or berry-like, containing 2–5 cartilaginous (thin and elastic) carpels, each with 2 seeds.—Trees or shrubs. Leaves simple or pinnate. Flowers showy, white or rose-colored, in cyme-like umbels. *May, June.*

§ Leaves pinnate. Fruit as large as peas, scarlet when ripe....6, 7
§ Leaves simple....*a*
    a Wild shrubs, 5–8f. high. Flowers small, in compound clusters....5
    a Trees wild or cultivated. Flowers large, in simple clusters....b
        b Flowers white. Pome bell-shaped, acute at base....1
        b Flowers rose-white. Pome with a pit at base....2–4
1 P. commu'nis. *Pear.* Leaves ovate-lanceolate. Styles 5, distinct. † (Fig. 280.)
2 P. malus. *Apple.* Leaves ovate, not lobed, the veinlets incurved. (Fig. 188.)
3 P. corona'ria. *American Crab.* Leaves ovate, often lobed, cut-serrate, straight-veined. (Fig. 454.)
4 P. angustifo'lia. *Narrow-leaved C.* Leaves lanceolate, scarcely veiny.
5 P. arbutifo'lia. *Choke-berry.* Leaves obovate or oval, with glands on mid-vein.
6 P. America'na. *Mountain-Ash.* Leaflets 13–15, lanceolate, pointed.
7 P. Aucupa'ria. *English M.* Leaflets lance-ovate, acute. Fruit larger.

4. CYDO'NIA. Quince.

Calyx urn-shaped, 5-cleft. Petals 5. Styles 5. Stamens many. Pome with 5 parchment-like cells, each with several seeds.—Shrubs. Leaves simple. Flowers solitary or few in a cluster.

1 C. vulga'ris. *Common Quince.* Leaves downy beneath, broadly ovate, acute, entire, with small, half-ovate stipules. Flowers roseate, solitary terminal. Fruit large, obovate, highly esteemed in preserves, &c. (Fig. 1)
2 C. Japon'ica. *Japan Quince.* Leaves glabrous, ovate-lanceolate, acute at each end, serrulate. Stipules reniform. Flowers red, side clusters, opening early.

5. RO'SA. Rose.

Calyx tube urn-shaped, fleshy, contracted at the throat, limb 5-cleft the sepals generally with a little leaf at tip Petals 5 (greatly multiplied by culture); achenia 00, bony, hispid, included in and attached to the in side of the fleshy calyx-tube.—Shrubby and prickly plants. Leaves unequally pinnate. Stipules attached to the petiole, or often free.

In the table, the first ten species are found growing wild in this country, and sometimes also cultivated. The other species never grow wild here.

§ Styles growing together into an inserted column. Climbers....h
§ Styles not cohering into a column....a
    a Stipules nearly free from the petiole and falling off....g
    a Stipules adhering to the petiole....b
      b Plant armed with curved or hooked prickles, erect....d
      b Plant armed with straight prickles....c
        c Wild, native Roses, 1-3 f., erect....5-7
        c Cultivated exotics climbing (No. 20) or erect....21-23
      d Leaflets glandular and fragrant beneath....f
      d Leaflets not at all glandular. Shrubs erect....e
        e Wild, native Rose, flowers single....8
        e Cultivated exotics, mostly double-flowered....13, 14
          f Flowers single. Wild....9, 10
          f Flowers double. Exotic, cultivated....15-17
      g Leaflets 5-9. Flower-stalk enveloped in bracts....4
      g Leaflets 3-5. Flower-stalk bractless, very smooth....2, 19
        h Leaflets 3-5, mostly 3. Native and cultivated....1
        h Leaflets 5-9....k Stipules and sepals mostly entire....11, 12
          k Stipules fringed, sepals entire....3
          k Stipules entire, sepals pinnatifid....18

1 R. setig'era. *Michigan R.* Flowers in corymbs, rose-colored, changeable. W. †
  2 R. læviga'ta. *Cherokee R.* Lfts. very smooth, ellip. Fls. solit., white. S. †
3 R. multiflo'ra. *Japan R.* Lfts. soft, wrinkled. Fls. corymbed, double. S. †
  4 R. bractea'ta. *Macartney R.* Fls. solitary, with large bracts beneath it. S.-W. †
5 R. lu'cida. *Shining R.* Lfts. 5-9, elliptic, shining. Prickles few. Calyx hispid.
6 R. nit'ida. *Wild R.* Leaflets 5-9, narrow-lance, shining. Prickles numerous.
7 R. blanda. *Bland R.* Lfts 5-7, oblong, dull. Prickles very few. Calyx smooth.
  8 R. Caroli'na. *Swamp R.* Stems 4-7f. high. Flowers in corymbs. Dull green.
9 R. rubigino'sa. *Sweet Brier.* Sepals persistent. Some of the prickles awl-shaped.
10 R. micran'tha. *Eglantine.* Sepals deciduous. All the prickles hooked alike. Fls. small.

  11 R. sempervi'rens. *Evergreen R.* Prickles alike. Lfts. evergreen, leathery. †
  12 R. arven'sis. *Ayrshire R.* Prickles unlike. Lfts. soft, deciduous. †
13 R. cinnamo'mea. *Cinnamon R.* Stipules broad, pointed, involute, wavy. †
14 R. cani'na. *Dog R.* Stipules broad, serrulate. Sepals fall off after flowering. †
  15 R. centifo'lia. *Cabbage R. Moss R.* Sepals spread in flower, often very glandular. †

16 R. damasce'na. *Damask R. Monthly R.* Sepals reflexed in flower. Flowers very double. †

17 R. alba. *White R.* Sepals pinnatifid, spreading. Fls. corymbed, large. †

18 R. moscha'ta. *Musk R.* Leaflets lanceolate, pointed. Fls. panicled, large, white. †

19 R. In'dica. *Chinese Monthly. Bengal R. Tea Rose, &c.* Lfts. ovate, pointed. †

20 R. Alpi'na. *Boursault R.* Lfts. 5–11, obovate, sharp-serrate. Stipules narrow. †

21 R. eglante'ria. *Yellow Rose.* Lfts. broad-oval. Petals obcordate, fugacious. †

22 R. Gal'lica. *French R.* Leaflets elliptical. Petals large, spreading. †

23 R. Pimpinellifo'lia. *Burnet R.* Lfts. small, roundish. Flowers small. †

6. GE'UM. Avens.

Calyx 5-cleft, usually with 5 alternate bractlets outside. Petals 5. Stamens many, collected on a dry receptacle, and bearing the long, persistent style.—♃ Leaves pinnate or lyrate.

§ Style bent and jointed near the middle....a
§ Style straight and not jointed, wholly persistent. Rare plants.... 6, 7
≃ Head of fruits quite sessile, with the styles finally hooked....b, 1
≃ Head of fruits stalked in the calyx more or less....4, 5
　b Petals yellow, longer than the calyx....2, 3
　1 G. Virginia'num. Petals white, as long as the calyx. Receptacle hairy.
2 G. macrophyl'lum. *Mountain A.* Lvs. ending with a very large roundish leaflet.
3 G. stric'tum. *Yellow A.* The end leaflet but little larger than the rest. Height 3–5f
　4 G. vornum. *Head-stalk A.* Petals yellow, small. Stalk as long as head. W.
　5 G. riva'le. *Water A.* Whole flower dark purple, large, nodding.
6 G. triflo'rum. Bractlets longer than the calyx or *purplish* petals. Fls. 3. W.
7 G. Peck'ii. *Peck's A.* Bractlets minute. Pet. yellow. Stem almost leafless. Mts.

7. RU'BUS. Bramble. Blackberries and Raspberries.

Calyx 5-parted, without bractlets. Petals 5, deciduous. Stamens ∞. Ovaries many, becoming many pulpy, drupe-like achenia (grains) united into a compound fruit.—Half-shrubby plants with ♃ roots and ⊚ stems, armed with prickles. Flowers mostly white. In the *Blackberries* the pulpy receptacle constitutes a part of the fruit, but in the *Raspberries* it does not.

＊ Leaves simple, 3–5-lobed. Flowers large....1–3
＊ Leaves compound, of 3–7 leaflets....a
a Stems stout, upright, often recurved at top....b
a Stems weak, trailing or prostrate....7
　b The side leaflets stalked. Prickles strong, recurved....8
　b The side leaflets sessile. Prickles weak, nearly straight....4

Raspberries.

1 R. odora'tus. *Rose Flowering.* Petals round, purple. Stalks hairy-clammy.
2 R. Nutka'nus. *White-flowering.* Petals broad-oval, white. Fls. several. N.-W.
8 R. Chamæmo'rus. *Cloud-berry.* Petals obovate, white. Flower only one. Mts.
  4 Petals as long or longer than the calyx....5, 6
  4 R. Idæus. *Garden Raspberry.* Petals shorter than the calyx.
5 R. strigo'sus. *Wild Red Raspberry.* Corolla cup-shaped, single.
6 R. rosæfo'lius. *Bridal Rose.* Corolla spreading, double. Cultivated.
  7 Stems prickly, shrubby, biennial. Fruit of many grains....11
  7 R. triflo'rus. Stems entirely unarmed, green, ⊙ Fruit of few grains.
8 R. occidenta'lis. *Thimble-berry.* Pl. glaucous. Petals shorter than sepals. Fr. dark.
8 Plants not glaucous. Petals much longer than the sepals....9, 10
    Blackberries.
  9 R. villo'sus. *High Blackberry.* Flowers in racemes. Leaflets ovate.
  10 R. cunsifo'lius. *Sand Bl.* Fls. 1-3 together. Lfts. wedge-obovate. M. S.
11 Prickles many. Flower-stalks without leaves or bracts....12, 13
11 R. Canaden'sis. *Dewberry.* Prickles few. Flower-stalks with leafy bracts.
  12 R. his'pidus. *Hispid, Running Bl.* Flowers small, with spreading sepals.
  13 R. trivia'lis. *Low Bush Bl.* Flowers large, with reflexed sepals. S.

8. DALIBAR'DA. False Violet.

Calyx deeply 5 or 6-parted, 3 of the segments larger. Petals 5. Stamens many. Styles 5-8, long, deciduous. Fruit 5-8 dryish, drupe-like achenia.—♃ Low herbs with creeping stems, simple leaves and 1-2 white flowers on each stalk. North.

D. re'pens. *Creeping F.* Found in damp woods. Creeping stems a few inches to a foot in length. Leaves roundish-cordate, crenate. Stipules very narrow-linear. Petioles 1-8' long. Scapes 1-flowered, about as long as the petioles. *June.*

9. WALDSTEI'NIA. Dry Strawberry.

Calyx 5-cleft, with 5 alternate, sometimes minute and deciduous bractlets. Petals 5 or more, sessile. Stamens many. Styles 2-6. Achenia few, dry, on a dry receptacle.—♃ Acaulescent herbs with lobed or divided radical leaves and yellow flowers on scapes.

W. fragario:'des. A pretty plant, in hilly woods, bearing some resemblance to the strawberry. Root-stock thick, scaly, blackish. Leaves trifoliate, on petioles 8-6' long; leaflets broad-wedge-shaped, cut-toothed, of a shining green above. Scapes about as high as the leaves, bearing 2-6 flowers, which are ½' across. *June.*

W. loba'ta. *Lobed D.* Along rivers, &c. Leaves simple, roundish-cordate, generally 8-5-lobed, &c. *April, June.* S.

10. POTENTIL'LA. Cinquefoil.

Calyx deeply 4–5-cleft, with an equal number of alternate bractlets outside. Petals 4–5, obcordate. Stamens ∞. Achenia ∞, collected in a head on a small, dry receptacle.—Herbs or shrubs with compound leaves and (mostly) yellow flowers. (Figs. 76, 77.)

```
    * Leaves palmately compound....a
    * Leaves pinnately compound....6-8
a Leaflets 3 only in each leaf....1
a Leaflets 5.  Stems prostrate or inclining....4, 5
    1 Flowers yellow.  Stems herbaceous....2, 3
    1 P. tridenta'ta.  Trident C.  Fls. white.  Lfts. wedge-obov., 3-toothed at end.  N.
2 P. Norve'gica.  Norway C.  Erect, many-flowered.  Petals short.  N. M.
3 P. min'ima.  Tiny C.  Low.  Stems 1-flowered.  Pet. longer than sepals.  Mts,
    4 P. Canaden'sis.  Canada C.  Leaflets green both sides, serrate, oblong.
    5 P. argen'tea.  Silver C.  Leaflets silvery-white beneath, pinnatifid.
6 P. frutico'sa.  Shrubby C.  Erect, shrubs with yellow flowers.  Height 1-2f.  N.
7 P. Anseri'na.  Goose-grass.  Stemless herbs.  Leaves and peduncles radical.
8 Herbs with leafy stems.  (3 rare species omitted.)
```

11. FRAGA'RIA. Strawberry.

Calyx deeply 5-cleft, with an equal number of alternate bractlets outside. Petals 5, obcordate. Stamens ∞. Achenia many, fixed to the surface of the large, conical, pulpy, scarlet or white receptacle.—Low 2f plants with trifoliate leaves. (Figs. 265, 455, 456.)

```
1 F. Virginia'na.  Common S.  Bractlets under the calyx entire.  Flowers white, on
    scapes.  Root-stock sending out runners which take root and form new plants.
2 F. In'dica.  Indian Strawberry.  Bractlets under the calyx 3-lobed.  Petals yellow.
    Stems trailing on the ground.  Fruit roundish, bright red, tasteless.  S. †  (272.)
```

12. SPIRÆ'A. Meadow-sweet. Hard-hack.

Calyx 5-cleft, persistent. Petals 5, roundish. Stamens 10–50, exserted. Carpels distinct, 3–12, forming little 1-celled, several-seeded pods. Styles terminal.—2f Beautiful, unarmed herbs or shrubs with alternate leaves and branches, and small white or rose-colored flowers. *May, Aug.*

```
    * Shrubs 4-9 f. high....a
    * Herbs with the leaves once or thrice pinnate....7
a Stipules present....1, 2
a Stipules none.  Leaves simple and undivided....b
```

9*

b Flowers in panicles. Leaves lance-ovate....3, 4
b Flowers in corymbs or little umbels. Leaves oval or ovate....5, 6
1 S. opulifo'lia. *Nine-bark.* Leaves simple, 3-lobed. Corymbs umbellate. N.
2 S. sorbifo'lia. *Sorb-leaved M.* Leaves odd-pinnate. Flowers in panicles.
  3 S. tomento'sa. *Hard-hack.* Lvs. with a rusty white dense wool beneath.
  4 S. salicifo'lia. *Willow-leaved.* Lvs. nearly smooth. Shrub 3 or 4f. high.
5 S. corymbo'sa. Corymb very large, terminal, flat-topped. Height 1-2f. S.
6 S. hypericefo'lia. *St. Peter's Wreath.* Little umbels many, lateral. Cultivated.
  7 Leaves once-pinnate. Inflorescence terminal, on a long stalk....8, 9, 10
  7 S. Arun'cus. *Goat's Beard.* Lvs. thrice-pinnate. Fls. in slender spikes. M.
8 S. loba'ta. *Queen of the Prairie.* Flowers purple. Side leaflets 3-lobed. W.
9 S. filipen'dula. *Dropwort.* Fls. white. Lfts. pinnatifid-serrate. Gardens.
10 S. Ulma'ria. *Meadow-sweet.* Flowers white. Lfts. doubly-serrate. Gardens.

Order LII. ONAGRA'CEÆ. Evening Primroses.

Herbs with alternate or opposite *leaves;* and with the parts of the
flowers generally in 4's, sometimes in 3's, 2's, or 1's; with the
sepals united below into a tube, valvate in the bud; the
petals and *stamens* inserted into the throat of the calyx;
ovary coherent with the tube of the calyx; becoming in the
fruit a 2–4-celled capsule or berry with many seeds.

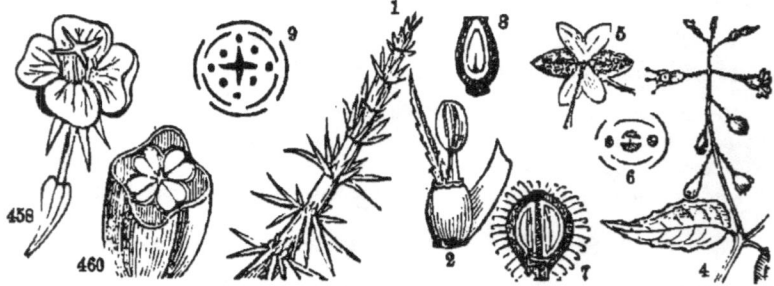

Fig. 458. Flower of Œnothera fruticosa. 9. Plan of the flower. *Fig.* 460. Section of the
4-celled capsule of Œ. biennis. 1. Hippuris vulgaris. 2. Its flower, with 1 stamen, 1 ovary,
2 style. 3. Vertical section of its 1-seeded fruit. 4. Circæa Lutetiana. 5. The flower enlarged. 6. Plan of the flower. 7. Vertical section of the 2-celled and 2-seeded fruit.

Analysis of the Genera.

* Flowers 4 or 5-parted (that is, with 4 or 5 petals, sepals, &c.)....2
* Flowers 3-parted, *i. e.*, with 3 sepals, 3 stamens, &c. (no petals)....g

* Flowers 2-parted, with 2 sepals, 2 petals, &c.....f
* Flowers 1-parted, with 1 stamen, 1 pistil, 1 seed (no petal)....h
  2 Flowers perfect (that is, having both stamens and pistils)....3
  2 Flowers monœcious (some with stamens, some with pistils)....e
    3 Stamens 8, twice as many as the sepals....4
    3 Stamens 4, same number as the sepals....d
      4 Calyx tube much prolonged above the ovary....5
      4 Calyx tube not prolonged above the ovary....a
        5 Garden exotics, with showy purple flowers....c
        5 Wild, native herbs, rarely cultivated....b
a Seed comous with a tuft of silky hairs. Fls. purplish. *Willow Herb.* EPILO'BIUM. 1
a Seed not comous, &c. Fls. large, yellow. Southern. *Yellow Jessie.* JUSSLÆ'A.
  b Petals equal, not clawed, yellow. Pods ∞-seeded.
                          Evening Primrose. ŒNOTHE'RA. 2
  b Petals hardly equal, clawed, red. Pods 1–4-seeded. *Gaura.* GAURA.
  c Herbs from California. Calyx tube short. Petals clawed. *Clarkia.* CLAR'KIA.
  c Shrubs from Chili. Cal. tube long, enlarged. Fls. hanging. *Ear-drop.* FUCH'SIA.
  d Petals yellow, sometimes minute or 0. Lvs. entire. *Seed Box.* LUDWIG'IA
  e Petals greenish or none. Leaves many-cleft. Water-plants.
                    Water Milfoil. MYRIOPHYL'LUM.
  f Delicate herbs with small, pale flowers. *Enchanter's Nightshade.* CIRCÆ'A. 3
g Small herbs in wet places, with pinnatifid lvs. *Mermaid Weed.* PROSERPINA'CA.
  h In water, rare. Leaves linear, whorled. *Mare's Tail.* HIPPU'RIS.

1. EPILO'BIUM. Willow-herb.

Calyx tube not prolonged above the ovary. Limb deeply 4-parted, deciduous. Petals 4. Stamens 8. Stigma often with 4 spreading lobes. Ovary and capsule linear, 4-cornered, 4-celled, 4-valved. Seeds crowned with a tuft of long hairs.—♃ Flowers purplish or white.

E. angustifo'lium. *Narrow-leaved Willow-herb. Rose-bay.* A tall, showy herb (4–6f. high), common at the North. Leaves narrow-lanceolate, nearly entire, with a vein running along the margin. Flowers large, all parts pale purple or white, in a long, terminal spike. Style and stamens declined. Stigma with 4 long lobes. *July, Aug.* Our four other species, with small flowers, and a club-shaped, undivided pistil, we omit.

2. ŒNOTHE'RA. Evening Primrose.

Calyx tube prolonged beyond the ovary, deciduous: segments 4, reflexed. Petals 4, equal, obcordate or obovate, inserted into the top of the calyx tube. Stamens 8. Capsule 4-celled, 4-valved. Stigma 4-lobed. Seeds not tufted.—Herbs with alternate leaves, and yellow flowers (in all the following species). *May, Aug.*

§ Flowers opening by night. Pods rounded at the corners, sessile....1, 2
§ Flowers opening by day. Pods club-shaped, sharply 4-cornered....a
a Stems erect, 1-3 feet high. Flowers large (1-2′ across)....3
a Stems half-erect, 6-16′ long. Flowers small (5-8″ across)....6, 7
  1 Œ. bien′nis. Lvs. slightly toothed. Pods oblong. Fls. 1′ or more wide.
  2 Œ. sinua′ta. Leaves sinuate-toothed or pinnatifid. Flowers ½′ wide. S.
  3 Pods scarcely winged on the 4 sharp angles. Leaves narrow....4, 5
  3 Œ. frutico′sa. Pods with the 4 angles distinctly winged. Leaves lanceolate.
    4 Œ. ripa′ria. Leaves linear-lanceolate. Flowers finally racemed. S. M.
    5 Œ. linea′ris. Leaves linear. Flowers on the ends of the branches. S. M.
  6 Œ. pum′ila. Flowers straw-yellow. Pods almost sessile. Common. N. M.
  7 Œ. chrysan′tha. Fls. orange-yellow. Pods distinctly stalked. Rare N.-W.

8. CIRCÆ′A. Enchanter's Nightshade.

Calyx tube a little prolonged above the ovary, lobes 2. Petals 2, obcordate. Stamens 2, opposite the sepals. Fruit reflexed, inversely egg-
shaped, with hooked hairs, 2-celled, 2-seeded.—♃ Small, tender herbs,
with opposite leaves and terminal racemes of small, reddish-white flowers.

C. Lutetia′na. (*See the figure.*) Stem 1-2f. high, sparingly branched, pubescent
  Leaves dark green, ovate, subcordate, acuminate, coarsely toothed. Pedicels
  without bracts, bent down after flowering. Fruit clothed with bristly hooks
  June, July.
C. alpi′na Stem 5-10′ high, very smooth. Leaves pale green, broad cordate, thin
  slightly dentate. Common in rocky woods at the North. *July.*

Order LV.—GROSSULACÆ. Currants.

Small *shrubs*, often prickly, with alternate,
  lobed, plaited *leaves;*
flowers in axillary racemes, regular, 4 or 5-
  parted, small;
petals inserted into the throat of the calyx,
  small, distinct, and the
fruit a 1-celled, many-seeded, 2-carpeled berry.

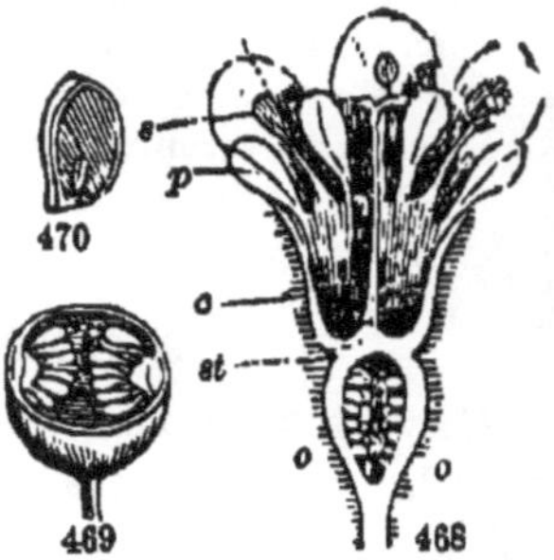

Fig. 468. A flower of the Red Currant cut open; *o*, the ovary and ovules; *st*, the style; *s*, the calyx tube; *p*, the petals; *a*, the stamens. *Fig.* 469. A berry cut open, showing the two placentæ and seeds. *Fig.* 470. A seed cut open, showing the little embryo.

RI'BES. Currants and Gooseberries.

The character of the genus is about the same as of the Order.

§ Currants. Stems without prickles or thorns....a
§ Gooseberries. Stems armed with prickles or spines....c
  a Leaves rolled in the bud (convolute). Fls. bright yel....1
  a Lvs. plaited (plicate) in the bud. Fls. not yellow....b
    b Fruit hairy....2, 3
    b Fruit smooth....4–6.
      c Fruit hispid....7, 8
      c Fruit smooth....d
        d Stalks of the flower or fruit long....11, 12
        d Stalks very short....9, 10

Fig. 471. Missouri Currant,—flower divided.

1 R. au'reum. *Missouri Currant.* Shrub 6–8f., with smooth, 8-lobed leaves (Fig. 471). W. †
2 R. sanguin'eum. *Oregon C.* Flowers bright red, showy. Leaves 8–5-lobed. †
8. R. prostra'tum. *Skunk C.* Fls. striped with red. Lvs. 5–7-lobed. Mts. N. M.
4 R. ru'brum. *Common Red C.* Leaves not dotted, downy beneath. Berries globular, red or white, in pendulous racemes as well as the fls. (Figs. 248, 261.)
5 R. flor'idum. *Flowering C.* Leaves yellow-dotted. Berries obovate, black.
6 R. nigrum. *Black C.* Leaves yellow-dotted. Berries roundish, black. Petiole shorter than the blade. Racemes loose, partly nodding, Gardens.
  7 R. Cynos'bati. *Prickly Gooseberry.* Racemes 2 or 8-flowered. Styles united. (Fig. 281.)
  8 R. lacus'tre. *Swamp G.* Racemes 5–8-flowered. Style 2-cleft. Berry small.
9 R. hirtil'lum. *Smoothish G.* Stems not prickly. Calyx tube bell-shaped. North.
10 R. oxycanthoi'des. *Hawthorn G.* Stems very prickly. Calyx tube cylindric. North.
  11 R. rotundifo'lium. *Round-leaved G.* Calyx cylindric. Stalk 1–8-flowered.
  12 R. Uva Cais'pa. *Garden G.* Calyx bell-shaped. Stalk hairy, 1-flowered. †

ORDER LX. CRASSULACEÆ. The Houseleeks.

Thick, juicy *plants*, with simple, mostly entire *leaves ;* with *flowers* perfectly symmetrical and regular; the *petals, sepals,* and *pistils* being of the same number (8–20); and the *stamens* either the same or twice as many; the *follicles* (as many as the *ovaries*) distinct or somewhat united.

Analysis of the Genera.

§ Pistils (follicles) entirely distinct and separate....2
§ Pistils 4 or 5, united into a 4 or 5-celled capsule....4
  2 Stamens twice as many as the pistils, petals, or sepals....3
  2 Stamens as many (3 or 4) as the pistils, &c. Herb 1-3' high. *r.* TILLÆ'A.
3 Flowers 5 (rarely 4)-parted. Stamens 10 or 8.   *Stone-crop.* SP'DUM. 1
3 Flowers 12 (or 6-20)-parted. Stamens 12-40.   *Houseleek.* SEMPERVI'VUM.
  4 ☉ Herb 2-4' high, fleshy, with 4-parted flowers. S.   DIAMOR'PHA.
  4 ♃ Herb 10-16' high, not fleshy, with 5-parted flowers. *c.* PENTHO'RUM.

Fig. 472. A flowering branch of Sedum acre. *Fig.* 473. A flower of S. acre, natural size. *Fig.* 474. A flower (12-parted, symmetrical, regular) of *Sempervivum* (Houseleek).

1. SE'DUM. Stone-crop. Orpine.

Sepals and petals 5, sometimes 4, distinct. Stamens 10 or 8. Pods 5, sometimes 4, distinct, many-seeded, with an entire scale at the base of each.—Mostly ♃ herbs, with 5-parted flowers in cymes, or in one-sided clusters.

1 Flowers white, or purplish, or rose-colored....2
1 S. a'cre. *Iceland Moss.* Fls. yellow. Plant in low tufts. Gardens.
  2 Leaves scattered, 1-3' long....3-5. (Figs. 472, 473.)
  2 S. terna'tum. *Stone-crop.* Leaves in whorls of 3's. Flowers white, in a 3-spiked cyme.
3 S. telephioi'des. *False O.* Leaves lanceolate or obovate, nearly entire. M. S.
4 S. Tele'phium. *Common O.* Leaves oval, serrate, obtuse. Flowers purplish.
5 S. pulchel'lum. *Handsome O.* Lvs. linear. Fls. in an umbel of spikes, purp. S.

ORDER LXI. SAXIFRAGACEÆ. Saxifrages.

Herbs or *shrubs* with the *pistils* fewer than the *sepals* of the flower; the *petals* as many as the *calyx sepals* (4 or 5), and together with the

5–10 *stamens* inserted on the calyx; the
styles 2, distinct, with their
2 *ovaries* more or less united below, and
  either free or adhering to calyx;
vods capsular, many-seeded;
embryo slender, in albumen.

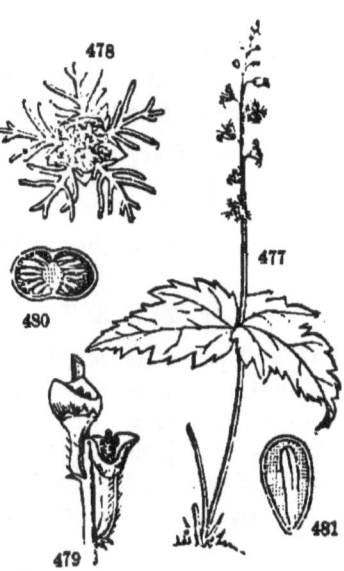

Fig. 475. Section of flower of Early Saxifrage
(Class Book, page 371). *Fig.* 476. Ovary and pistils,
cut across to show the two cells. *Fig.* 477. Mitella
diphylla; 8, a flower, magnified; 9, the fruit pods
open, showing the black seeds. *Fig.* 480. Cross-section of the ovary; 1, seed cut open, showing the long
embryo.

Analysis of the Genera.

§ Herbs. Petals imbricated in the bud....a
§ Shrubs. Petals valvate or convolute (twisted) in bud....e
  a Flowers with 10 stamens....b
  a Flowers with 5 stamens....d
b Petals 4–6, usually 5, entire....e
b Petals 5, all pinnatifid. Stamens 10.   *Mitrewort.* MITEL'LA. 1
b Petals 0. Low, prostrate, in wet places.   *Water Carpet.* CHRYSOSPLE'NIUM.
  c Pods 2-celled. Leaves simple, mostly radical.   *Saxifrage.* SAXIF'RAGA.
  c Pods 2-celled. Leaves bi-ternately compound, cauline. S   ASTIL'BE.
  c Pods 1-celled. Leaves palmately lobed.   *False Mitrewort.* TIAREL'LA. 2
d Styles 2, pod 2-celled. Scape reclined, 8–12' long. W.   SULLIVAN'TIA.
d Styles 2, pod 1-celled. Scape erect, a foot or more. M. W.   HEU'CHERA.
d Styles 3, pod 1-celled. Herb in tufts ½' high. S.   LEPUROPET'ALON.
  e Leaves opposite, simple....f
  e Leaves alternate. Shrub 4–8f. erect. Racemes white. M. S.   ITE'A.
f Shrub climbing trees, &c. Flowers white, fragrant. S.   DECUMA'RIA.
f Shrubs erect. Cymes not radiate—all the flowers perfect.   PHILADEL'PHUS. 3
f Shrubs erect. Cymes radiate. Stamens 8–10.   HYDRAN'GEA. 4

1. MITEL'LA. Mitrewort.

Calyx 5-cleft, bell-shaped. Petals 5, pinnatifid with linear divisions, inserted on the throat of the calyx. Stamens 5 or 10, included. Styles 2, very short. Capsule short, 2-beaked, 1-celled, 2-valved.—♃ Small, slender herbs, with roundish, lobed, and cordate leaves, mostly from the root. Flowers small, in a slender raceme. N.

1 **M. diphyl'la.** Scape 12-20′ high, with 2 opposite leaves nearly sessile, and many white flowers above with curiously cleft petals. *May, June.* (See Fig. 477.)

2 **M. nu'da.** Scape leafless, thread-like, 5-7′ high, few-flowered. *May, June.*

Both species send out runners from the base.

2. TIAREL'LA. False Mitrewort. Gem-fruit.

Calyx 5-parted, lobes obtuse· Petals 5, entire, the claws inserted on the calyx. Stamens 10, exserted. Styles 2. Capsule 1-celled, 2-valved, 1 valve much larger.—♃ Fls. white. N. M.

T. cordifo'lia. Scape about 10′ high, sometimes bearing a leaf, the flowers white in all their parts, forming a cylindrical raceme. In rocky woods, with the *Mitrewort*, very common at the North. *May, June.*

3. PHILADEL'PHUS. False Syringa.

Calyx 4-5-parted, tube adherent to the ovary, persistent. Corolla 4-5-petaled. Styles 4, more or less united. Stamens 20-40, shorter than the petals. Capsule 4-celled, 4-valved, many-seeded. — Handsome flowering shrubs, with opposite leaves. Petals convolute in the bud.

Fig. 482. "Radiant" panicle of Oak-leaved Hydrangea; the larger flowers neutral.

P. grandiflo'rus. *Large-flowered Syringa.* A very showy shrub, 6f. high. Leaves ovate, acuminate, 3-veined. Stigmas 4, styles united into 1. Flowers large, in umbels of 2-7, white nearly inodorous. Cultivated, but wild at the South. *June.*

P. corona'rius. *Mock Orange.* Stems 5-8f. high. Leaves oval and ovate, short-pointed, feather-veined. Styles and stigmas 4, distinct. Flowers numerous, white, handsome, very fragrant. Cultivated. *June.*

4. HYDRAN'GEA. Hydrangea.

Flowers in cymes, the marginal ones generally barren, with the sepals much enlarged (that is, the cymes are *radiant*). The fertile flowers are small, calyx about 4-toothed, petals 4, stamens 8 or 10; capsule 2-beaked, many-seeded.

1 **H. arbores'cens.** *Big Wild H.* Leaves ovate, obtuse or cordate at base, nearly smooth. Cymes flat. Shrub 4 to 6 feet high. M. W. Cultivated.
2 **H. quercifo'lia.** *Oak-leaved H.* Leaves deeply siunate-lobed. Cymes in the form of a panicle. South. Cultivated. (See *Fig.* 482.)
8 **H. radia'ta.** *Silver-leaved H.* Leaves ovate, clothed with a silvery-white down beneath. Cymes flat. Shrub 6-8f. high. S. †
4 **H. horten'sis.** *Changeable H.* Leaves elliptical, narrowed at each end, smooth. Cymes mostly all barren, changing from green to white, pink, blue, &c.

ORDER LXIII. UMBELLIFERÆ. The Umbelworts.

Herbs with hollow, furrowed stems, simple or compound *leaves;*
no *stipules,* but with a broad sheathing base to the *petioles;*
the small *flowers* in umbels, and the *calyx* wholly adherent to the ovary;
the *petals* and *stamens* 5, standing on the top of the ovary;
the *styles* 2, and the *fruit* dry, its 2 *carpels* seed-like and separating,
marked outside by ribs and furrows running lengthwise.

Analysis of the Genera.

* Plants growing wild, some of them cultivated for the eatable root....2
* Plants never wild, but cultivated for their fruit, &c.....q
   2 Flowers white, rarely rose-colored or cream-colored....3
   2 Flowers yellow, or (in one instance) dark purple....4
      3 Umbels simple, leaves simple. Little creeping wet plants....a
      3 Umbels regularly compound, the flowers not sessile....c
      3 Umbels irregular, flowers in crowded heads, sessile....b
  4 Fruit decidedly flattened on the back....p
  4 Fruit flattened on the sides or not at all....o

a Fruit flattened. Leaves roundish. *Pennywort.* Hydroco'tyle.

a Fruit globular. Lvs. linear. Fls. pedicelled. Height 1–2′. *r.* Crant'zia.

b Fruit clothed with hooked prickles. Heads small, 2–4. *c. Sanicle.* Sanic'ula. 1

b Fruit clothed with scales. Heads often near 1′ thick. W. S. *c.* Eryn'gium.

 c Umbels not radiate (§ 255, *u*, outer flowers not larger than the rest)....d

 c Umb. rad., very large. Huge herbs, 4–8f. high. *c. Cow Parsnip.* Herac'leum.

d Leaves simple linear petioles without blades. S. Tiedman'nia.

d Leaves only once divided, pinnately or ternately....*u*

d Leaves twice or thrice compounded....g

 e Fruit flattened or contracted, more or less, on the sides....f

 c Fruit much flattened on the back. M. S. *Archemore.* Archemo'ra.

f Leaflets 3, ovate, doubly serrate. Stem 1–2f. high. *Honewort.* Cryptotæ'nia. 2

f Leaflets 3, long, linear, grass-like. ·Rare. S. *Nerveleaf.* Neurophyl'lum.

f Leaflets 5–11, lanceolate or lance-linear. 2–6f. *Water Parsnip.* Si'um.

f Leaflets 5–9, oblong. Stem procumbent. S. *Marsh Umbel.* Helosciad'ium.

 g Bracts of the involucre (not involucel) entire....h

 g Bracts of the involucre cleft and divided....k

 g Bracts of the involucre none or almost none....m

h Fruits bristly, club-shaped, few. Stem 1–2f. high. *Cicely.* Osmorhi'za. 3

h Fruits smooth, flattened on the sides, ribs wavy. *Poison Hemlock.* Coni'um. 4

h Fruits smooth, flattened on the back, ribs winged, straight. *r.* Conioseli'num.

h Fruit smooth, terete, not flattened, ribs straight. *Lovage.* Ligus'ticum.

 k Fruits bristly, short, numerous. Often cultivated. *Carrot.* Dau'cus.

 k Fruits smooth. Stems and leaflets thread-like. Rare. Discopleu'ra.

 k Fruits smooth. Stem 3–6′ erect, bulbous. W. *Pepper-and-Salt.* Erigeni'a.

m Fruit flattened on the back. Stems large. *c.* *Angelica.* Archangel'ica.

m Fruit flattened on the sides....n

m Fruit terete, not flattened. Poison. N. Rare. *Fool's Parsley.* Æthu'sa.

 n Calyx 5-toothed. Stems diffuse, slender. W. *Chervil.* Chærophyl'lum.

 n Cal. 5-toothed. Umbels stalked. Sts. erect, very slender. S. Leptocau'lis.

 n Calyx teeth none, fruit strongly ribbed. Poison. *Water Hemlock.* Cicu'ta. 6

 n Calyx teeth none, fruit scarcely ribbed. W. Rare. *Crest Umbel.* Eu'lophus.

o Involucels leafy. Leaves perfoliate, simple, entire. *Modesty.* Bupleu'rum.

c Involucels minute. Seed with 5 winged ribs. *Golden Alexanders.* Thas'pium. 7

o Involucels minute. Seed with 5 ribs not winged. *Alexanders.* Ziz'ia. 8

 p Involucels minute. Fruit corky. Leaves bi-pinnatifid. Polytæ'nia.

 p Involucels none. Fruit thin. Leaves pinnate. *Parsnip.* Pastina'ca.

q Flowers white. Involucre 0 or of 1 entire bract....r

q Flowers white. Involucre of a few cleft bracts. *Parsley.* Petroseli'num.

q Flowers yellow. Leaf segments very narrow and many. *Fennel.* Fœnic'ulum.

 r Umbellets radiate. Fruit round. Lvs. finely cut. *Coriander.* Corian'drum.

 r Umbellets not radiate (the flowers all similar)....s

s Fruit flattened on the sides, roundish. Lf. segm. wedge-form. *Celery.* A'pium.

s Fruit flattened on the sides, oval. Leaf segments linear. *Caraway.* Ca'rum.

s Fruit egg-shaped, not flattened. Leaf segments linear. *Anise.* Pimpinel'la.

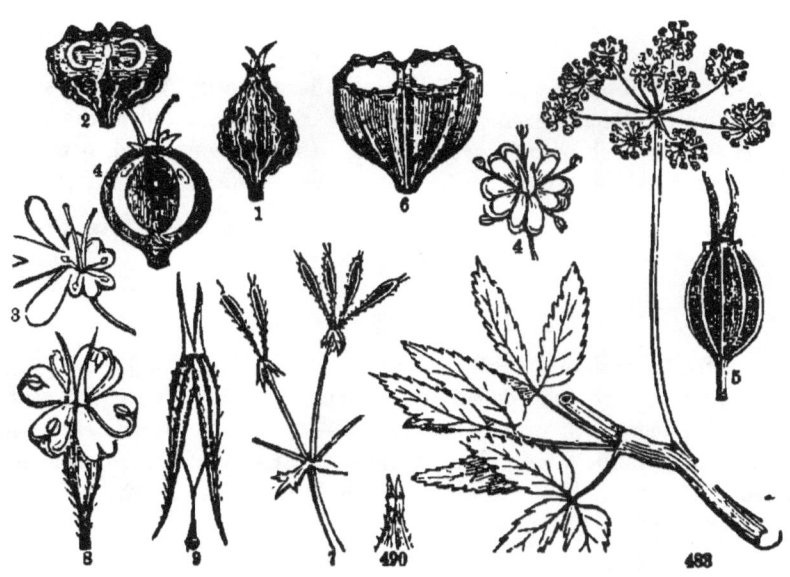

Fig. 483. Gnlden Alexanders, with its compound, naked umbel, &c. 4. A flower enlarged.
5. The fruit with its thread-shaped ribs and two persistent styles. 6. Cross-section, showing the
two carpels with the oil-tubes and flat inner face. 7. Umbel of Sweet Cicely, in fruit. 8. A
flower enlarged. 9. The fruit with the two carpels separating from the base and supported by a
two-cleft stalk. *Fig.* 490. Summit of the fruit of Bitter Cicely. 1. Fruit of Poison Hemlock,
with the undulate-crenulate ribs. 2. Cross-section, showing the grooved inner face and involute
albumen. 3. Radiate flower of Coriandrum. 4. Vertical section of the globose fruit, showing
the minute embryo.

1. SANIC'ULA. **Sanicle.**

Flowers polygamous. Calyx teeth leafy, tube bristly. Petals obovate,
erect, with the point inflected. Fruit roundish, armed with hooked
prickles. Carpels without ribs.—♃ Plants 1–2f. high. Umbel with a few
capitate umbellets. Involucre of few cleft bracts, involucel of several,
entire. *June–Aug.*

1 S. **Marylan'dica.** *Long-styled S.* Sterile flowers many, pedicellate; fertile flowers
    sessile. Styles slender, conspicuous, recurved. Leaves 5–7-parted. Common.
2 S. **Canaden'sis.** *Short-styled S.* Sterile flowers few, much shorter than the fertile.
    Styles shorter than the prickles. Leaves 5-parted, upper 8-parted. Umbels
    (or heads) small. Woods. Common.

2. CRYPTOTÆ'NIA. Hone-wort.

Calyx teeth obsolete. Petals with an inflexed point. Fruit linear-oblong or ovate-oblong. Seeds with 5 obtuse ribs, contracted at the sides. —♃ A smooth herb with 3-parted leaves. Umbels compound, with very unequal rays, white flowers, no involucre, aud few-leaved involucels.

C. Canaden'sis. St. 1–2f. high, erect. Leaflets large, the side ones often 2-parted or lobed. Common in moist woods. *July.*

3. OSMORHI'ZA. Cicely.

Calyx teeth obsolete. Fruit linear-oblong, club-shaped, tapering to the base, crowned with the conical styles; carpels each with 5 equal, acute, bristly ribs, and a deep groove on the face.—♃ Leaves bi-ternately divided, with the umbels opposite. Involucre few-leaved; involucel 4–7-leaved. Flowers white. Fruit an inch in length. Height about 2f. *May, June.* (Figs. 244, 487–9.)

O. longis'tylis. *Sweet C.* Styles thread-like, nearly as long as ovary. Plant downy. The root has an agreeable spicy flavor.

O. brevis'tylis. *Bitter C.* Styles conical, 5 times shorter than ovary. Plant hairy. Less interesting than No. 1. (See Fig. 490.)

4. CONI'UM. Poison Hemlock.

Calyx teeth obsolete. Fruit ovate, flattened on the sides, each carpel with 5 wavy-crenulate ribs on the back, and a deep narrow groove on the inner face.—ⓑ Herbs with large, decompound leaves, with very many leaflets. Involucre and involucels of 3–5 leaves, the latter one-sided. Flowers white. (Figs 65, 491, 492.)

C. macula'tum. Stem spotted with purple, glaucous, about 4f. high. Leaves bright green, leaflets small, lanceolate, pinnatifid. Umbels terminal, the involucels with the inner half wanting. *June, July.*

5. ERIGENI'A. Pepper-and-salt.

Calyx limb wanting. Petals flat, entire. Carpels (half-fruits), 3-ribbed, contracted on the face, forming together a fruit much broader than long. —♃ Root tuberous. See Fig. 333.

£. bulbo'sa. A small, early-flowering herb, Western N. Y. to Mo. Stem arises from a roundish tuber deep in the ground. The root leaf is thrice ternate. The involucrate leaf twice ternate. The dark-brown stamens with the little white petals suggest its common name.

6. CICU'TA.   Water Hemlock.

Calyx 5-toothed. Petals with the point inflected. Fruit roundish, a little contracted on the sides so as to appear somewhat double. Seeds with 5, flattish, equal ribs, 2 of them on the margin.—♃ Poisonous herbs with compound leaves and perfect umbels of white flowers. Involucre few-leaved or 1. Involucels many-leaved.

1 C. macula'ta. *Spotted Water-Hemlock.* Stem streaked with purple, 3–6f. high, smooth, striate, hollow. Lower leaves triternate and tripinnate, segments lanceolate, serrate. Umbels 2–4' broad. Fruit 10-ribbed. Involucels of 5 or 6 short, slender, acute bracts. Common in wet meadows. *July, Aug.*

2 C. bulbi'fera. *Narrow-leaved Water-Hemlock.* Stem green, striate, slender, with little bulblets in the axils of the branches. Leaves bi-ternately divided. Leaflets linear or lance-linear, 2–4' long, with distant teeth. In wet meadows and swamps. *Aug.*

7. THAS'PIUM.   Alexanders.

Calyx minutely 5-toothed. Fruit elliptical, roundish across, not flattened either way, seeds each with 5 winged ribs.—♃ Leaves divided. Involucre none, involucels few-leaved. The species resemble the *Zizias* except in their fruit. *May, June.* (Figs. 483–6.)

1 Root leaves simple, cordate, stem leaves once-ternately divided....2
1 T. barbino'de. Leaves bi- or tri-ternate, lfts. cut-serrate. St. hairy at joints.
2 T. au'reum. *Golden A.* Fruit oval. Flowers yellow. Stem 2–3f. high.
2 T. atropurpu'reum. *Purple A.* Fruit roundish. Flowers dark purple. Stem 2–3f. high. S. M.

8. ZIZIA.   Alexanders.

Calyx minutely 5-toothed. Fruit oval or ovate, flattened at the sides so as to appear somewhat double. Seeds each with 5 ribs which are not winged, but thread-like.—♃ Smooth, with divided leaves and yellow flowers. Umbels compound, with no involucre or involucels.

Z. integer'rima. *Entire-leaved A.* Root and stem leaves bi- and tri-ternate, leaflets entire. Plant 1–2f. high, in rocky woods. *May–July.*

Order LXV. CORNACEÆ. Cornels.

Trees and *shrubs*, seldom *herbs*, with sim-
    ple, mostly opposite *leaves ;* with
flowers 4-parted, arranged in cymes; the 4
petals valvate in the bud; and with the 4
stamens standing on the top of the 2-
    celled
ovary, which is adherent to the calyx-
    tube; *styles* united ;
fruit a 1 or 2-seeded drupe.

Fig. 495. Low Coroel; *b*, the 4-leaved involucre
surrounding the head of flowers.

Analysis of the Genera.

§ Flowers perfect, 4-parted. Petals 4. Drupe 2-celled.   *Corneil.* CORNUS. 1
§ Flowers imperfect, 5-parted. Petals often 0. Drupe 1-celled. Trees
    with small, green flowers in side clusters. Fruit plum-like. *Tupelo.* NYSSA.

CORNUS. Cornel. Dogwood.

Trees, shrubs, or perennial herbs. Flowers in cymes. Sepals, petals,
and stamens each 4, with a double pistil.

1 C. Canaden'sis. *Low Cornel.* A small herb, with a creeping, woody root-stock
    sending up annually its stems 4–6' high. Some stems bear only 4 whorled
    leaves, others bear 6 leaves at top and an umbellate cyme of small white flowers
    supported by a large, white, 4-leaved involucre. The whole resembles a single
    white flower. *May, June.* (Fig. 495.)
2 C. flor'ida. *Flowering Dogwood.* A small tree, 20–30f. high, with opposite, ovate,
    pointed, entire leaves. The cyme of small greenish flowers is supported by a
    very large, white, 4-leaved involucre in May.

Six other species, without involucres, grow in the country

COHORT II.

THE GAMOPETALOUS EXOGENS.

Essential Character. — Flowering plants (PHÆNOGAMIA) with their stems growing by additions to the outside in layers (EXOGENS), their seeds inclosed in a seed-vessel or pericarp (ANGIOSPERMS), their flowers with a double perianth and their petals united (MONOPETALÆ).

ORDER LXVI. CAPRIFOLIACEÆ. Honeysuckles.

Shrubs and *herbs*, often twining, with opposite *leaves ;* with *flowers* clustered and often fragrant, 5-parted and often irregular; *corolla* monopetalous, tubular or rotate; *stamens* on the tube of the corolla, often one less than its lobes; *ovary* adherent to the calyx; *style* 1; *fruit* a berry, drupe, or capsule; *embryo* small, in fleshy albumen.

Analysis of the Genera.

1 Corolla tubular. Stigma capitate, on a slender style....2
1 Corolla rotate, deeply 5-lobed. Stigmas 8, rarely 5, sessile. Shrubs....c
  2 Herbs....a
  2 Shrubs....b
a Stamens 4, capsule 3-celled. A trailing evergreen.   *Twin-flower.* LINNÆ'A. 1
a Sta. 5, drupe bony, 3-5-celled. Erect, unbranched.   *Fever-root.* TRIOS'TEUM.
  b Cor. bell-shaped, reg'r. Berry glob., 4-celled, 2-seeded. SYMPHORICAR'PUS. 2
  b Cor. tubular, lobes unequal. Berry 2-3-celled, few seeded. c. LONICE'RA. 3
  b Corolla funnel-shaped. Capsule 2-celled, many-seeded. c. DIERVIL'LA. 4
c Leaves pinnate. Berry globose, pulpy, 8-seeded.   *Elder.* SAMBU'CUS. 5
c Leaves simple. Drupe flattish, 1-seeded. Handsome shrubs. VIBUR'NUM.

1. LINNÆ'A. Twin-flower.

Calyx tube ovate, limb 5-parted, deciduous, with 2 bractlets at base. Corolla bell-shaped, limb a little irregular, 5-lobed. Stamens 4, 2 longer than the other 2. Capsule 3-celled, but only 1-seeded, 2 of the cells being empty.—A trailing evergreen herb, dedicated to Linnæus, the first and greatest of botanists.

L. borea'lis. The only species, a fine little plant, found in moist woods in cool climates. It has long, thread-like, creeping stems, rooting at the joints, the upright branches about 3' high. Leaves small, roundish. Flowers in pairs, rose-colored, nodding, at the top of the slender stalk. *June.*

2. SYMPHORICAR'PUS. Snow berry.

Calyx tube globose, limb 4–5-toothed. Corolla bell-shaped, 4–5-lobed, regular. Stamens 4–5, short. Fruit a globose berry, 4-celled but only 2-seeded, 2 cells being empty.—Small erect shrubs with oval, entire leaves, rose-white flowers in short clusters.

    * Stamens and style included (*i. e.*, not longer than the corolla)....1, 2
    * Stamens and bearded style exserted (extending out of the corolla)....3
1 S. racemo'sus. *Cult.* Fls. in loose, leafy racemes. Berries snow-white, large.
2 S. occidenta'lis. *Wolf-berry.* Fls. in dense, nodding spikes. Berries white. N.-W.
3 S. vulga'ris. *Coral-berry.* Fls. in axillary heads. Berries red. M. S. W.

3. LONICE'RA. Honeysuckle.

Calyx tube globular, limb 5-toothed, very short. Corolla tubular or funnel-form, limb 5-cleft, irregular or almost regular. Stamens 5. Ovary 2 or 3-celled. Berry few-seeded.—Climbing or erect shrubs, with opposite and often connate leaves (that is, their bases growing together around the stem), entire on the margins.

§ Stem climbing, flowers sessile, whorled (in pairs in one species)....a
§ Stem mostly erect, leaves never connate, flowers in pairs....2
    a Upper pair or pairs of leaves united (connate) at base....b
    ∷ Leaves all distinct, corolla ringent, (In gardens only.)....7, 8
b Corolla tube gibbous (swelled out on one side) at base, limb ringent....5, 6

Fig. 496. Trumpet Honeysuckle. Flowers and the connate leaves.

b Corolla tube equal and slender (not gibbous) at the base....1
    1 Corolla ringent, lower lip linear, upper 4-lobed....2–4
    1 L. sempervi'rens. *Trumpet H.* Cor. trumpet-shaped, nearly regular, scarlet.

2 **L. flava.** *Wild-yellow H.* Flowers in a terminal, 2 (or more)-whorled spike, pale yellow. Leaves glaucous. W. S.

3 **L. grata.** *Wild-sweet H.* Fls. in terminal and axillary whorls, reddish white.

4 **L. Caprifo'lium.** *Italian H.* Fls. in a single, terminal whorl, red, yel., and white.

5 **L. parviflo'ra.** *Small-fl. H.* Leaves oblong, smooth and glaucous beneath. Flowers 1' long, yellowish and purplish, or crimson.

6 **L. hirsu'ta.** *Hairy H.* Leaves broad-oval, hairy and downy, green (not glaucous). Flowers sulphur-yellow. N.

7 **L. Periclym'enum.** *Woodbine H.* Fls. whorled, capitate, red and yellow, sweet-scented. Leaves deciduous. Berries red. †

8 **L. Japon'icum.** *Japan H.* Flowers in pairs, axillary, sweet-scented, deeply two-lipped, reddish. Leaves evergreen. †

9 Corolla gibbous at base, lobes more or less irregular. Wild....10–12

9 **L. Tartar'icum.** *Tartarian H.* Corolla scarcely gibbous, lobes spreading, equal, rose-color, handsome. Leaves cordate, obtuse. †

10 **L. cilia'ta.** *Fly H.* Corolla lobes short, erect, nearly equal. Berries red.

11 **L. oblongifo'lia.** *Swamp Fly H.* Corolla deeply ringent. Pedicels long. Berries double, purple. Shrub 3–4f. high, swamps. N.

12 **L. cœru'lea.** *Blueberried H.* Corolla lobes short, subequal, yellow. Pedicels very abort. Berries double, blue. N.

4. DIERVILLA. Bush Honeysuckle.

Calyx tube oblong, limb 5-cleft. Corolla twice as long, limb 5-cleft and nearly regular. Stamens 5. Capsular fruit 2-celled, many-seeded.— Small erect shrubs with opposite leaves and axillary flowers.

D. trif'ida. Stem about 2f. high, branching. Leaves ovate, serrate, ending in a long, narrow point. Peduncles 1–3-flowered, the ovaries slender, about half as long as the greenish-yellow corolla. Hedges and woods. *June.*

5. SAMBU'CUS. Elder.

Calyx small, 5-parted. Corolla regular, rotate, 5-cleft into obtuse lobes. Stamens 5. Stigmas sessile. Berry globose, pulpy, 3-seeded.—Shrubs (5–6f. high) or perennial herbs with pinnate or bi-pinnate leaves. Flowers (white) in cymes.

S. Canaden'sis. *Sweet E.* Leaflets 7–11. Cymes flat. Berries dark-purple. *June.*
S. pu'bens. *Red E.* Leaflets 5–7. Cymes oblong, panicled. Berries red. *May.*

10

Order LXVII. RUBIACEÆ. The Madderworts.

Plants with opposite, sometimes whorled, entire *leaves;* the *stipules* between the petioles; the *calyx* adherent to the ovary; *corolla* regular, inserted on the calyx tube; *stamens* inserted on the corolla and as many as its lobes; *ovaries* 2, united; with the 2 *styles* more or less united.

Analysis of the Genera.

§ Leaves whorled. Herbs with square stems....a
§ Leaves opposite, with small stipules between the petioles....2
    2 Herbs, with the flowers habitually 4-parted....3
    2 Shrubs or trees....d
        3 Fls. twin (always in pairs)....b
        3 Flowers single (not twin)....c
a Flowers 5-parted. Fruit twin, fleshy, berry-like. *Madder.* Ru′bia.
a Flowers 4-parted. Fruit twin, dry, separable nuts. *Bedstraw.* Ga′lium.
  b Two fls. on one ovary. Creeping stems. *Partridge-berry.* Mitchel′la. 1
c Carpels 2, 1-seeded, both never opening. Fls. axillary, solitary. Dio′dia.
c Carp. 2, 1-seeded, one never opening. Fls. axillary, clustered. Spermaoo′ce.
c Carpels 2, few-seeded. Corolla much exserted. *Bluets.* Housto′nia. 2
c Carpels 2, many-seeded. Cor. scarce exserted. *Greenhead.* Oldenlan′dia.
d Flowers 4-parted, in globular hds. *Button-bush.* Cephalan′thus.
d Fls. 5-parted, cymes radiant with scarlet sepals. S. Pincene′ya.

1. MITCHELLA. Partridge-berry.

Flowers 2 on each double ovary. Calyx 4-parted. Corolla funnel-shaped, hairy within. Stamens 4, short, inserted on the corolla. Stigmas 4. Berries composed of the 2 united ovaries. *Jn.*

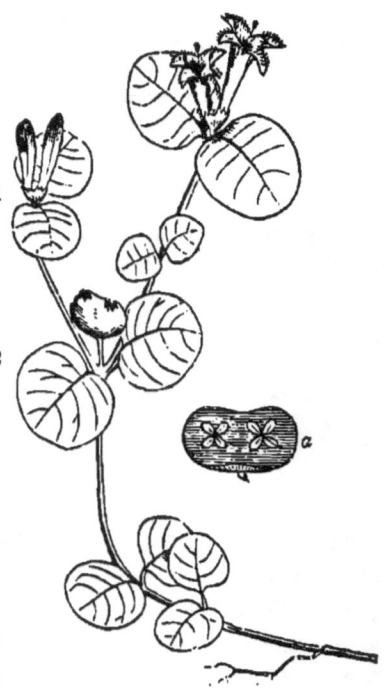

Fig. 497. Mitchella repens, whole plant, with flowers and fruit. *a*, cross-section of the double fruit, showing the two ovaries.

M. re'pens. Common in woods. Leaves round-ovate. Flowers white or pinkish. Berries red, remaining through the winter.

2. HOUSTO'NIA. Bluets.

Calyx tube round egg-shaped, 4-cleft, persistent. Corolla tubular, much exceeding the calyx, limb 4-lobed, spreading, filaments 4, on the corolla. Style 1. Capsule 2-lobed, half-free.—Herbs. Stipules connected to th petiole. Flowers never yellow.

§ Corolla salver-form, glabrous. Peduncles 1-flowered....a
§ Corolla funnel-form. Peduncles many-flowered, cymous....b
  a Flowers terminal. Small, delicate herbs....1, 2
   a Flowers axillary. Small, delicate herbs....3, 4
    b Leaves lance-ovate. Cymes terminal....5
    b Leaves lance-linear. Cymes terminal....6, 7.
1 H. cœru'lea. *Dwarf Pink.* Stems very numerous, upright, 3–6'. Root leaves ovate-spatulate. Flowers pale blue. *May, June.*
2 H. serpyllifo'lia. *Thyme-leaved B.* Stems thread-form, decumbent, 6–12'. Leaves round-ovate, petiolate, fringed. Flowers on long stalks, pale. S.
3 H. min'ima. *Tiny B.* Leaves linear-spatulate. Stems 1–3' high. Prairies.
4 H. rotundifo'lia. *Round-leaved B.* Lvs. roundish. Stems 2–5'. S. Mts.
5 H. purpu'rea. *Prairie Innocence.* Stems upright, much branched, 1f., with numerous clusters of roseate or white, very delicate flowers. W. S.
6 H. longifo'lia. *Long-leaved I.* Stems 4–10', erect. Leaves oval-elliptic, narrowed to end.
7 H. angustifo'lia. *Narrow-leaved I.* Stems 1–2f. erect. Lvs. linear. Flowers numerous. W. S.

ORDER LXX.—COMPOS'ITÆ. Asterworts.

An immense family of *herbs* or *shrubby plants*, with *compound flowers*, that is, the *flowers* (or *florets*) collected into close *heads* upon a common receptacle, and surrounded by an *involucre* of many bracts (called *scales*), with 5 *stamens* which have their anthers united into a tube around the style, with the *calyx tube* closely adhering to the 1-celled ovary (an achenium in fruit), and the *calyx limb* crowning the ovary in the form of a *pappus* consisting of scales, awns, bristles, or hairs, or else entirely wanting; the *corolla* consisting of 5 united petals, either strap-shaped (ligulate or tubular, and the *style* 2-cleft at the top.

In this Order the pupil will remember that the *heads* are called *radiate*, when the outer florets only have *rays* or are *ligulate* (see Fig. 498); *radi-*

ant, when all the florets are ligulate (Fig. 504) ; *discoid*, when all the florets
are tubular, there being no rays (Fig. 509). The *receptacle* is the broad
top of the stalk on which the florets sit (Fig. 499). It is *chaffy* when there
are scales or bracts growing among the florets, and *naked* when none.

The tubular florets constitute the *disk*, and the ligulate, if any, the *ray ;*
the disk is generally *yellow*, while the ray is about as often *cyanic* (that is,
blue, red, white, or *any* color except yellow) as yellow.

Fig 498. A Sunflower,—head radiate. 9. Vertical section of the head, showing the scales of
the involucre, and a single disk-flower remaining upon the convex receptacle. *Fig*. 500. A perfect disk-flower magnified, showing the achenium, the 2 awns of the pappus, the 5-toothed tubular corolla, the 5 stamens united around the branched style, and the chaff-scale at base. 1. Head
(radiate) of Solidago cæsia. 2. A pistillate, ligulate flower of the ray. 3. A perfect disk-flower.
4 A (radiant) head of Dandelion. 5. A perfect, ligulate flower. 6. Achenium, with its long
beak and feathery pappus. 7. A (radiant) head of Nabalus altissimus. 8. A flower. 9 *Lappa*
(Burd ck), head discoid. 10. A flower. 11. One of the hooked scales. 12. A (discoid) head of
Eupatorium purpureum. 13. A flower. 14. *Ambrosia* (Pigweed). 15. Staminate head enlarged.
16. Pistillate involucre enlarged. 17. The fertile flower.

Analysis of the Genera.

Sub-order First, TUBULIFLORÆ,
having all the perfect flowers tubular (§ 95), the ligulate flowers, if any, imperfect.

§ Heads of flowers radiate, with yellow rays....2
§ Heads of flowers radiate, the rays not yellow....3
§ Heads of flowers discoid (no rays). These genera, about 50 in number, such as the *Tansy, Wormwood, Boneset, Ironweed* (Figs. 248-250), *Everlasting, Burdock* (Fig. 509), *Thistle, Hogweed* (Fig. 514), and even *Bachelor's-button,* are all, for want of room, omitted. (See Class Book of Botany, p. 410, &c.)

2 Leaves alternate or scattered on the leafy stems....4
2 Leaves opposite or whorled on the stems, or all radical....6
   3 Leaves alternate or scattered on the leafy stem....7
   3 Leaves opposite or whorled on the leafy stem....x
   3 Leaves all radical and the flowers on a scape....y
4 Receptacle chaffy (with bracts growing among the florets)....5
4 Receptacle with deep, horny cells, like a honeycomb....e
4 Receptacle not chaffy, flat or merely convex....a
4 Receptacle not chaffy, conical or globular....f
   5 Rays sterile, disk fertile. Receptacle conical or columnar....g
   5 Rays sterile, disk fertile. Receptacle flattish. Fruit flattened on the sides....h
   5 Rays fertile, disk sterile. Receptacle flat. Fruit flattened same way as scales....k
6 Receptacle chaffy. Rays sterile, disk fertile....u
6 Receptacle chaffy. Rays fertile, disk sterile....p
6 Receptacle chaffy. Rays fertile, disk perfect....q
6 Receptacle naked or destitute of chaffy scales....m
   7 Receptacle not chaffy, naked of scales....8
   7 Receptacle chaffy with scales among the florets. Lvs. finely divided....v
   7 Receptacle chaffy with scales, &c. Lvs. undivided, merely toothed....w
8 Pappus of numerous bristly hairs....9
8 Pappus of 2 or 3 awns and minute hairs. Glabrous plants....t
8 Pappus wholly wanting, or only a membranous margin....u
   9 Involucre of unequal scales, imbricated in several rows....10
   9 Involucre scales nearly equal, narrow, and almost in one row....t
10 Pappus simple, the bristly hairs abundant and about equal....r
10 Pappus double, the outer row of hairs extremely short....s
   a Involucre scales imbricated in several rows....b
   a Involucre not imbricated, the outer scales very short or none   SENE'CIO.
   a Invol. not imbr., outer scales equal to the inner. *Marigold.* CALEN'DULA.
   a Involucre not imbricated, outer scales longer than inner. S. GAILLAR'DIA
b Pappus simple, the bristles all equal and of one kind....o
b Pappus double, the outer very short and chaffy. Lvs. entire. W. S. CHRYSOP'SIS
b Pappus double in the disk, none in the rays. Lvs. toothed. S. HETEROTHE'CA

c Heads small, rays few (2-15)....d
   ᴐ Heads quite large, rays narrow, about 80. Tall. *c.* *Elecampane.* In'ula.
d Pappus scaly, very short. Root lvs. cordate. Rays 4 or 5. S. Brachychæ'ta.
d Pappus abundant bristly hairs. Root lvs. not cordate. *Goldenrod.* Solida'go. 1
d Pap. of a single row of equal bristly hairs. Ped. long, slender. S. Isopap'pus.
   e Involucre about 4-rowed. Rays 20-30. Head solitary. S. Baldwin'ia.
   e Involucre about 2-rowed. Rays 8-10. Heads corymbed. S. Actinosper'mum.
   f Ray florets pistillate. Leaves decurrent.   *Sneezewort.* Hele'nium.
   f Ray florets neutral.   *False Sneezewort.* Leptop'oda.
     g Fruit (achenia) 4-angled. Heads large, showy. *Cone-flower.* Rudbeck'ia.
     g Fruit flattened, winged. Heads showy. Rays droop. W. S. Lep'achys.
  h Achenia wingless. Pappus of 2 deciduous scales. *Sunflower.* Helian'thus. 2
  h Achenia winged. Pappus of 2 persistent awns. Lvs. often decurrent. W.
                        Rag-Sunflower. Actinom'eris.
   k Achenia wingless, in more than 1 row. Coarse herbs with large heads.
                    M. W. *Leaf-cup.* Sil'phium.
   k Ach. winged, in only 1 row. Small, with middling hds. S. Berlandie'ra.
m Stems leafy, erect, about 2f. (or 1-3f.) high....n
m Stemless plants, leaves radical, appearing after heads. *Colt's-foot.* Tussila'go.
   ᴜ Scales 5, united in 1 row. Leaves pinnate. *French Marigold.* Tage'tes.
   ᴜ Scal. in 2 rows, the out. united. Lvs. pinn. W. S. *False Dog-fennel.* Dyso'dia.
   ᴜ Scales in 1 or 2 rows, all distinct. W. S.   *Arnica.* Ar'nica.
o Involucre imbricated in 3 or more rows of scales. *Sunflower.* Helian'thus. 2
 ᴐ Invol. 2-rowed. Pappus of downwardly hispid awns. *Burr Marigold.* Bi'dens. 3
o Invol. 2-rowed. Pappus upwardly hispid if at all. *Tick Sunflower.* Coreop'sis. 4
   p Achenia wingless. Rays 5-12. Herbs viscid, 2-10f. high. S. Polym'nia.
   p Achenia wingless. Rays 5. Herbs 2-10' high, at first stemless. Flowers
     early in Spring. W. S.   Chrysog'onum.
   p Achenia broadly winged. Rays 12-25. Coarse, tall herbs. M. S. W.
                 Resin Weed, Polar Plant. Sil'phium.
  q Herbs 3-6f. high. Rays 1-5. Recept. flat. S.-W. *Crown Beard.* Verbesi'na.
  q Herbs 2-3f. high. Rays 6-9. Receptacle convex. S.-W. Tetragonothe'ca.
  q Herbs 2-6f. high. Rays 10-15. Recept. conical. *False Sunflower.* Heliop'sis.
  q Shrubs 3-10f. high, with solitary heads. S.   Borrich'ia.
    r Ach. very silky, biggest at top. Rays about 5. *False Aster.* Sericocar'pus.
    r Achenia smooth or smoothish, flattened. Rays 6-100. *Starwort.* Aster. 5
 s Wild plants 1-4f. high, with middle-sized heads (about 1' broad). Diplopap'pus.
 s Garden plants 1-2f. high, with very large heads. *China Aster.* Callis'tephus.
   t Herbs 2-3f. high, very smooth. Leaves lanceolate, entire. W. Bolto'nia.
   t Herbs ½-9f. high, hairy or rough. Rays 20-200. *White-weed.* Erig'eron. 6
u Involucre broad and flattish. Pappus 0. Rays white. *Ox-eye.* Leucan'themum.
u Involucre hemispherical. Pappus a membranous margin. Cult. Pyre'thrum.
ᴜ Involucre hemispherical. Pappus 0. Lvs. lobed. Cultivated. Chrysan'themum.
ᴜ Inv. bell-shaped. Pappus 0. Lvs. entire. Rays violet-purp. W. † *Daisy.* Bel'lis.

v Disk florets yellow, perfect. Rays pistillate. *Camomile.* ANTHEM'IS.
v Disk florets yellow, perfect. Rays neutral. *May-weed.* MARU'TA.
v Disk florets white, perfect. Rays pistillate. *Yarrow.* ACHIL'LEA. 7
w Rays short, white, 3 or 4. W. S. *Crown-beard.* VERBESI'NA.
w Rays very short, white, 5, ear-shaped. W. M. PARTHE'NIUM.
w Rays very large, purple, pendulous. *Purple Cone-flower.* ECHINA'CEA.
x Leaves pinnately divided. Inner involucre of 8 united scales. † DAH'LIA.
x Leaves simple. Receptacle conical with large chaff. † ZIN'NIA.
x Leaves simple. Receptacle flat. Rays rose-color. *Tick-seed.* COREOP'SIS.
x Leaves simple. Receptacle flat. Rays white, short. W. ECLIP'TA.
y Heads in corymbs. Disk florets regularly 5-toothed. r. NARDOS'MIA.
y Heads solitary. Disk florets regularly 5-toothed. S. *Daisy.* BEL'LIS.
y Hds. solitary. Disk flts. 2-lipped, outer lip 8-toothed, inner 2. S. CHAPTA'LIA.

Sub-order Second, LIGULIFLORÆ,

having *all* the florets ligulate (§ 96) and perfect, *i. e.*, the heads radiant.

2 Flowers bright yellow....3
2 Flowers cream-color or purplish....5
2 Flowers blue. Stems leafy, erect....e
8 Pappus none. Involucre of about 8 equal scales....a
8 Pappus double, the outer of scales, inner of bristles....b
8 Pappus wholly of feathery bristles....f
8 Pappus wholly of hair-like bristles, generally abundant....4
4 Fruit bearing the pappus on its slender beak....c
4 Fruit not lengthened into a beak, pappus sessile....d
5 Pappus consisting of equal, feather-like bristles....f
5 Pappus of simple, hair-like bristles, abundant....g
= Leaves all alternate. Heads panicled. r. *Nipplewort.* LAMPSA'NA.
a Leaves partly opposite. Heads solitary or umbeled. Pappus 0. S. APO'GON.
b Leaves all radical, pinnatifid-toothed. Pappus scales 5, with
5 bristles. c. *Dwarf Dandelion.* KRIG'IA.
b Lvs. all or mostly rad., seldom pinn. Pap. scal. and brist. many. CYN'THIA.
c Stemless leaves runcinate. Pappus white. *Dandelion.* TARAX'ACUM. 8
c Stems leafy or not. Pappus reddish or tawny. S. PYRRHOPAP'PUS.
e Stems leafy, leaves runcinate. Pappus silky-white. c. *Lettuce.* LACTU'CA.
d Pappus brownish. Stems mostly leafy, with many heads. c.
Hawkweed. HIERA'CIUM. 9
d Pap. silky white. Stemless; scapes each with one head. W. TROX'IMON.
d Pappus silky white. Stems bear prickly leaves. c. *Sow Thistle.* SON'CHUS.
e Pappus of many small scales. Branched stems 2f. high. Heads axillary,
large. Common. Eastward. *Succory.* CICHO'RIUM.
e Pappus of many hair-like bristles. 3–8f. *Blue Lettuce.* MULGE'DIUM.
f Leaves on the stem linear, entire. Purpl. † *Vegetable Oyster.* TRAGOPO'GON.
f Leaves all radical, toothed. Flowers yellow. Fruit taper-beaked.
Hawkbit. LEON'TODON.

g Ach. with a long beak, pap. silk-white. Heads erect. *c. Wild Lettuce.* Lactu'ca.
g Achenia not beaked, pappus dull-white. Hds. nod. *c. Drop-flower.* Nab'alus. 10
g Achenia not beaked, pap. dull-white. Hds. erect, purple. S. *r.* Lygones'mia.

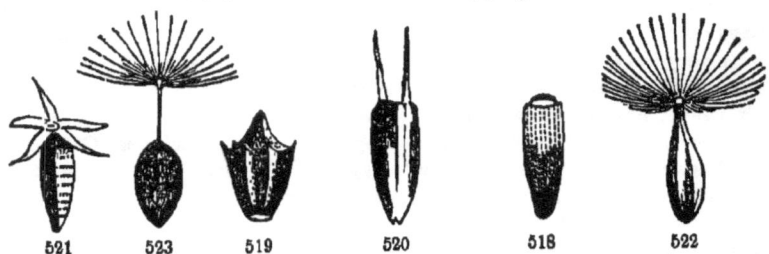

Achenia of Asterworts, showing the varying pappus. *Fig.* 518. Achenium of Eclipta, w pappus. *Fig.* 519. Horseweed (*Ambrosia trifida*). *Fig.* 520. Sunflower; pappus 2 teeth *Fig.* 521. Ageratum,—5 scales. *Fig.* 522. Blue Lettuce,—many hair-like bristles. *Fig.* 523. Wild Lettuce, pappus raised on a beak.

1. SOLIDA'GO. Goldenrod.

Heads few-flowered, the rays 1–15, pistillate, disk florets perfect. Involucre oblong, imbricate, with close-pressed scales. Receptacle alveolate, narrow. Pappus simple, of equal, hair-like, rough bristles.—Herbs, very abundant in the United States. Stem erect, branching near the top. Leaves alternate. Heads small, florets all yellow (in **S. bicolor**, whitish), opening from August to October. (See Figs. 501–503.)

¶ Shrub woody, 1–3f. high. Heads with 1–3 rays. S....1
¶ Herbs. Heads without rays (discoid). S....2, 8
¶ Herbs. Heads with rays (1–15, generally small)....a
    a Scales of the involucre with recurved, leafy, green tips....4, 5
    a Scales of the involucre erect, tips scarcely at all green....b
b Herbs (white or yellow) in axillary, close clusters or short racemes....6–9
b Heads in terminal racemes forming a close or a spreading panicle....c
b Heads in terminal compound corymbs....n
    c Racemes erect, not one-sided. Leaves feather-veined....d
    c Racemes spreading or recurved, the flowers all on one side....f
d Alpine species (growing only on mountains). Heads quite large....10–12
d Not alpine—growing in plains or low grounds. Heads not large....e
    e Plants very smooth, at least the stem and leaves. Rays 4–7....13–15
    e Plants downy or hoary with very close soft hairs. Rays 9–12....16, 17
    Leaves 3 or 1-veined. Very smooth salt-marsh herbs....18, 19
f Leaves evidently 3-veined. Herbs inland, &c....g
f Leaves not veiny, thick, subentire....27–29
f Leaves evidently feather-veined, mostly serrate....k

g Leaves entire or very nearly so....20, 21
g Leaves serrate. Stem smooth and glabrous....22–24
g Leaves serrate. Stem roughish-pubescent....25, 26
k Stem downy or hairy. Leaves rough or not....30–32
k Stem smooth and glabrous. Leaves smooth or rough....m
   m Rays 6–12. Racemes close, forming a compact panicle....38–40
   m Rays 6–12. Racemes distant, loosely or scarcely panicled....36, 37
   m Rays 2–5. Racemes, or the panicle, long and slender....33–35
n Leaves lanceolate, large. Stem smooth....44–46
n Leaves lanceolate, large. Stem rough-downy....41–43
n Leaves linear, entire. Stems much branched, smoothish....47, 48
   1 S. paucifloscullo'sa. *Shrubby Goldenrod.* Bush 2f. high, very smooth, with lanceolate leaves and the 5-flowered heads in erect, panicled racemes. S.
   2 S. discoi'dea. *Rayless G.* Disk florets 10–15. Racemes erect, panicle slender. S.
   8 S. brachyphyl'la. *Chapman's G.* Florets 5–7. Racemes spreading, one-sided. S.
   4 S. squarro'sa. *Ragged S.* Rays 10–15. Scales stiff, with spreading, green tips. Heads large. N.
   5 S. squarrulo'sa. *Rough S.* Rays 6–10. Scales awl-shaped, with slender, loose tips. S.
   6 S. bi'color. *Creamy S.* Rays about 8, creamy-white. Plant hairy. Lva. elliptic.
   7 S. Buck'lyi. *Buckly's S.* Rays 4–6, yellow. Plant woolly. Leaves oblong. S.
   8 S. latifo'lia. *Broad-leaved S.* Rays yellow. Plant smoothish. Leaves broad lanceolate, coarse-toothed. Seed downy. *c.*
   9 S. cæ'sia. *Polished S.* Rays yellow. Plant smooth and glaucous. Lvs. lin.-lanceolate. Stem flexuous, tall, slender. A beautiful Goldenrod. Woods. *c.*
   10 S. thyrsoi'dea. *Thyrse G.* Leaves ovate, long-stalked. Tall, 1–3f. high. Heads large. Coarse, showy. In mountain woods.
   11 S. Virgau'rea. *True G.* Leaves oval, short-stalked. Low, 2–3' high. Heads about 30-flowered, few, often only one.
   12 S. hum'ilis. *Mountain G.* Lvs. oblanceolate. High 6–12'. Heads about 12-flwd.
13 S. virga'ta. *Virgate G.* Heads all in one raceme at top.
14 S. stric'ta. *Upright G.* Heads in a panicle, which is narrow and erect.
15 S. specio'sa. *Showy G.* Heads in a thyrse-like panicle, large and very showy. Pedicels shorter than the involucre, pubescent. Leaves very broad.
   16 S. verna. *Early S.* Whitish-downy. Lower leaves ovate. *May, June.* S.
   17 S. puber'ula. *Dusty S.* Dusty-puberulent. Lower leaves oblanceolate. Panicle long, compound, dense. Scales acute. *Aug.* N.
18 S. sempervi'rens. *Evergreen S.* Lvs. lanceolate, thick, obscurely 8-veined. E.
19 S. angustifo'lia. *Narrow-lv. S.* Lvs. lance-lin. 1-veined, thick. Hds. small. S
   20 S. nemora'lis. *Wood S.* Plant dusty, roughish. Lvs. acute. Rays showy. *c*
   21 S. rupes'tris. *Rock S.* Plant smooth. Lvs. acuminate. Rays very short. W
22 S. Missourien'sis. *Missouri G.* St. 1–2f. All glabrous. Panicle dense. W.
23 S. sero'tina. *Late G.* Stem 3–6f. Leaf veins hairy beneath. Panicle loose.
24 S. gigan'tea. *Giant G.* Stem 8–8f. Branchlets hairy. Leaves lanceolate.

10*

25 S. Canaden'sis. *Canada G.* Leaves pointed, rough. Panicle broad. *c.*

26 S. Short'ii. *Short's G.* Leaves acute, very smooth. Panicle long, narrow. W.

27 S. pilo'sa. *Hairy S.* Hairy, 4–7f. high. Leaves remotely serrulate. N.–J. S.

28 S. odo'ra. *Sweet S.* Stem downy in lines, slender, 2–3f. high. Leaves very entire, smooth, punctate with pellucid dots. Fragrant. *c.*

29 S. tortifo'lia. *Twist-lv. S.* Stem rough. Lvs. often twisted, not punctate. S.

30 S. altis'sima. *Tall S.* Stem hairy, 4–6f. Lvs. veiny, rough. Scales acute. *c.*

31 S. Drummon'dii. *Drummond's S.* Stem 1–2f. Lvs. velvety. Scales obtuse. W.

32 S. rad'ula. *Rasp-lv. S.* Stem rough-downy. Lvs. oblong-spatulate. W.

33 S. ulmifo'lia. *Elm S.* Branchlets hairy. Scales acute. Rays 3 or 4, disk flowers 3 or 4. N. W.

34 S. Boot'ii. *Boott's S.* Branchlets hairy. Scales obtuse. Rays 2–5, disk flowers 8–12. S.

35 S. linoi'des. *Flax S.* Smooth all over. Scales obtuse. Rays 1–4. 12–20′. N.

36 S. Muhlenber'gii. *Muhlenberg's S.* Lvs. large, thin, notched, smooth both sides. Heads 15-flowered. N.

37 S. pat'ula. *Spreading S.* Lvs. large, thick, very rough on the upper side. Stem 2–4f, branches leafy. Heads 20-flowered. N.

38 S. ellip'tica. *Marsh S.* Very leafy. Lvs. elliptic. Panicle dense, pyramidal.

39 S. argu'ta. *Saw-lv. S.* Leaves few, elliptic, sharply serrate. Panicle spreading.

40 S. neglec'ta. *Neglected S.* Leaves few, serrate, lin.-lanceolate. Panicle narrow.

41 S. Ohien'sis. *Ohio S.* Smooth all over. Lvs. obtuse, flat. Corymbed. W.

42 S. Riddel'lii. *Riddell's S.* Branches, &c., dust-downy. Lvs. acute, concave. Heads corymbed. W.

43 S. corymbo'sa. *Corymbed S.* Branches corymbed, hirsute. Outer secund.

44 S. Houghto'nii. *Houghton's S.* Hds. few, very large. Otherwise like No. 41. N.–W.

45 S. rig'ida. *Stiff S.* Lvs. rigid. Heads very large. Scales obtuse. Height 3–5f.

46 S. Spithamæ'a. *Dwarf S.* Leaves thin, sharp-serrate. Scales acute. Height 6–12′. Mountains. S.

47 S. lanceola'ta. *Lance-lv. S.* Leaves linear-lanceolate, 3–5-veined. Rays minute, about 17. Corymbs crowded, fragrant. *c.*

48 S. tenuifo'lia. *Linear-lv. S.* Leaves narrow-linear, one-veined. Rays short, about 10. More slender, with thinner clusters. *c.*

2. HELIANTHUS. Sunflower.

Heads many-flowered, rays neutral, disk-florets perfect. Scales of the involucre in several rows, more or less imbricated. Torus flat or convex, the chaff persistent, embracing the 4-sided, flattened achenia. Pappus of 2 chaffy awns, deciduous.—Herbs, mostly ⅔, rough. Leaves opposite, the upper often alternate, mostly 3-veined. Heads mostly large, the disk from half an inch to 1f. broad. Rays yellow, disk yellow or purple. *July–Oct.* (Figs. 498, &c.)

§ Disk with its corollas and pales dark purple....a
§ Disk with its corollas and pales yellow....o
      a Herbs annual. Leaves chiefly alternate....1, 2
      a Herbs perennial. Leaves opposite....b
        b Scales of the involucre acuminate....3-5
        b Scales of the involucre obtuse....6, 7
      o Leaves chiefly alternate and feather-veined....8-11
      o Leaves chiefly opposite and 3-veined or triple-veined....d
        d Scales of involucre erect, closely imbricated....e
        d Scales loosely spreading. Heads large, 9-15-rayed....f
        d Scales loosely spreading. Heads small, 5-8-rayed....22-25
          e Plants green, rough....12, 13
          e Plants whitish, downy....14, 15
            f Scales lance-linear, longer than disk. Leaves thin....16, 17
            f Scales lance-ovate, as long as the disk. Leaves thick....18-21

1 H. an'nuus. *Common S.* Stout and tall (3-10f.). Heads large 6-10 across, nodding. Achenia (seeds) glabrous. A variety has all the flowers ligulate.

2 H. deb'ilis. *Slender S.* Slender, decumbent. Heads small. Seeds downy. S.

3 H. Rad'ula. *Rasp-lv. S.* Leaves roundish, rough, obtuse. Rays 7-10 or none. S.

4 H. heterophyl'lus. Leaves oval, lanceolate, &c. Rays 12-18. Pales acute. S.

5 H. angustifo'lius. Leaves lance-linear, pointed. Pales 3-toothed. N.-J. S.

6 H. rigidus. *Rigid S.* Lvs. lanceolate, pointed. Scales ovate, acute. Rs. 12-20. W.

7 H. atrorubens. *Livid S.* Leaves ovate, obtusish. Scales oblong, obtuse. S.

8 H. gigan'teus. *Tall S.* Hairy, rough. Lvs. lanceolate, pointed, serrate. c.

9 H. tomento'sus. *Velvet S.* Very downy. Lower lvs. ovate, nearly entire. W. S.

10 H. grosse-serra'tus. *Coarse-toothed S.* Stem smooth. Leaves lance-pointed, sharp-serrate. Rays 15-20. W.

11 H. tubero'sus. *Artichoke.* Cultiv. Lvs. 3-veined, lower cordate-ovate.

12 H. lætiflo'rus. *Laughing S.* Branched. Leaves lance-oval, short petioled.

13 H. occidenta'lis. *Western S.* Stem slender, simple, nearly leafless above.

14 H. mol'lis. *Soft S.* Leaves ovate, cordate, sessile. Plant woolly. W.

15 H. cine'reus. *Ashy S.* Lvs. ovate-oblong, tapering to base. Ashy-downy. O.

16 H. decapet'alous. *Ten-rayed S.* Rays 9-12. Leaves all opposite. Stem 3-4f. N. M.

17 H. tracheliifo'lius. *Trach-leaved S.* Rays 12-15. Branch lvs. alternate. 4-8f. W.

18 H. doronicoi'des. *False Tiger-bane.* Leaves petiolate, ovate, and lance-ovate, upper alternate. Scales longer than disk. Rays 12-15. W. S.

19 H. strumo'sus. *Warty S.* Leaves short-petioled, lance-ovate, all alike. Scales equalling the disk. c. A double-flowered variety is cultivated.

20 H. hirsu'tus. *Hairy S.* Leaves petiolate, hairy beneath. Scales hairy. W.

21 H. divarica'tus. *Forked S.* Leaves sessile, very rough, opposite or ternate. c.

22 H. microceph'alus. *Small S.* Stem smooth, much branched. Lvs. narrow. W.

23 H. Schweinit'zii. *Schweinitz's S.* Stem downy, rough. Leaves white, downy. Car.

24 H. læviga'tus. *Polished S.* Stem and leaves very smooth. Not branched. S. Mts.

25 H. longifo'lius. *Long-lvd. S.* Leaves lance-linear, acute, smooth. Rays 8-10. Ga.

3. BI'DENS. Burr-Marigold.

Involucre scales nearly equal, double, the outer generally large and leafy. Rays few (3–8, or sometimes none), neutral, disk perfect. Receptacle chaffy, flat. Achenia flattened or 4-sided, crowned with 2–4 awns which are hispid backwards.—Leaves opposite. *July–Oct.*

* Rays inconspicuous or none....a
* Rays quite showy, yellow....4, 5
    a Achenia flattened, broadest at top....1–3
    a Achenia slender, 4-sided....6, 7
1 **B. frondo'sa.** *Leafy B.* Leaves pinnately 3–5-fol., divisions distinct. Rays 0.
2 **B. conna'ta.** Leaves simple, lower ones sometimes 3-parted. Rays 0.
3 **B. cer'nua.** *Nodding B.* Leaves simple, scarcely connate. Rays few or 0.
    4 **B. chrysanthemoi'des.** *Mud B.* Lvs. narrow-lance., equally serrate, connate.
    5 **B. Beck'ii.** *Beck's B.* Lvs. mostly under water and very finely divided. M.
    6 **B. leucan'tha.** *White B.* Heads small, with white rays. Lvs. pinnate. S.
    7 **B. bipinna'ta.** *Spanish Needles.* Rays very short, yellow. Lvs. bi-pinnate.

4. COREOP'SIS. Tick-seed.

Involucre many-flowered, double, each of 8–18 scales, the outer leafy, the inner membranous. Receptacle flat, the chaff falling with the fruit. Achenia flattened, often winged, emarginate, each commonly with 2 teeth or awns which are not hispid downwardly as in BIDENS.—Leaves generally opposite. Heads showy (rarely without rays).

* Heads discoid (without rays)....1, 2
* Heads radiate, rays showy....a
    a Disk yellow, rays also yellow, mostly entire....b
    a Disk yellow, rays rose-colored, 3–5-toothed at the end....20, 21
    a Disk purple, rays yellow with a purple base, toothed....18, 19
    a Disk purple, rays wholly yellow, toothed at the end....14–17
b Leaves petiolate, compound, with lanceolate, toothed divisions....3–5
b Leaves petiolate, compound, with linear, entire divisions....6, 7
b Leaves petiolate, simple, or some of them eared at base....8–10
b Leaves sessile, 3-parted, divisions entire or not often, seeming whorled....11–13
    1 **C. discoi'dea.** *Rayless T.* Leaves on long petioles, ternately divided. W.
    2 **C. bidentoi'des.** Leaves on short petioles, toothed, lance-linear. Penn. *r.*
3 **C. au'rea.** *Golden T.* Leaflets 3–5. Outer scales about 8. Achenia 2–4-toothed. S.
4 **C. trichosper'ma.** Leaflets 5–7. Outer scales about 8. Ach. slender, 2-toothed.
5 **C. aristo'sa.** Leaflets 5–7. Outer scales 10–13. Achenia 2 or 4-awned. W.
    6 **C. trip'teris.** Stem 4–8f. high. Hds. on short stalks. Rays 1' long, entire. W. S
    7 **C. grandiflo'ra.** St. 1–2f. high. Heads on long stalks. Rays 1' long, 4–5-cleft. S

8 C. latifo'lia. Stem 4–6f. high. Rays entire. Leaves ovate, serrate. S.

9 C. auricula'ta. Stem 1–3f. high. Rays 2–5-toothed. Lvs. often eared at base. S.

10 C. lanceola'ta. Stem 2–3f. high. Rays 4–5-toothed. Lvs. lanceolate, entire. S.

11 C. senifo'lia. Leaf divisions all entire, appearing in 6-leaved whorls. S.

12 C. verticilla'ta. Leaf divisions all again divided into narrow-linear lobes. W.

13 C. palma'ta. Lvs. deeply 3-cleft, wedge-shaped, lobes linear, not whorled. W.

14 C. delphinifo'lia. Leaves sessile, 3-parted, the divisions often lobed. S.

15 C. gladia'ta. *Sword-lv. C.* Leaves petioled, lanceolate, sometimes divided. Stem round. S.

16 C. angustifo'lia. *Narrow-lv. C.* Leaves petioled,
narrow-spatulate, entire. Stem square. S.

17 C. Œm'leri. *Œmler's C.* Leaves petioled, lance-
ovate, entire. Stem round below. S.

18 C. Drummon'dii. *Drummond's C.* Lvs. pinnately 3–5-foliate, divisions oblong-oval,
entire. †

19 C. tincto'ria. *Dyer's C.* Leaves pinnately
much divided, divisions linear, entire. †

20 C. ro'sea. *Rose C.* Stem leafy, leaves narrow-
linear, entire. Rays rose-white. E.

21 C. nuda'ta. *Leafless C.* Stem few-leaved, leaves
awl-shaped, entire. Rays rose-red. S.

Fig. 524. Aster patens.

5. ASTER. Starwort.

Heads many-flowered. Scales of the involucre generally imbricated in two or more
rows, and with green tips. Disk florets tubular, perfect, rays fertile, in one row, oblong,
revolute when old. Receptacle flat, marked
with pits. Pappus simple, hair-like, rough. Achenium usually flattened.
—A large genus of ♃ herbs, very abundant in the United States, flowering in late summer and autumn. Leaves alternate; disk florets yellow,
changing to purple; rays blue, purple, or white, never yellow.—The species are very variable, and many of them are hard to distinguish.

* Radical and lower leaves cordate and petiolate....a

* Radical leaves never cordate....o

    a Heads in loose corymbs. Rays white or whitish....1, 2

    a Heads in racemes or panieles, blue or bluish....b

b Leaves evidently serrate; rays light blue, about 12, spreading ½' ...3, 4

b Leaves entire or nearly so; rays bright blue, spreading near 1'....5–7

    o Involucre scales tipped with green, or the outer ones wholly green....d

    o Involucre scales with scarious margins or wholly scarious....f

d Stem leaves clasping, with a cordate or auricled base,...e
d Stem leaves sessile, rarely clasping, never cordate or auricled....19
 e Involucro scales close, in several rows, outer ones gradually shorter....8, 9
 e Involucre scales loose, nearly equal, outer ones often wholly green....10–12
f Leaves lanceolate and linear-lanceolate, more or less rough....13–15
f Leaves linear, fleshy, very smooth, entire. Salt-marsh herbs....16–18
 1 A. corymbo'sus. *Corymbed S.* Slender, with thin, serrate leaves.
 2 A. macrophyl'lus. *Big-lvd. S.* Stout, with large, thick, ser., rough lvs. 13-rayed.
 3 A. cordifo'lius. *Heart-leaved S.* Involucre scales close, obtuse. Lvs. sharp-serrate.
 4 A. sagittifo'lius. *Arrow-leaved S.* Scales awl-shaped, long, loose. Lvs. blunt-serrate.
 5 A. azu'reus. *Azure S.* Stem leaves sessile, rough, lanceolate, and linear.
 6 A. undula'tus. Stem lvs. on winged stalks, with rounded clasping bases, wavy.
 7 A. Shor'tii. *Short's S.* Stem leaves on naked stalks, all cordate, pointed, entire.
 8 A. patens. *Patent S.* Plant rough-downy. Leaves entire. Scales pointed.
 9 A. lævis. *Polished S.* Plant smooth and glaucous. Scales broad, acute.
 10 A. prenanthoi'des. Lvs. sharply cut-serrate, with a long, slender, entire base.
 11 A. punic'eus. *Red-st. S.* Lvs. sparingly serrate, lance. Stem hairy, often red.
 12 A. Novæ-Angliæ. *New-England S.* Leaves entire, rough, numerous. Rays nearly 100, ¾' long. Stems 4–6f. high. A fine species, often cultivated.
13 A. acumina'tus. *Dell S.* Leaves coarsely-toothed, broad-lanceolate, long-pointed, often clustered. Rays white. In dark woods. N.
14 A. nemora'lis. *Wood S.* Leaves narrow-lanceolate, nearly entire, acute, with edges revolute. Heads 1–3. In damp woods. N. M.
15 A. ptarmicoi'des. *Sneezewort S.* Leaves entire, stiff, acute. Heads corymbed.
 16 A. flexuo'sus. *Zigzag S.* Heads large, with showy rays. Stem flexuous.
 17 A. linifo'lius. *Flax S.* Heads numerous, with very short rays in 2 rows.
 18 A. subula'tus. Heads with showy blue rays. Scales in 2 or 3 rows. S.
19 Many species, very variable, here omitted. (See p. 420, Class Book.)

6. ERIG'ERON. Fleabane. Whiteweed.

Heads many-flowered, mostly hemispherical, rays very numerous (20-200), narrow, linear, pistillate ; disk flowers perfect. Receptacle flat, naked (no chaff or pits). Scales of the involucre nearly equal and in one row. Pappus generally simple.—Herbs with alternate leaves. Rays white, blue, or reddish. Flowering from May to September.

* Rays showy, longer than the involucre. Heads large (½–1' broad)....a
* Rays obscure, shorter than the involucre, whitish. Heads very small....1, 2
 a Rays purple, very numerous. Heads loosely corymbed....3–5
 a Rays white or whitish. Heads loosely panicled....6–8
1 E. Canaden'se. *Canada F.* Erect, hairy. Leaves lanceolate. Heads panicled.
2 E. divarica'tum. *Prostrate F.* Low, diffuse. Lvs. linear. Heads corymbed. W.

3 E. bellidifo'lium. *Daisy F.* Leaves nearly entire. Rays 50–80, bluish-p.
4 E. Philadel'phicum. Leaves nearly entire. Rays 150–200, reddish-purple.
5 E. qoercifo'lium. *Oak-lv. F.* Lvs. sinuate-pinnatifid-toothed. Rays 100–200. **S.**
6 E. an'nuum. *Annual F.* Stem leafy, 3–5f. high. Leaves coarse-toothed.
7 E. strigo'sum. *Rough F.* Stem leafy, 2–3f. high. Leaves nearly entire.
8 E. nudican'le. *Naked F.* Stem leafless, 1–2f. high. Rays about 30. **S.**

7. ACHILLE'A. Yarrow. Millfoil.

Heads many-flowered, rays few, fertile; receptacle flat, chaffy; achenia flattened, margined, without a pappus.—4 European herbs with small, 4–12-rayed heads in corymbs. *June–Sept.*

1 A. millefo'lium. Leaves twice pinnatifid with fine segments. Rays 4 or 5. *c.*
2 A. Ptar'mica. *Sneezewort.* Leaves undivided, lance-linear, serrate. Rays 8–12. *r.*

8. TARAX'ACUM. Dan'delion.

Involucre many-flowered, double, the outer of small scales much shorter than the close, erect row of the inner. Receptacle naked. Achenia produced into a long beak crowned with copious white, hair-like bristles of the pappus.—Acaulescent herbs with runcinate leaves. (Figs. 504–506.)

T. Dens-leo'nis. *Dan'delion.* Outer scales of the involucre reflexed. Leaves runcinate, smooth, dentate.—In all open situations, blossoming at all seasons except winter. Scape round, hollow, lengthening after flowering, and bearing a globular head of seeds and seed-down, whose light and airy form is a very familiar sight to all.

9. HIERA'CIUM. Hawkweed.

Involucre more or less imbricated, egg-shaped, many-flowered. Achenia not prolonged into a beak, striate. Pappus of rough, brittle, numerous tawny bristles in a single row.—4 Leaves alternate, entire, or toothed. Florets yellow. *July–Sept.*

* Involucre and stalks smooth or nearly so....a
* Involucre, stalks, &c., rough with glandular hairs....b
    a Heads with 50 to 60 florets....1
    a Heads with 10 to 20 florets....2, 3
      b Heads with 40 to 50 florets....4
      b Heads with 20 to 30 florets....5, 6
    1 H. Canaden'se. *Canada H.* Stem leafy, corymbed at top. Leaves sharp-toothed. **N.**

2 H. panicula'tum. *Panicled H.* Stem leafy, widely panicled. Leaves fine-toothed.
3 H. veno'sum. *Robin's Plantain.* Stem almost leafless, corymbed. Lvs. entire.
    4 H. scabrum. *Rough H.* Heads corymbed. Plant stiff, rough-hairy.
5 H. longip'ilum. *Long-haired H.* Plant clothed with straight bristles 1' long. W.
6 H. Grono'vii. *Gronovius' H.* Plant slender, quite hairy below.

10. NAB'ALUS. Lion's-foot.

Involucre cylindrical, double, the inner of many linear scales in one row, the outer of a few short scales at base. Receptacle naked. Achenia smooth, striate, not beaked, crowned with a copious, straw-colored or brownish hair-like pappus.—Erect herbs, with a thick, tuberous, bitter root. Heads 5–18-flowered, white or straw-colored, often purplish. *Aug.–Oct.*

* Heads glabrous, pendulous. Leaves multiform in the same plant....<
* Heads hairy, erect or nodding. Leaves reniform, undivided....7–9
    a Tall (2–4f. high). Heads (8–12-flowered) in a corymb-like panicle....1, 2
    a Tall (2–6f. high). Heads in a long, raceme-like panicle....3, 4
    a Low (5–10' high). Heads racemed. Found only on high mountains....5, 6
1 N. al'ba. *White L.* Pappus cinnamon-color. Leaves hastate, often lobed.
2 N. Fra'seri. *Fraser's L.* Pappus straw-color. Leaves deltoid, often cleft.
3 N. altis'simus. *Tall L.* Heads 5-flowered. Leaves divided, or cleft, or entire.
4 N. virga'tus. *Rod L.* Heads 8–12-flowered. Lowest leaves pinnatifid.
5 N. na'nus. *Dwarf L.* Outer involucre of short-ovate, close scales.
6 N. Boot'tii. *Boott's L.* Outer involucre of linear, loose scales.
    7 N. racemo'sus. *Racemed L.* Heads nodding, 9–12-flowered. W. M.
    8 N. crepidin'eus. *Crepis L.* Heads nodding, 25–35-flowered. W. S.
    9 N. as'per. *Rough L.* Heads erect, 11–14-flowered. Panicle racemed. W.

Order LXXI. LOBELIACEÆ. Lobeliads.

Herbs with alternate *leaves*, scattered *flowers*, and often milky *juice*; *calyx* superior; *corolla* irregular, 5-lobed, tube split down to the base; *stamens* 5, united into a tube both by the filaments and anthers; *ovary* adherent to the calyx tube; *styles* united into one; *stigma* fringed; *fruit* a 2–3-celled, many-seeded capsule.

LOBE'LIA. Cardinal-flower. Indian Tobacco.

The two upper lobes of the irregular corolla are smaller than the three lower. Stamens united into a curved tube. Stigma 2-lobed. Capsule opening at top. Seeds very small.—Flowers axillary, generally forming leafy or bracted racemes. *July–Sept.*

§ Stems leafy....a
§ Stems leafless, leaves nearly all crowded at the root, under water....11, 12
    ⚹ Flowers bright red or scarlet, large and showy....1, 2
    a Flowers blue, varying to bluish-white....b
b Stem stout, 2-3 or 4f. high. Flowers large, about 1' long....3-5
b Stem slender, 6'-2f. high. Flowers small ($\frac{1}{4}$-$\frac{1}{2}$' long)....c
    c Stem branched, racemes several, loose, or flowers scattered....6, 7
    c Stem generally simple, bearing a single raceme....8-10
1 L. cardina'lis. *Cardinal-flwr.* Stem smooth. Leaves oblong-lanceolate, acute.
2 L. ful'gens. *Mexican.* Stem downy. Leaves linear-lanceolate, long-pointed. †
    3 L. puber'ula. Leaves obtuse, denticulate. Raceme one-sided. Plant downy.
    4 L. syphilit'ica. *Blue C.* Lvs. acute, slightly toothed. Racemes equal, hairy.
    5 L. amœ'na. *Pretty C.* Leaves acuminate, toothed. Racemes one-sided, smoothish. S.
6 L. infla'ta. *Indian Tobacco.* Hairy. Lvs. ovate-lanceolate, toothed. Pod inflated.
7 L. Kal'mii. *Kalm's C.* Smooth. Leaves linear-spatulate, entire. Fls. blue-white.
    8 L. Nuttal'lii. *Nuttall's L.* † Pedicels twice as long as the flowers. Leaves linear, extremely slender. S. M.
    9 L. spica'ta. *Spiked L.* Pedicels as long as the flowers. Racemes dense. Leaves oblong.
    10 L. leptostach'ya. *Slender-spiked L.* Pedi. none. Lvs. lance-oval, smooth. W
11 L. Dortman'na. *Water L.* Root leaves linear, terete, hollow, fleshy. Scape long.
12 L. palndo'sa. *Marsh L.* Root leaves linear-oblong, flat. Stem tall. S.

ORDER LXXII. CAMPANULACEÆ. Bellworts.

Herbs with a milky juice, alternate *leaves;*
flowers mostly blue and showy, with a superior
calyx; a regular and mostly campanulate 5-lobed *corolla;* with the 5
stamens usually separate, and *ovary* adherent to the calyx tube; and with
the 2–5-celled *pod* crowned with the remains of the calyx.

Analysis of the Genera.

Calyx tube very short (below the flower).	CAMPAN'ULA. 1
Calyx tube long and three-angled.	SPECULA'RIA.

CAMPAN'ULA. Bell-flower. Harebell.

Calyx 5-cleft. Corolla bell-shaped, funnel-shaped, or wheel-shaped, its 5 lobes valvate in the bud, closed at the base inside by the valve-like bases of the 5 stamens. Pod opening on the sides.—♃ Herbs with axillary or terminal flowers. *June–October.*

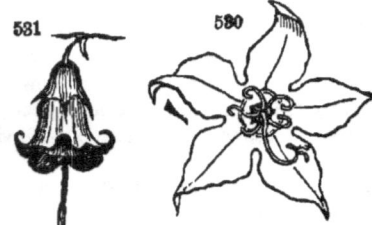

Fig. 526. The Harebell, the whole plant. 7. Ovary of Canterbury Bells, with *f*, a broad filament, *a*, an anther, and *p*, the hairy style. 8. A cross-section of the curious 5-celled seed-vessel, 2 placentæ in each cell. 9. Seed cut open, showing the large embryo. *Fig.* 530. Flower of American Bellwort. *Fig.* 531. Flower of Patent Bellwort.

§ Corolla wheel-shaped, flat, in leafy spikes....1, 2
§ Corolla bell-shaped, &c., broadly or narrowly....a
a Flowers on slender pedicels, solitary or panicled....b
a Flowers sessile or nearly so. Stem erect. Gardens....7–9
  b Flowers large (6–12″ broad). Root leaves unlike the stem leaves....3, 4
  b Flowers small (2–5″ broad). Leaves all similar in form....5, 6
1 C. America'na. *American B.* Stem tall (2–4f.). Leaves pointed at ends, smooth.
2 C. planiflo'ra. Stem low (7–12′), simple. Lvs. thick, shining, obtuse, or acute. †
  3 C. rotundifo'lia. *Harebell.* Stem weak. Root lvs. roundish, stem lvs. linear.
  4 C. persicifo'lia. *Peach B.* Stem erect. Leaves lance-linear. Flowers very broad. †
5 C. aparinoi'des. *Bedstraw B.* Stem reclining, rough backwards. Flowers white.
6 C. divarica'ta. *Patent B.* Erect. Panicle wide. Leaves toothed. Fla. blue. S.
7 C. glomera'ta. Flowers crowded above, funnel-shaped. Plant smooth. †
8 C. me'dium. *Canterbury B.* Flowers distant, very large, obtuse at base. †
9 C. lanugino'sa. *Woolly B.* Flowers scattered, rather large, acute at base. †

ORDER LXXIII. ERICACEÆ. **Heathworts.**

Herbs, or more generally *shrubs*, with simple, often evergreen *leaves ; flowers* regular or nearly so, 4 or 5-parted ; *petals* rarely almost distinct ; *stamens* as many or twice as many as the lobes of the corolla, and the *anthers* oddly appendaged and generally opening by two terminal pores ; the *style* 1, and the *ovary* 4–10-celled, with albuminous *seeds.*

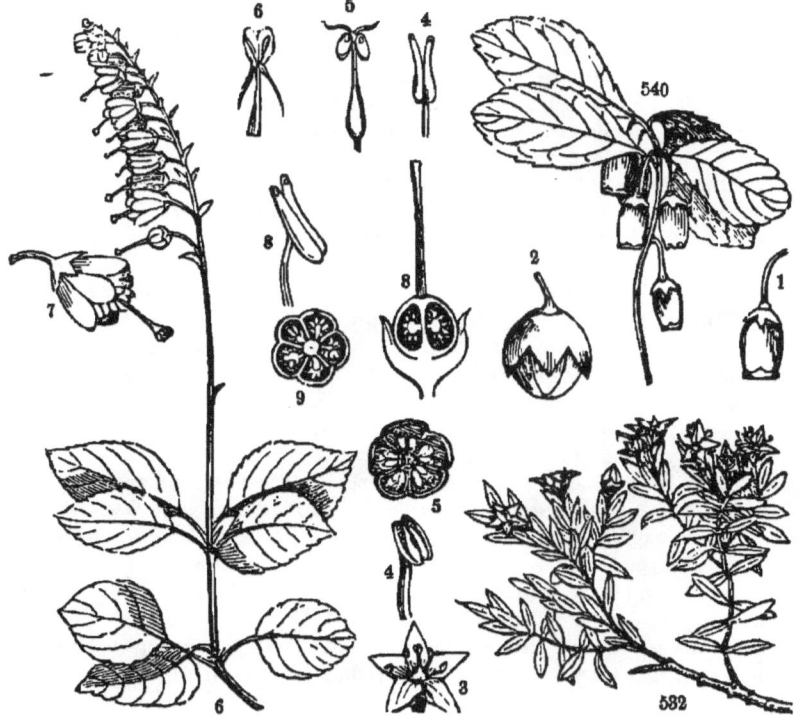

Fig. 532. Azalea procumbens. 3. A flower enlarged. 4. A stamen, much enlarged, showing the lengthwise opening of each of the cells. 5. Cross-section of a 5-celled capsule of Rhododendron, showing the inflexed margins of the valves. 6. Pyrola secunda. 7. A flower enlarged. 8. A stamen enlarged, showing the terminal tubes and pores. 9. Cross-section of a 5-celled, many-seeded capsule. *Fig.* 540. Checkerberry (*Gaultheria*). 1. A flower enlarged. 2. A berry. 3. Vertical section of the ovary, showing the free, fleshy calyx. 4. Anther of the Vaccinium Vitis-Idea. 5. Stamen of Bearberry (*Arctostaphylos*). 6. Awned stamen of a Blueberry (*Vaccinium*).

Analysis of the Genera.

§ Shrubs or trees, or shrublets....2
§ Herbs evergreen, with green herbage and leaves....m
§ Herbs leafless, without verdure. Bracts scale-like....n
  2 Calyx adherent, crowning the berry in fruit....a
  2 Calyx free from the ovary, or inferior....3
    3 Petals united into a gamopetalous corolla....4
    3 Petals entirely or very nearly separate and distinct....5
      4 Flowers 4-parted. Stamens 8....b
      4 Flowers 5-parted. Stamens 5 or 10....
      5 Pods 2 or 3-celled, cells only 1-seeded. Southern....k
      5 Pods 3-celled, cells many-seeded....g
      5 Pods 5 or 7-celled, cells many-seeded....h
        6 Corolla urn-shaped (oval or globular), lobes small....u
        6 Corolla not urn-shaped, open or spreading....e
a Erect shrubs with 5-parted flowers and 10-seeded berries.
                         Huckleberries. GAYLUSSA'CIA.
a Erect shrubs with 5-parted flowers and ∞-seeded berries.
                         Blueberries. VACCIN'IUM.
a Trailing shrublets. Corolla 4-cleft, reflexed. Fr. red. *Cranberry.* OXYCOC'CUS. 1
a Trailing shrublets. Corolla 4-cleft, spreading. Fruit white.
                         Boxberry. CHIOG'ENES. 2
b Leaves linear-accrose, whorled or crowded. Cultivated.   *Heath.* ER'ICA.
b Leaves oval-lanceolate. Shrub, 4f. high. Penn. S.   } *Mountain Heath.* MENZIE'SIA.
  c Pod dry, opening bet. the cells. Lvs. lin. N.   }
  c Pod dry, opening into the cells....d
  c Drupe fleshy, 5-seeded. Shrubs trailing. *Bearberry.* ARCTOSTAPH'YLOS.
  c Berry fleshy, many-seeded. Little shrublets. *Checkerberry.* GAULTHE'RIA. 3
d Shrublet moss-like, on high Mts. Leaves linear. *Moss Andromede.* CASSI'OPE.
d Shrubs with ample leaves. Pod-valves entire. *Andromede.* ANDROM'EDA.
d Tree with ample leaves and slender racemes. *Sorrel-tree.* OXYDEN'DRUM.
  e Corolla saucer-form, holding the anthers in 10 pits. *Laurel.* KAL'MIA. 4
  e Corolla salver-form, very fragrant. Trailing. *May-flower.* EPIGÆ'A. 5
  e Corolla funnel or bell-form, with spreading lobes....f
f Stamens 5, included. Plant and leaves very small. Mts. N. H. LEIOSELEU'RIA.
f Stamens 5 (rarely more), long-exserted. Corolla funnel-form. AZA'LEA. 6
f Stamens 10 (rarely fewer), exserted. Corolla bell-form. *Bay.* RHODODEN'DRON. 7
  g Leaves alternate, deciduous, serrate. Flowers racemed. CLE'THRA.
  g Leaves mostly opposite, evergreen, entire. Flowers umbeled.
                         Sand Myrtle. LEIOPHYL'LUM.
h Flowers 5-parted. Corolla regular. *Labrador Tea.* LE'DUM.
h Flowers 5-parted. Corolla irregular. RHODO'RA.
h Flowers 7-parted, regular. Stamens 14. S. BEFA'RIA.

k Flowers 4-parted, with 8 stamens and a 3-seeded pod. 8.   ELLIOT'TIA.
k Flowers 5-parted, with 5 stamens. Leaves lanceolate, entire. S. CYRIL'LA.
k Flowers 5-parted, with 10 stamens. Lvs. lanccol., entire. S. MYLOCA'RYUM.
m Flowers racemed, many. Perennial, low, smooth, erect.   PYR'OLA. 8
m Flower solitary, one only. Perennial, small. N. r.   MONE'SES.
m Flowers corymbed, few. Leaves evergreen, thick. *Pipsissiwa.* CHIMAPH'ILA. 9
  n Corolla polypetalous. Plant white, reddish, &c. *Indian Pipe.* MONOT'ROPA.10
  n Corolla gamopetalous, bell-shaped, in a short spike. S.   SCHWEINIT'ZIA.
  n Corolla gamop., egg-shaped, in a loose rac. *Albany Beechdrops.* PTEROS'PORA.

1. OXYCOC'CUS. Cranberry.

Calyx superior, 8-cleft. Corolla 4-parted, with lance-linear, reflexed segments. Stamens 8, anthers tubular, 2-parted, opening by oblique pores. Berry globular, 4-celled, many-seeded.—Trailing and very slender, with woody stems, alternate, thick, narrow, entire leaves, and acid, eatable fruit. Flowers purplish. *June.*

1 O. palus'tris. *Bog C.* Stems thread-form, trailing. Leaves ovate, 2-4" long. Pedicels terminal, 1-flowered.

2 O. macrocar'pus. *Market C.* Stems thread-form, trailing. Leaves oblong, 4-6" long. Pedicels axillary, 1-flowered.

8 O. erythocar'pus. *Bush C.* Stems 1-3f. high, erect. Leaves oval, pointed, serrulate. Petals not reflexed at first. S. Mountains.

2. CHIOG'ENES. Boxberry.

1 C. hispid'ula. *Running B.* A little woody creeper, 4 to 6' long, in old woods, northward. Leaves many, small, oval. Flowers white, 4-parted. Berry white. Plant tastes like Checkerberry. (Fig. 547.)

Fig. 547. Boxberry, the entire plant.

3. GAULTHE'RIA. Checkerberry.

Calyx 5-cleft, with 2 bractlets at base. Corolla urn-shaped, the limb of 5 short, revolute lobes. Stamens 10. Capsule 5-celled, invested by the calyx, which becomes a pulpy berry.—Little shrubby or half-shrubby plants, with alternate, evergreen leaves. (Figs. 540-543.)

G, procum'bens. *Common Checkerberry*, or *Wintergreen.* Branches ascending 3' from the prostrate, slender root-stock, which is usually concealed. Leaves obovate, and few nodding flowers, all clustered at top of the stem, and spicy in flavor. Berries scarlet. Flowers in Summer, white.

4. KAL'MIA. Calico-bush. Mountain Laurel.

Calyx 5-parted. Corolla with 10 prominences beneath, and 10 corresponding pits within, holding the 10 anthers. Filaments recurved. Border with 5 shallow lobes. Capsule 5-celled, many-seeded.—Beautiful shrubs, with entire, evergreen, leathery leaves. Flowers white and red, in racemed corymbs. *May–June.*

> 1 Flowers in terminal corymbs. Leaves smooth, thick....2, 8
> 1 Flowers in lateral corymbs. Leaves rusty or downy beneath....4, 5
> 1 **K. hirsu'ta.** *Hairy L.* Flowers axillary, solitary, stalked, red. Plant hairy. Leaves mostly scattered, acute, sessile. 1–2f. S.
> 2 **K. latifo'lia.** *Great L.* Leaves scattered, green both sides. Corymbs large, rose-white, numerous and very showy. 3–20f.
> 8 **K. glau'ca.** *Polished L.* Leaves opposite, glaucous-white beneath, revolute on the margin. Corymbs small, lilac. 2–3f.
> 4 **K. cunea'ta.** *Wedge-leaved L.* Leaves scattered, wedge-oblong. Corymbs small, roseate, each of 4–8 flowers. Plant 3–5f. S.
> 5 **K. angustifo'lia.** *Sheep-poison.* Leaves opposite and in 3's, blunt at each end. Corymbs small, deep purple. 3–4f.

5. EPIGÆ'A. May-flower.

Calyx large, 5-parted, with 3 bracts at base. Corolla salver-form, tube hairy within, limb of 5 spreading lobes. Stamens 10. Anthers open by slits. Capsule 5-celled, 5-valved.—Little trailing shrubs.

> **E. repens.** *Trailing Arbutus.* Leaves cordate-ovate, entire. Corolla tube cylindrical. Stems slender, flat on the ground, 10–15' long. Leaves evergreen, rounded at the end, 2' or more long. Flowers tinged with red, very fragrant. *April, May.*

6. AZA'LEA. Azalea.

Calyx small, 5-parted. Corolla funnel-form, somewhat irregular, with 5 spreading lobes. Stamens 5, and, with the 1 style, long exserted, curved toward the lower side, Anthers open by pores. Capsule 5-celled, 5-valved.—Erect shrubs. Leaves alternate, deciduous, entire. Flowers large, showy, fragrant, clustered. *April–July.*

§ Lobes of the calyx all (rarely 1 excepted) very short or minute....1, 2
§ Lobes of the calyx all oblong, and of conspicuous length....3–5

1 A. visco'sa. *Clammy Swamp Pink.* Flowers very viscid, appearing with the full-grown leaves, the tube much longer than the segments. Shrub 4–7f. White or roseate.

2. A. nudiflo'ra. *Pinxter Bloom.* Clusters naked, appearing with or before the young leaves. Corolla tube downy, scarcely longer than the segments. Branches often whorled. Colora pink, purple, white, buff, &c. †

3 A. calendula'cea. *Flaming Pinxter.* Young branchlets downy, corymbs nearly or quite leafless. Tube of the corolla hairy, shorter than the ample lobes. Common. Penn. S. & W. Flowers very many, flame-color, bright red, saffron-yellow, &c. †

4 A. arbores'cens. *Tree Azalea.* Branches smooth. Leaves glaucous beneath. Corymbs leafy with full-grown leaves. Corolla tube longer than the lobes, not viscid. Height 10–20f. Mountains. S.

5 A. Pon'tica. *Asiatic A.* Flowers viscid, with full-grown leaves. Tube wide-mouthed, as long as segments. All colors. †

7. RHODODEN'DRON. Rose Bay.

Calyx 5-parted. Corolla broad, campanulate, regular or slightly irregular, 5-lobed. Stamens 10, mostly declined, anthers opening by pores. Capsule 5-celled, 5-valved.—Shrubs with alternate, entire, evergreen leaves. Flowers umbeled, splendid.

§ Calyx lobes large, leaf-like. Exotic....7
§ Calyx lobes small, scale-like....a
    a Leaves small, obtuse at each end. Mountains....1, 2
    a Leaves large, acute, rusty or silvery beneath....5, 6
    a Leaves large, acute, glabrous beneath....3, 4

1 R. Lappon'icum. *Lapland R.* Shrub 5–10' high. Lvs. scaly, elliptic. N.
2 R. Catawbien'se. *Catawba R.* Shrub 3–5f. high. Leaves smooth, oval. S.
    3 R. max'imum. *Great R.* Leaves oblanceolate, acute, paler beneath. Flowers in large umbels, white, with yellow dots. Rocky woods.
    4 R. Pon'ticum. *Asiatic R.* Leaves lanceolate, acuminate, not paler beneath. Flowers large, purple, variegated.
5 R. puncta'tum. *Dotted-lf. R.* Lvs. with rusty, resinous dots beneath. Mts. S.
6 R. arbo'reum. *Tree R.* Leaves with silvery spots beneath. Asia. †
    7 R. In'dicum. *Indian R.* Leaves rough, wedge-lance. Fls. few together. †

8. PYR'OLA. False Wintergreen.

Calyx 5-parted. Petals 5, equal. Stamens 10, anthers large, turned outwards, opening by 2 pores at the obtuse top. Style thick, long; stigmas 5, often projecting like rays. Pod 5-celled, 5-valved, opening into the cells, many-seeded.— ♃ Low, evergreen herbs, almost woody, with the leaves generally radical, and the scape bearing a raceme of nodding flowers. Mostly northern. *June, July.*

§ Stamens ascending, style declined and curved....a
§ Stamens and style straight and erect....5, 6
   a Leaves thick and shining. Flowers white or rose-colored....1, 2
   a Leaves green, not shining. Flowers greenish-white....3, 4
1 P. rotundifo′lia. *Round-leaved P.* Lvs. orbicular. Mostly white petals. (Fig. 14.)
2 P. asarifo′lia. *Heart-leaved P.* Leaves round-cordate. Rose-colored petals.
8 P. ellip′tica. *Pear-leaved P.* Leaves large, thin, elliptical, on short petioles.
4 P. chloran′tha. *Green-fl. P.* Lvs. small, thick, roundish, shorter than petioles.
5 P. secun′da. *One-sided P.* Raceme with the green-white flowers all on one side.
6 P. minor. *Lesser P.* Raceme spike-form, with small, globular, white fls. Mts.

9. CHIMAPH′ILA. Pipsissiwa.

Calyx 5-parted. Petals 5, spreading, round. Stamens 10. Anther cells lengthened above into tubes. Style very short, thick. Capsule 5-celled.—Small evergreens, with oblong, serrate, clustered leaves, and terminal flowers. *June, July.*

1 C. umbella′ta. *Prince's Pine.* Leaves wedge-lanceolate, in 4's-6's. Umbel 4-7-flowered, on an erect stalk. *July.*
2 C. macula′ta. *Spotted P.* Leaves lanceolate, acuminate, marked with whitish streaks along the midvein. Flowers 2 or 8. (See Fig. 548.)

10. MONOT′ROPA. Pine Sap.

Calyx of 1-5 bract-like sepals. Petals 4 or 5, connivent in a bell-shaped corolla. Stamens 8-10. Capsule 4-5-celled, 4-5-valved.—Low herbs growing on the juices of other plants, all white or tawny, with scales instead of leaves.

1 M. uniflo′ra. *Indian Pipe.* Sepals 1-8. Flower solitary, scentless. Stem 6′ high, common in woods. Whole plant white. *Summer.*
2 M. Hypop′itis. *Pine Sap.* Downy, tawny. Sepals 4, 5. Flowers racemed, fragrant. Stem 6-8′ high. Root a tangled ball of fibres. *Aug.*

548

ORDER LXXVIII. PRIMULACEÆ. Primworts.

Fig. 549. Primula Mistassinica, the whole plant. *Fig.* 550. The corolla cut open, showing the stamens on the tube. 1. The plan of the flower, showing the stamens opposite the petals. 2. The calyx and ovary. 3. The fruit cut open, showing the seeds on the central placenta. *Fig.* 554. Dodecatheon Meadia, whole plant. 5. A single flower, natural size. *Fig.* 556. Fruit (pyxis) of Anagallis, with its lid open, showing the seeds.

Herbs low, with the *leaves* either radical or mostly opposite; with the *flowers* 5 (rarely 4 or 6)-parted; the *corolla* monopetalous, regular; the *stamens* inserted on the corolla-tube and opposite to its lobes; the *ovary* 1-celled, with a free, central placenta; *style* 1; *stigma* 1; the *capsule* 1-celled, many-seeded; *seeds* with fleshy albumen.

11

Analysis of the Genera.

* Stemless. Leaves all radical, scape bearing an umbel....**a**
* Stems leafy. Flowers yellow, corolla wheel-form (tube none)....**b**
* Stems leafy. Flowers white, red, &c., never yellow....2
2 Leaves whorled, at least those near the flowers. Corolla white....c
2 Leaves opposite, entire. Flowers axillary, solitary....d
2 Leaves alternate, entire. Flowers white....e
   a Cor. tube egg-shaped, lobes short, spread. *Dwarf Primrose.* ANDROSA'CE.
   a Corolla tube cylindrical, lobes spreading. *Primrose.* PRIM'ULA. 1
   a Corolla tube cylindrical, lobes reflexed. *American Cowslip.* DODECATH'EON. 2
   b Corolla 5-parted, without intermediate teeth. *Loose-strife.* LYSIMACH'IA. 3
   b Corolla 6-parted, with 6 intermediate teeth. Racemes axillary. NAUMBER'GIA.
   c Fls. 7-part. Lvs. entire, in a single whorl. *Chick Wintergreen.* TRIENTA'LIS. 4
   c Fls. 5-parted. Leaves finely pinnatifid, in water. *Feather-foil.* HOTTO'NIA.
   d Plant prostrate, with scarlet corollas. *Pimpernel.* ANAGAL'LIS. 5
   d Plant erect, with no corolla, but white calyxes. *Black Saltwort.* GLAUX.
   e Fls. 5-parted, panicled. Plant 8–15' high. *Water Pimpernel,* SAM'OLUS.
   c Fls. 4-parted, axillary. Plant 1–2' high. *Dwarf Pimpernel.* CENTUN'CULUS.

1. PRIM'ULA. Primrose. Auricula.

Calyx angular, 5-cleft. Corolla salver-shaped, or often rather funnel-shaped, with 5 entire, or notched, or bifid lobes. Stamens 5, included. Pod opening at the top, many-seeded.— ♃ Herbs with the leaves all radical, and the flowers showy, in an umbel on a scape.

   * Corolla salver-form, limb abruptly spreading. Plants wild, rare....1, 2
   * Corolla salver-form, limb abruptly spreading. Plants cultivated....3, 4
   * Corolla funnel-form, limb gradually spreading. Cultivated....a
a Leaves hairy, rugose, toothed, or crenate, or wavy at edge....5, 6
a Leaves smooth, plane, entire, or sometimes crenate....7, 8
   1 P. Mistassin'ica. *Mistassins P.* Smooth, green, 3–8' high. Flowers 1–8, flesh-colored. On lake shores. N. First seen on L. Mistassins.
   2 P. farino'sa. *Bird's-eye P.* Mealy, 8–10' high. Flowers 8–20, lilac-yellow Shores of the great lakes. N.
8 P. grandiflo'ra. *Common P.* Petals obcordate, notched, yellow, purple, &c. †
P. purpu'rea. *Purple P.* Petals obtuse, entire, dark-violet, never yellow. †
   5 P. officina'lis. *Cowslip P.* Lvs. hairy. Outer fls. nodding, border concave. †
   6 P. ela'tior. *Oxlip P.* Leaves smooth above. All fls. nodding, border flat. †
7 P. Auric'ula. *Auricula.* Lvs. and calyx mealy-glaucous. Bracts very short. †
8 P. calyci'na. *Cup P.* Lvs. white-edged, calyx inflated. Bracts long. Purpl. †

2. DODECATH'EON. American Cowslip.

Calyx 5-parted, reflexed. Corolla tube very short, limb rotate, 5-parted, with the limb reflexed. Stamens 5, inserted into the throat of the corolla, filaments short, anthers long, acute connivent at apex, but shorter than the style.— ♃ Leaves all radical, oblong, scape erect, bearing an umbel of nodding rose or white flowers. *May, June.* (Fig. 554.)

D Mea'dia. *Pride of Ohio.* A striking and elegant plant, in prairies throughout the Western States. Scape 1–2f. high. Petals white or pink. Stamens yellow. †

3. LYSIMACH'IA. Loose-strife.

Calyx 5-parted. Corolla tube very short, limb 5-parted, spreading. Stamens 5, on the base of the corolla, filaments often united. Pods 5–10-valved. Seeds several or many.— ♃ Leaves opposite or whorled, entire. Flowers mostly yellow. *June, July.*

§ Erect Peduncles several-flowered, or flowers panicled....a
§ Erect. Pedicels 1-flowered, flowers racemed....8, 9
§ Erect. Pedicels 1-flowered, flowers axillary....1
§ Prostrate, creeping. Pedicels (or umbels) axillary....10, 11
= Leaves thick, rather obtuse, with the edges rolled back....4, 5
a Leaves thin, acuminate, with the edges not rolled....6, 7
1 Leaves mostly opposite, on petioles fringed with hairs....2, 3
1 L. quadrifo'lia. *Whorled L.* Leaves whorled in 3's, 4's, and 5's, sessile.
2 L. cilia'ta. *Fringe-lf. L.* Leaves ovate, often cordate. Stems mostly branched.
8 L. hib'rida. *Hybrid L.* Lvs. lance-oblong, opposite or whorled. Stems branched.
4 L. asperifo'lia. *Rough-lf. L.* Leaves oblong-lanceolate. Panicle bracted. S.
5 L. longifo'lia. *Long-lf. L.* Lvs. lance-linear. Fls. large, scarcely pan. W.
6 L. lanceola'ta. *Lance-lf. L.* Lvs. whorled in 4's, lance. Upper fls. racemed. S.
7 L. Fra'seri. *Fraser's L.* Leaves opposite, ovate, often cordate. Panicle large. S.
8 L. stric'ta. *Strict L.* Leaves nearly opposite, narrow-lance., with bulblets.
9 L. Herbemon'ti. *Herbemont's L.* Lvs. whorled, in 4's or 5's, lance., acuminate. S.
10 L. rad'icans. *Rooting L.* Branches rooting at the end. Leaves lanceolate.
11 L. Nummula'ria. *Moneywort.* Stem simple. Leaves roundish, very obtuse

4. TRIENTALIS. Chick-wintergreen.

Calyx and corolla 7-parted. Stamens 7. Pod many-seeded.— ♃ Stem low, simple. Pedicels 1-flowered.

T. America'na. *American C.* A pretty little plant, common in woods northward. Stem 3–5' high, bearing several lanceolate leaves in a sort of whorl at top, and from their midst, 1 or more white, starlike flowers. *May, June.*

5. ANAGAL'LIS. Pimpernel.

Calyx and corolla 5-parted, wheel-shaped. Stamens 5. Pod globular, opening by a lid all around (*i. e.*, a pyxis).—Herbs with square stems and opposite leaves. (Fig. 556.)

A. arven'sis. *Scarlet P. Poor-man's-weather-glass.* A small, trailing plant, in fields, roadsides, &c. Leaves sessile, broad-ovate. Pedicels 1-flowered, axillary. Flower red, rarely blue. Opening at 8 A. M., closing at 2 P. M., and in damp weather not opening at all. (See the figure, 557.)

Order LXXXIII. BIGNONIACEÆ. Trumpets.

Plants with opposite *leaves*, destitute of stipules, often climbing;
flowers gamopetalous, irregular, 5-parted, showy;
stamens 5, but only 2 or 4 of them perfect, and didynamous;
ovary 2-celled, with 1 style, forming a dry pod with winged seeds.

Analysis of the Genera.

Stamens 4.	Pod valves and partition contrary. Leaves pinnate.	Teco'mia. 1
Stamens 4.	Pod valves and partition parallel. Leaves binate.	Bigno'nia.
Stamens 2.	Pod straight, cylindric. Trees. Leaves simple.	Catal'pa. 2

1. TECO'MA. Trumpet-flower.

Calyx bell-shaped, 5-toothed. Corolla trumpet-shaped, with a 5-lobed, nearly regular limb. Stamens didynamous, 4, with the 5th a small rudiment. Pod with the partition contrary to the valves.—Trees or shrubs, often climbing. Leaves digitate or pinnate. Flowers red.

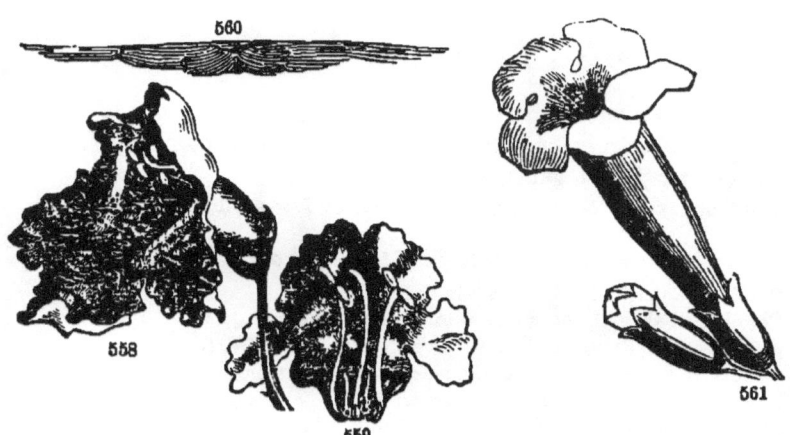

Fig. 558. Flower of Catalpa.
Fig. 559. The Corolla cut open, showing the 2 perfect stamens and the 3 rudiments of stamens.
Fig. 560. A 2-winged seed of Catalpa.   *Fig.* 561. Flower of the Trumpet Creeper.

1 T. rad'icans. *Trumpet Creeper*. Climbing by radicating tendrils. Leaflets 9–11,
ovate, acuminate, toothed. Corolla tube thrice longer than the calyx. Stamens
included. A well-known, splendid climbing vine. *Summer*.

2 T. Capen'sis. *Cape T.* Climbing. Leaflets 7–9, round-ovate, serrate. Stamens
and style exerted. Corolla tube curved. † S. Africa.

3 T. grandiflo'ra. *Chinese T.* Climbing. Leaflets 9–11, pointed, ovate, toothed.
Two glands on the nodding pedicels. Corolla tube scarce longer than calyx. †

2. CATALPA. Catalpa.

Corolla unequally bell-shaped, 4 or
5-lobed. Stamens 2 perfect, with 3
rudiments. Capsule long, cylindric,
with a thick partition.

C. bignonioi'des. Trees with large, broad-
ovate, cordate, velvety leaves, and terminal panicles of showy, white, variegated flowers. Common.

Fig. 562. A panicle (size much diminished)
of Catalpa.

Order LXXXV. LOGANIACEÆ.

Herbs or *shrubs* with opposite *leaves,* with the
stipules small or mere ridges connecting the base of the petioles;
flowers 4 or 5-parted, gamopetalous, regular;
ovary free; *fruit* 2-celled, many-seeded, or few-seeded.

Analysis of the Genera.

§ Corolla tubular, lobes 5, valvate in the bud....a
§ Corolla bell-shaped, lobes 4 or 5, imbricate in the bud....b
  a Styles wholly united into 1. Corolla tube long. W. S.   Spige'lia. 1
  a Styles distinct, with the stigmas united. Fls. small, white. S. Mitre'ola.
 b Flowers 4-parted. Diffuse, low herbs. M. S.   *Polyprenum.* Polypre'mum.
 b Flowers 5-parted. Slender climbing shrubs. S.   Gelsem'inum. 2

1. SPIGE'LIA. Pink-root.

Calyx segments linear-subulate. Corolla narrowly funnel-form. Stamens
5. Capsule 2-celled, few seeded.—Herbs with the flowers sessile in a
terminal one-sided coiled spike.

S. Maryland'ica. *Maryland P.* Stem square, erect. Leaves sessile, ovate-lanceolate. Corolla 4 or 5 times longer than the calyx, scarlet. *June.*

Fig. 563. Spigelia; the spike uncoils as the flowers open.   *Fig.* 564. A flower, natural size.

2. GELSEMI'NUM.   Yellow Jessamine.

Calyx lobes oblong.   Corolla funnel-bell-form, with 5 short, roundish lobes.   Filaments 5, on the corolla.   Style thread-form with 2 double stigmas.

G. sempervi'rens.   A shrub very common, South ; with long, wiry, twining stems, ascending bushes and hedges.   Leaves evergreen, shining, lanceolate.   Corolla tube 1 inch long, golden-yellow.

ORDER LXXXVI. SCROPHULARIACEÆ.   Figworts.

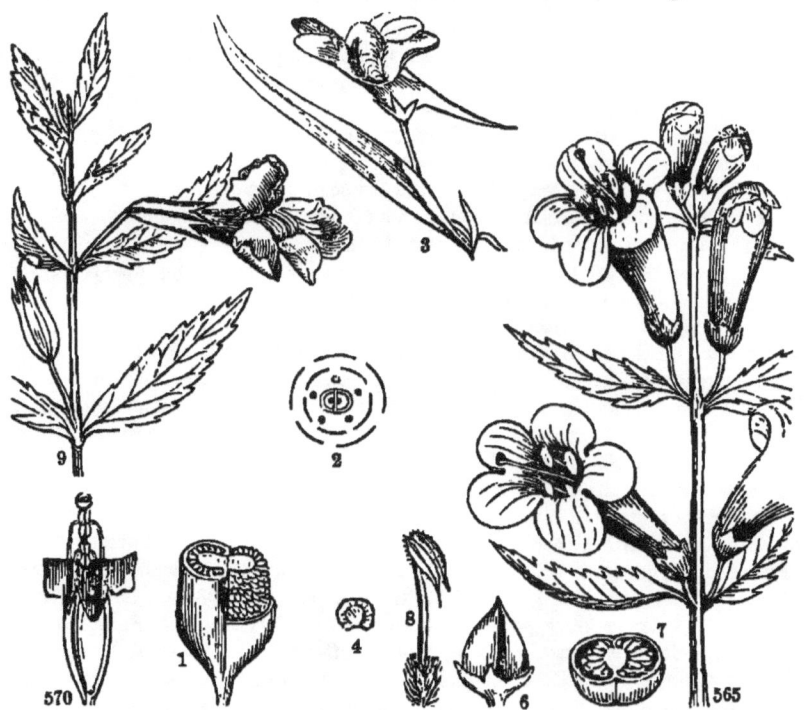

Fig. 565. The Yellow Foxglove (*Dasystoma pubescens*).   6. Mature fruit.   7. Cross-section of the 2-celled capsule.   8. A stamen enlarged.   9. Monkey-flower (*Mimulus ringens*).   *Fig.* 570. Calyx with the corolla partly removed, showing the didynamous stamens in pairs, with the stigma above the highest pair.   1. Sections of the 2-celled, many-seeded capsule.   2. Plan of the flower, showing the position of the fifth rudimentary filament.   3. Linaria vulgaris, leaf, and personate, bi-labiate, spurred flower.   4. A winged seed.

Plants mostly herbaceous, with unsymmetrical *flowers*, without fragrance;
calyx mostly 5-parted, free from the ovary, persistent;
corolla bi-labiate or otherwise irregular, lobes imbricated in the bud;
stamens on the corolla tube, 1 or 3 of the 5 usually imperfect or minute;
ovary 2-celled; *style* 1; *stigma* 2-lobed; *capsule* 2-celled, many-seeded.

Analysis of the Genera.

* Herbs with the leaves alternate or all radical....2
* Herbs with the leaves opposite or sometimes whorled....4
* Trees with large cordate leaves and large blue panicles....a
  2 Flowers diandrous, having but 2 perfect stamens....c
  2 Flowers didynamous, having 4 stamens, 2 of them longer....8
  2 Flowers pentandrous, having the 5 stamens all perfect....b
    8 Corolla bi-labiate, with the throat closed (personate)....d
    8 Corolla bi-labiate, throat open, upper lip arched....e
    8 Corolla rather bell-shaped, with 5 nearly equal lobes....f
  4 Flowers with only 2 perfect stamens....g
  4 Flowers with 4 perfect stamens, the 5th scarcely appearing....5
  4 Flowers with 4 perfect stamens and a 5th sterile distinct filament....n
    5 Corolla 2-lipped, the limb quite irregular....6
    5 Corolla limb nearly regular, with 4 or 5 plain, spreading lobes....o
      6 Stamens included in the tube of corolla, generally in pairs....k
      6 Stamens ascending beneath the arched upper lip....m
      6 Stamens descending into the sack-shaped lower lip....h

a Corolla trumpet-shaped, stamens arched downwards. Fragrant. † PAULOW'NIA.
b Corolla wheel-shaped, stamens declinate. Scentless. *Mullein.* VERBAS'CUM. 1
c Corolla 4-lobed, minute, white. Plant small. Lvs. radical. S. AMPHIAN'THUS.
c Corolla 4-lobed. Fls. spiked. Lvs. mostly radical. Scape 1f. N.-W. SYNTHI'RIS.
c Corolla deeply many-cleft, variously colored. Lvs. cleft. † SCHIZAN'THUS.
  d Corolla protracted into a spur behind. Racemes leafy. *Toad-flax.* LINA'RIA. 2
  d Cor. swollen into a sack behind. Rac. leafy. † *Snap-dragon.* ANTIRRHI'NUM.
e Bracts lobed, generally colored. Anth.-cells unequal. *Painted-cup.* CASTILLE'JA.
e Bracts and leaves entire, green. Flowers purplish. *Chaff-seed.* SCHWAL'BEA.
e Bracts and leaves serrate, green. Flowers yellow. *Lousewort.* PEDICULA'RIS.
  f Tall, erect, with large, nodding flowers. Gardens. *Foxglove.* DIGITA'LIS.
  f Low and minute. Corolla equally 5-cleft. In mud. *Mudwort.* LIMOSEL'LA.
  f Climbing, slender. Corolla large, gibbous at base. † Mexico. MAURAN'DIA.
  f Climbing, slender. Corolla large, equal at base. † Mexico. LOPHOSPER'MUM.
g Corolla labiate. Calyx 5-parted. Sterile filaments minute or 0. GRATI'OLA. 3
g Corolla labiate. Calyx 5-parted. Sterile filam. forked. *Mud-flower.* ILYSAN'THUS.
g Corolla labiate. Calyx 4-parted. Flowers very small. *Semi-flower.* HEMIAN'THUS.
g Corolla rotate, with 4 nearly equal lobes, lower smallest. *Speedwell.* VERON'ICA. 4

h Handsome herbs, 1–2f. high, with flowers blue and white.

Innocence. COLLIN'SIA. 5

k Leaves serr. Sts. square. Palate of lower lip prominent. *Monkey-fl.* MIM'ULUS. 6

k Leaves many-cleft into fine divisions. W. *Conobea.* CONO'BEA.

k Leaves entire. Corolla protracted into a spur behind. *Toad-flax.* LINA'RIA. 2

k Leaves entire. Cor. not spurred. Small, obscure weeds. W. M. S. HERPES'TIA.

  m Fls. yellow, in a terminal, one-sided spike. *Yellow-rattle.* RHINAN'THUS.

  m Fls. white, small, in a term. one-sided spike. Mts. *Eye-bright.* EUPHRA'SIA.

  m Fls. yellowish, axil., or in a leafy, equal spike. *Cow-wheat.* MELAM'PYRUM.

n Sterile filament shorter than the rest, smooth. *Snake-head.* CHELO'NE. 7

n Sterile filament long, projecting, bearded. *Beard-tongue.* PENTSTE'MON.

n Sterile filament a scale on the brown corolla. *Figwort.* SCROPHULA'RIA.

  o Corolla purple, in a long, slender spike. Leaves lance-ovate.

Blue-hearts. BUCHNE'RA.

  o Cor. purp. or rose-white, axillary. Lvs. narrow-lin., entire. GERAR'DIA. 8

  o Corolla yellow, and 5-lobed as well as the calyx....p

p Stamens scarcely longer than the tube of the corolla....q

p Stamens long-projecting, with very large anthers. S. MACRAN'THERA.

  q Stamens quite unequal in length. Sepals very short. DASYS'TOMA. 9

  q Stamens about equal in length, anths. all perfect. Sep. long. W. SEYME'RIA.

1. VERBAS'CUM. Mullein.

Calyx 5-parted. Corolla rotate, 5-lobed, slightly irregular. Stamens 5, all perfect, filaments woolly, at least the three upper ones. Pod roundish egg-shaped, 2-valved, many-seeded.—Mostly ② herbs. Flowers in spikes, or panicles, or racemes. Leaves alternate. *June–August.*

1 V. **Thap'sus.** *Common M.* Tall, woolly. Leaves decurrent. Flowers spiked, 2 filaments smooth.

2 V. **Blatta'ria.** *Moth M.* Branched, smooth. Leaves serrate. Flowers racemed. Filaments violet-woolly.

3 V. **Lych'nitis.** *White M.* White-downy, branched. Leaves crenate. Flowers panicled. Filaments white-woolly.

2. LINA'RIA. Toad-flax.

Calyx 5-parted. Corolla personate with the throat closed by the prominent palate, upper lip reflexed, lower 3-cleft, tube inflated and spurred behind. Pod 2-celled, bursting below the top.—Herbs with the lower leaves generally opposite, the upper alternate. Flowers solitary, axillary, often forming leafy racemes. *June–September.*

1 L. **vulga'ris.** *Butter and Eggs.* Leaves lance-linear. Flowers large, yellow and orange, in a close raceme. Erect.

2 L. Canaden'se. *Canada T.* Leaves linear, obtuse. Flowers small, blue, loosely racemed. Stem erect.
3 L. Elat'ine. *Pointed T.* Leaves ovate-hastate. Flowers small, yellow, and purple. Stem prostrate.

3. GRATI'OLA. Hedge-hyssop.

Calyx nearly equally 5-parted. Corolla upper lip entire or slightly 2-cleft, lower 3-cleft. Fertile stamens 2, mostly with 3 sterile filaments. Pod 2-celled, 4-valved, many-seeded.—Low herbs with opposite leaves. Peduncles axillary, 1-flowered, usually with 2 bracts near the calyx. *June–August.*

§ Flowers on peduncles. Plants weak, smooth, or viscid....a
§ Flowers sessile or nearly so. Plants rigid, bristly-hairy. S....8, 9
  a Sterile filaments thread-like, tipped with a small head....b
  a Sterile filaments none, or very minute and pointed....5–7
    b Leaves entire or nearly so. Plants smooth....1, 2
    b Leaves toothed. Plants generally viscid-downy. Flowers white....3, 4
  1 G. officina'lis. *Officinal H.* Stem erect. Leaves clasping. Fls. whitish. S.
  2 G. au'rea. *Golden H.* Stem ascending, branched. Leaves sessile. Flowers yellow, showy.
3 G. visco'sa. *Viscid H.* Leaves ovate-lanceolate. Sepals and bracts lanceolate. S.
4 G. ramo'sa. *Branching H.* Lvs. linear-lance. Bracts minute. Sepals linear. S.
5 G. sphærocar'pa. *Round-fruited H.* Peduncles not longer than calyx. Pod globular. W.   [calyx. S.
6 G. Florida'na. *Florida H.* Peduncles long. Corolla four times longer than the
7 G. Virginia'na. *Virginian H.* Peduncles long. Cor. twice longer than calyx.
8 G. pilo'sa. *Hairy H.* Leaves ovate, toothed. Corolla scarce longer than calyx. S.
9 G. subula'ta. *Awl-lv. H.* Leaves narrow, entire. Cor. thrice longer than calyx. S.

4. VERON'ICA. Speedwell.

Calyx 4-parted. Corolla with a wheel-shaped, spreading, 4-cleft border, the lower segment smallest. Stamens 2, inserted into the tube, projecting. Sterile filaments 0. Pod flattened, mostly obtuse or notched at the apex, 2-celled, few or many-seeded.—Mostly herbs, with opposite leaves. Flowers small, solitary, axillary, or racemed, blue, flesh-color, or white. *March–September.*

§ Erect, tall (1½–4f.). Flowers in dense terminal spikes....1, 2
§ Low, weak (3–12'). Leaves opposite. Corolla tube very short....a
  a Racemes mostly opposite, from the axils of the leaves, sky-blue....3, 4
  a Racemes mostly alternate, from the axils of the leaves, light-colored....5, 6
  a Racemes terminal, or the flowers axillary and not racemed....b

' Floral leaves like the rest, not longer than the recurved peduncles....7-9
' Floral leaves bract-like, longer than the erect flower-stalks....c
  c Perennial. Flower-stalks equalling or exceeding the calyx....10, 11.
  o Annual Flower-stalks shorter than the calyx, or none....12, 13

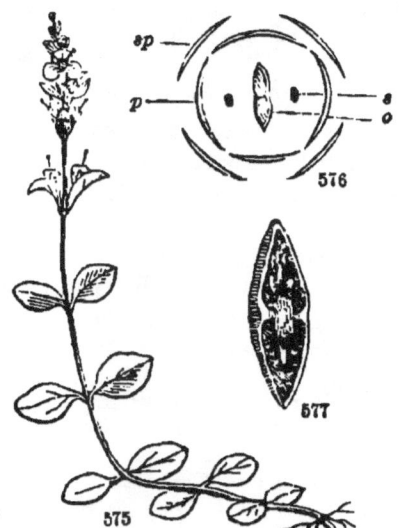

Fig. 575. Speedwell (*Veronica serpyllifolia*), whole plant. Fig. 576. Plan of the flower: *o*, is the 2-celled ovary; *s*, the 2 stamens; *p*, the 4 petals; *sp*, the 4 sepals. Fig. 577. Cross-section of the pod, showing its 2 cells, &c.

, V. Virginia'na. *Culver's Physic.* Leaves whorled. Corolla tube longer than limb.
$ V. spica'ta. *Spike-flowered S.* Leaves opposite. Corolla limb longer than tube. †
  3 V. Anagal'lis. *Water S.* Leaves sessile, cordate-clasping, ovate-lance.
  4 V. America'na. *Brooklime.* Leaves petiolate, oblong-ovate, base roundish or cordate.
5 V. scutella'ta. *Marsh S.* Leaves linear. Racemes very slender, few-flowered.
6 V. officina'lis. *Common S.* Lva. obovate-elliptical, finely serrate. Racemes dense.
  7 V. Buxbaum'ii. *Buxbaum's S.* Lva. roundish-ovate. Pod triangular-obcordate. Fields. E.
  8 V. agres'tis. *Neckweed.* Lvs. cordate-ovate. Pod roundish, acutely notched. Fields. E.
  9 V. hederæfo'lia. *Ivy-leaved S.* Leaves cordate, roundish, 3-5-lobed. Capsule 4-seeded. M. Rare. [than long. c.
10 V. serpyllifo'lia. *Thyme-leaved S.* Flower-stalks longer than calyx. Pod broader
11 V. alpi'na. *Alpine S.* Fl.-stalks as long as the calyx. Pod obov. Hairy. White Mts.
  12 V. peregri'na. *Purslane S.* Smoothish. Leaves petiolate, oblong, few-toothed, fleshy. c.
13 V. arven'sis. *Corn S.* Hairy. Lower leaves ovate, crenate, petiolate; upper lanceolate, sessile, entire. Stem 2-6' high. In fields. Common.

5. COLLIN'SIA. Innocence.

Calyx 5-cleft. Corolla 2-lipped, throat closed, upper lip bifid, lower lip trifid, with the middle segment keel-like, holding the style and 4 stamens in a kind of sack. Capsule roundish.—Annual herbs.

1 C. verna. *Early Collinsia,* or *Innocence.* Corolla 2 or 3 times longer than the calyx. Plant 8 to 18' high, tender and delicate. Leaves lance-ovate, dentate, opposite. Flowers variegated with blue and white, singular and pretty. M. W.

2 O. parviflo'ra. *Small-flowered I.* Corolla scarcely longer than the calyx, blue. Plant small. N.-W.

6. MIM'ULUS. Monkey-flower.

Calyx prismatic, 5-angled and 5-toothed. Corolla tubular, upper lip reflexed or erect, 2-lobed, lower lip spreading, with a prominent palate, 3-lobed. Pod '2-celled, many-seeded.—Herbs prostrate or erect, with square stems, opposite leaves, and axillary solitary flowers. *July.*

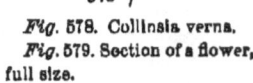

* Species from California, cultivated in gardens..:..3, 4
* Species growing wild, in fields, road-sides, &c. Fls. blue....1, 2

  1 M. rin'gens. *Ringent M.* Stem not at all winged. Leaves sessile. Peduncles longer than the flower. *o.*

  2 M. ala'tus. *Wing-stem M.* Stem slightly winged. Leaves petiolate. Peduncles shorter than the calyx.

3 M. lu'tea. *Yellow M.* Flowers yellow, often spotted. Leaves round-ovate. †
4 M. cardina'lis. *Cardinal M.* Fls. scarlet, large and brilliant. Leaves ovate. †

Fig. 578. Collinsia verna.
Fig. 579. Section of a flower, full size.

7. CHELO'NE. Turtle-head.

Calyx deeply 5-parted, or the sepals distinct. Corolla inflated, upper lip broad, concave, lower 3-lobed, bearded in the throat. Stamens 4, woolly, with a 5th sterile filament shorter than the others. Seeds many, broadly wing-margined.—♃ Plants about 2f. erect, with opposite serrate leaves. *Aug.–Sept.*

1 O. purpu'rea. *Purple T.* Leaves lanceolate, petiolate. Flowers purple. Probably a variety of the next. W. M.
2 O. gla'bra. *White T.* Leaves lanceolate, sessile or nearly so. Flowers white or purplish. By brooks and wet places.
3 O. Lyo'ni. *Lyon's T.* Lvs. ovate, petiolate, rarely cordate. Fls. purple or white. S.

8. GERAR'DIA. Purple Gerardia.

Calyx bell-shaped, 5-toothed, valvate in the bud. Corolla tubular, swelling above, with 5 unequal, spreading lobes, which are shorter than the tube. Stamens 4, quite unequal by pairs, included, hairy. Pod ovate, pointed, many-seeded.—① Erect and branching herbs, with opposite, slender leaves, and large, showy, purple or rose-colored flowers. *July–Sept.*

§ Calyx segments longer than its tube. Two anthers very small. W. (Omitted.)
§ Calyx segments short, equal. Anthers all equal....a
   a Corolla bi-labiate, upper lip very short, erect. S. (Omitted.)
   a Corolla lobes subequal, all spreading; throat usually hairy....b
      b Leaves almost none; opposite scales instead. S. (Omitted.)
      b Leaves all alternate, filiform. S. (Omitted.)
      b Leaves opposite....c
         c Peduncles not longer than the calyx. Leaves linear....1, 2
         c Peduncles much longer than the calyx. Leaves linear, long....d
            d Flowers large, about 9″ long....3, 4
            d Flowers small, about 6″ long....5, 6

1 G. mariti′ma. *Sea-side G.* Leaves linear, fleshy. Fls. small, their stalks scarce as long as the truncate calyx. Cor. upper lobes fringed. Salt marshes. E.
2 G. purpu′rea. *Purple G.* Leaves linear. Peduncles shorter than the calyx, which is a truncate tube with setaceously acute teeth. Flowers 1′ long. Common.
3 G. as′pera. *Rough-lv. G.* Pedunc. twice longer than calyx, which has teeth. W.
4 G. linifo′lia. *Flax G.* Peduncles many times longer than the toothless calyx. S.
5 G. tenuifo′lia. *Slender G.* Leaves linear, 1′ long. Peduncle 1′, longer than the corolla, which is purple, with spots inside. Slender, branched, 6–12′ high. c.
6 G. seta′cea. *Bristle-lv. G.* Leaves linear-setaceous, the floral ones much shorter than the very long peduncles. Plant 12–18′ high. Flowers rose-color. W.

9. DASYS′TOMA. Woolmouth.

The characters are the same as in GERARDIA, except that the calyx is 5-cleft, and imbricated in the early bud; the corolla yellow, with tube longer than the lobes, and woolly inside; the leaves rather large, and mostly pinnatifid, and the root ♃. Flowers very showy. Plants 2–4f. high. *July–Sept.* (Figs. 565–568.)

§ Sepals finely toothed. Leaves all pinnatifid, with toothed lobes....4, 5
§ Sepals entire. Leaves entire or mostly once pinnatifid-toothed....1
   1 Glabrous. Leaves acute at apex, lanceolate in outline....2, 3
1 D. flava. *Downy W.* Downy. Leaves obtuse, entire, except the lower. Sepals obtuse. Common in woods.
2 D. quercifo′lia. *Oak-leaved W.* Glaucous. Leaves mostly pinnatifid. Corolla 2′ in length. Calyx segments lance-acuminate, longer than its tube.
3 D. integrifo′lia. *Entire-leaved W.* Green. Leaves lanceolate, entire. Stalks shorter than calyx. Flowers 1′ long. In woods. Ohio, W.
4 D. pedicula′ria. *Lousewort W.* Smoothish or downy, branched. Flower-stalks longer than calyx. Leaves lance-ovate, twice pinnatifid.
5 D. pectina′ta. *Combed W.* Very hairy. Leaves lanceolate, pectinate-pinnatifid. Stalks shorter than calyx.

Order LXXXIX. LABIATÆ. Labiate Plants.

Herbs with square stems, and opposite, aromatic *leaves;*
flowers axillary, in verticils, sometimes as if in spikes or heads;
corolla labiate (rarely regular), the upper lip 2-cleft or entire, overlapping
in the *bud* the lower 3-cleft lip; *stamens* 4, didynamous, or 2;
ovary deeply 4-lobed, forming in *fruit* 4 hard nuts or achenia.

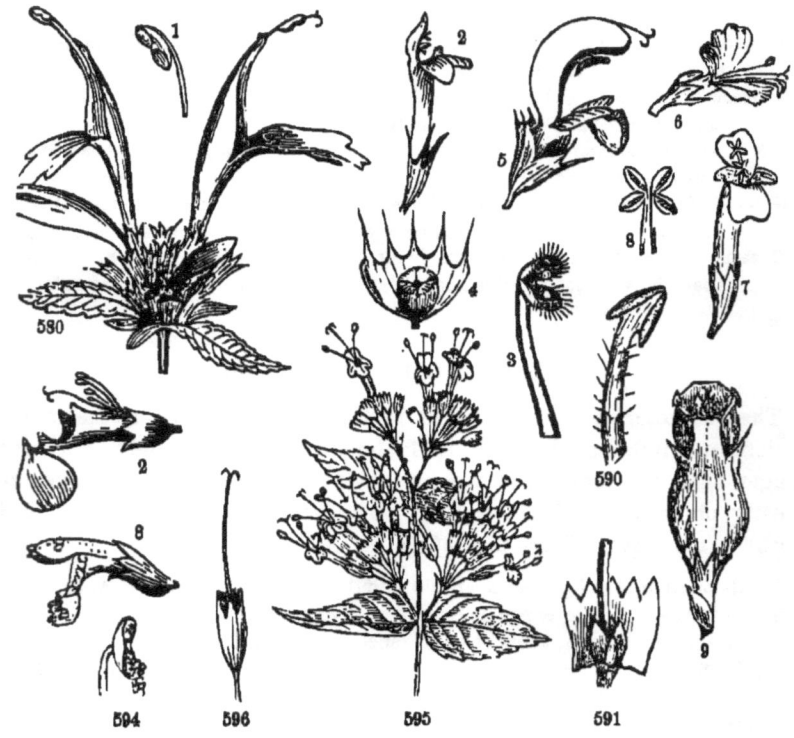

Fig. 580. Monarda didyma. 1. An anther enlarged. 2. Flower of Hemp Nettle (*Galeopsis*). 3. One of its stamens much enlarged. 4. The calyx opened, showing the 4 achenia. 5. Flower of Sage (*Salvia*). 6. Flower of Ocimum basilicum. 7. Flower of Nepeta Glechoma. 8. A pair of the anthers forming a cross. 9. Flower of Physostegia Virginica seen from beneath. *Fig.* 590. One of its stamens. 1. Its ovaries with the rudimentary filament. 2. Flower of Teucrium Canadense. 3. Flower of Catnep (*Nepeta Cataria*). 4. One of its anthers. 5. Dittany (*Cunila Mariana*). 6. A calyx and style.

Analysis of the Genera.

§ Flowers with only 2 perfect stamens....7

§ Flowers with the 4 perfect stamens all declining to the lower lip....a

§ Flowers with the 4 perfect stamens erect or ascending to the upper lip....2

    2 Stamens of equal length.   Corolla almost regular, 4 or 5-lobed....c

    2 Sta., the upper pair longer than the lower (outer).   Calyx 13-15-veined....k

    2 Stamens, the lower pair longer than the upper (interior) pair....3

        3 Stamens diverging apart, mostly straight and exserted....6

        3 Stamens parallel, ascending and long-exserted from the upper side....b

        3 Stamens parallel, ascending in pairs beneath the upper lip....4

            4 Calyx 13-veined, 5-toothed, and somewhat 2-lipped....g

            4 Calyx 5-10-veined or irregularly netted....5

                5 Calyx strongly 2-lipped, upper lip truncate, closed in fruit....m

                5 Calyx not labiate, 3 or 4-lobed, open in fruit....n

                5 Calyx subequally 5-toothed, teeth not spinescent....o

                5 Calyx subequally 5-toothed, teeth ending in sharp spines....q

                5 Calyx unequally 8-10-toothed....s

        6 Calyx hairy in the throat, mostly labiate....f

        6 Calyx naked in the throat, mostly equal, 5-toothed....e

    7 Stamens ascending beneath the galea (upper lip).   Anthers 1-celled....h

    7 Stamens exserted, distant.   Anthers 2-celled. ..d

a Corolla upper lip 4-lobed, lower entire, flattish.   †   *Sweet Basil.* O'CYMUM.

a Corolla upper lip 4-lobed, lower saccate, deflexed.   S.   *Hyptis.* HYPTIS.

a Corolla upper lip 2-lobed, lower 3-lobed, long, lilac.   † *Lavender.* LAVAN'DULA.

    b Stamens exserted through a fissure in the tube. *Blue Curls.* TRICHOS'TEMA.

    b Stamens very long, involute, arching the corolla. *Germander.* TEU'CRIUM.

c Corolla limb equally 5-lobed.   Stamens short.   *Blue-false-Gentian.* ISAN'THUS.

c Corolla limb 4-lobed, the broadest lobe notched.   *Peppermint,* &c. MENTHA. 1

    d Cor. nearly reg., 4-cleft. Calyx naked in throat. *Water Hoarhound.* LYC'OPUS. 2

    d Corolla labiate, cyanic, throat naked. Stam. straight. *Dittany.* CUNI'LA. 3

    d Cor. labiate, cyanic, throat naked. Stam. ascend. *Pennyroyal.* HEDE'OMA.

    d Corolla labiate, yellow, throat hairy.   Stamens

        long-exserted.   *Horse-balm.* COLLINSO'NIA.

e Fls. yel.   Coarse herbs not fragrant, with large lvs.

e Fls. bright blue.   Handsome herbs.   Calyx 15-veined.   † *Hyssop.* HYS'SOPUS.

e Fls. pale blue, in dense hds.   Calyx 10 or 13-striate. *Wild Basil.* PYCNAN'THEMUM.

e Fls. pink-colored, axillary.   Lvs. linear, small.   † *Summer Savory.* SATURE'JA.

    f Corolla exserted, pink-color, racemed.   Leaves linear, smooth. Stem 1f. S.

                DIOSRAN'DRA.

    f Corolla short as calyx, pale-purple.   Bracts roundish, colored.

                Marjoram. ORIG'ANUM.

    f Corolla short as calyx, blue-purp.   Bracts minute, green. *Thyme.* THYMUS.

g Cor. tube straight. Lvs. small, suberenate or entire. *Calaminth.* CALAMIN'THA. 4
g Corolla tube curved upwards. Leaves large, coarse-crenate. *Balm.* MELIS'SA.
    h Anthers halved, the halves widely separated, each 1-celled. *Sage.* SAL'VIA. 5
    h Anthers halved, one half present, 1 celled. Filaments toothed. Shrub. †
                                 Rosemary. ROSMARI'NUS.
    h Anthers whole, 2-celled. Calyx subsequally 5-toothed.
                         Mountain Mint. MONAR'DA. 6
    h Anthers whole, 2-celled. Calyx labiate, teeth bristle-shaped. BLEPHIL'IA.
k Leaves serrate. Stamens diverging. Fls. spiked. *Tall Hyssop.* LOPHAN'THUS.
k Lvs. serrate. Stam. all ascend. Fls. capitate. *Dragonhead.* DRACOCEPH'ALUM.
k Lvs. crenate, cordate or reniform. Corolla smooth inside. *Catmint.* NEP'ETA. 7
k Leaves crenate, cordate. Corolla tube very broad, hairy inside. CEDRONEL'LA.
    m Calyx lips toothed. Filam. forked. Fls. spiked. *Self-heal.* BRUNEL'LA. 8
    m Calyx lips entire, the upper appendaged on back. *Skullcap.* SCUTELLA'RIA. 9
n Calyx 3-lobed. Anthers all distinct. Corolla large, purplish. S. MACBRI'DEA.
n Calyx 4-lobed. Anthers, upper pair, connate. White. W. SYNAN'DRA.10
    o Cor. tube inflated in the midst, whitish. Tall. *Lion's-heart.* PHYSOSTE'GIA.11
    o Cor. tube inflated at the throat, purple. Lvs. roundish. *Henbit.* LAM'IUM.
    o Corolla inflated in the broad concave upper lip. *Jerusalem Sage.* PHLOMIS.
    o Corolla not inflated, short....p
       p Calyx salver-form, 10-veined. *Black Hoarhound.* BALLO'TA.
       p Calyx broad-campanulate, netted. *Molucca Balm.* MOLLUCCEL'LA.
q Lvs. serrate. Anth. open crosswise. Nuts truncate. *Hemp Nettle.* GALEOP'SIS.
q Leaves serrate. Anth. open lengthwise. Nuts obtuse. *Hedge Nettle.* STACHYS.
q Leaves lobed. Nuts truncate at top, 3-angled. *Motherwort.* LEONU'RUS.
    s Cor. white, upper lip flattish. Style equally bifid. *Hoarhound.* MARRU'BIUM.
    s Corolla white, upper lip concave. Style unequally bifid. S. LEUCAS.
    s Corolla scarlet, exserted. Calyx upper tooth longest. *Lion's-ears.* LEONO'TIS.

1. MEN'THA. Mint.

Calyx equally 5-toothed. Corolla nearly regular, tube included in the calyx, border 6-cleft, the upper lobe mostly notched. Stamens 4, equal, straight, erect, distant.—Aromatic herbs, with the pale purple or white flowers in close axillary clusters, or forming spikes.

    * Whorls of flowers remote, axillary, not in spikes. Leaves petiolate....1, 2
    * Whorls of flowers approximate, forming terminal spikes....3, 4
1 M. Canaden'sis. *Wild Mint.* Plant grayish, fragrant. Lvs. acute at each end.
2 M. arven'sis. *Field M.* Plant green, ill-scented. Lvs. frequently obtuse at base.
    3 M. piperi'ta. *Peppermint.* Leaves petiolate, ovate, serrate, smooth. Spikes few, thick, short. Stems 2-3f. high.
    4 M. vir'idis. *Spearmint.* Leaves sessile, lance-oblong, acute, serrate. Spikes many, slender, long. Stems 1-2f. high.

2. LYC'OPUS. Water Hoarhound.

Calyx tubular, 4–5-cleft. Corolla nearly regular, 4-cleft, tube as long as the calyx, stamens 2, distant, diverging the length of the straight style. — ♃ Low herbs, with deeply toothed or pinnatifid leaves, and remote axillary whorls of small, whitish flowers. *July, Aug.*

L. **Virgin'icus.** Stem obtuse-angled. Leaves sharp-toothed. Calyx 4-cleft, blunt.
L. **Europæ'us.** Stem sharp-angled. Leaves sinuate-toothed. Calyx 5-cleft, spiny.

3. CUNI'LA. Dittany. (Figs. 595, 596.)

C. **Maria'na.** *Maryland D.* Stem branched, 1–2f. high. Leaves ovate, serrate, nearly sessile. Cymes axillary and terminal, corymbous, stalked. Corolla nearly twice us long as the calyx, pale-red. In rocky woods. N. Y. to Ga. *Summer.*

4. CALAMIN'THA. Calaminth.

Calyx 13-veined, tubular, throat mostly hairy, upper lip 3-cleft, lower 2-cleft. Corolla tube straight, exserted, throat enlarged, upper lip. erect, subcutive, lower lip spreading, its middle lobe largest. Stamens 4, lower pair longer.— ♃

1 C. **Clinopo'dium.** *Wild Basil.* Herb hairy, 1–2f. high. Leaves ovate, subserrate. Flowers many, in dense, axillary whorls, with subulate bracts. Calyx bent.
2 C. **Nep'eta.** *False Catmint.* Herb hairy, 2f., much branched below. Lvs. broad-ovate, petiolate. Whorls few-flowered above. Calyx straight. Hills. Va.
3 C. **glabel'la.** *False Pennyroyal.* Herb smooth, half erect, 6–12', branched. Lvs. oblong, those of the runners ovate. Cor. pale-violet. Fragrant. *June.* M.
4 C. **canes'cens.** *Hoary C.* Shrub 10' high. Lvs. linear. Fls. opposite, roseate. S.
5 C. **coccin'ea.** *Scarlet C.* Shrub with narrow obovate leaves, large scarlet fls. S.
6 C. **Carolinia'na.** *Carolina C.* Shrub 1f. Leaves ovate, serrate-crenate. Flowers rose-purple. S.

5. SAL'VIA. Sage.

Calyx striate, labiate, throat not hairy. Corolla ringent, upper lip straight or falcate, lower spreading, 3-lobed. Stamens 2. The connectile is placed transversely on the filament like the letter T, bearing at each end 1 lobe of the halved anther. (See Figs. 585, 176.)

§ Herbs native, in woods, &c....a
§ Herbs or shrubs in gardens, with blue flowers....7, 8
§ Shrubs from Mexico, cultivated, with large scarlet flowers....9, 10

⚹ Calyx slightly 2-lipped, obscurely 3-toothed, equal. South....1-3
a Calyx deeply 2-lipped, 5-toothed, lower lip longer....4-6

1 S. azu′rea. *Azure S.* Leaves linear-oblong. Fls. downy, azure-blue. *Summer.*
2 S. urticifo′lia. *Nettle-lv. S.* Lvs. rhombic-ovate. Corolla smooth, blue. *May.*
8 S. coccin′ea. *Scarlet S.* Lvs. ovate, cordate, hoary. Corolla red, smooth. *July.*
   4 S. Clayto′ni. *Clayton's S.* Lvs. lanceolate, pinnatifid, cauline. Fls. small. S.
   5 S. obova′ta. *Obovate S.* Lvs. broadly obovate, entire. Flowers blue. S.
   6 S. lyra′ta. *Lyrate S.* Leaves all radical, oblong, lyrate, erose-dentate, 1 or 2 on the scape, bract-like. Fls. 1′ long, violet-purple. M. S. *Spring.*
7 S. Scla′rea. *Clarry S.* Lvs. ample, broad-ovate. Corolla upper lip large, high-arched. † (Fig. 585.)
8 S. officina′lis. *Common S.* Lvs. not large, lance-oblong, rugous. Corolla upper lip scarce longer than the lower, some vaulted. Shrubby.
   9 S. ful′gens. Stem weak. Lvs. lance-ovate, long-stalked. Calyx scarcely colored.
   10 S. splen′dens. Stem erect. Leaves broad-ovate, stalked. Calyx scarlet also.

6. MONAR′DA. Mountain-mint.

Calyx tubular, lengthened, 15-ribbed, nearly equally 5-toothed. Corolla tubular, long, the lips linear or oblong, lower reflexed, 3-lobed, upper erect, entire, involving the filaments. Stamens 2, with rudiments of more. —Erect, fragrant herbs, with rather large flowers in bracted whorls or heads, the bracts generally tinged with the color of the flowers. *July–Sept.* (Figs. 580, 581.)

§ Calyx densely hairy in the throat. Corolla purple or whitish....1, 2
§ Calyx naked in the throat. Corolla scarlet or yellow....3, 4
   1 M. fistulo′sa. *Wild Bergamot.* Stem acutely angular, 2–4f. Leaves lance-ovate, petiolate. Heads of flowers large, dense, terminal. *b. p. w.* M. W.
   2 M. Bradburia′na. *Bradbury's M.* Stem simple, 3f. Leaves lance-oblong, subsessile, hairy both sides. Heads few, large, purple. W.
8 M. puncta′ta. *Horsemint.* Stem branched, 2–3f. high. Leaves lance-oblong, tapering to a petiole. Bracts longer than the pale yellow flowers. M. W. S.
4 M. did′yma. *Mountain Balm.* Stem branched, 2–3f. Leaves broad-ovate, acuminate. Heads large, with long crimson flowers and bracts. †

7. NEP′ETA. Catmint.

Calyx striate, obliquely 5-toothed. Upper lip of the corolla notched or 2-cleft, lower 3-lobed, middle lobe largest, throat naked and widened. Stamens ascending beneath the upper lip.—⚇ Lvs. crenate. (Figs. 587, 588.)

1 N. Cata′ria. *Catnep.* Tall. Cymes dense, terminal spikes. Leaves cordate.
2 N. Glecho′ma. *Gill.* Trailing. Cymes loose, axillary. Leaves round-reniform.

8. BRUNELLA. Blue-curls.

B. vulga'ris. *Common B.* Stem simple, ascending 8–18′. Leaves oblong-ovate, stalked, toothed. Whorls close together, forming an oblong, dense spike. Corolla blue, upper lip truncate, with 3 awns.

9. SCUTELLA'RIA. Skullcap.

Calyx campanulate, lips entire, with an appendage on the back and closed after flowering. Corolla with a long, ascending tube, the upper lip vaulted, nearly entire, middle lobe of the lower lip wide, spreading. Stamens approximate in pairs, ascending beneath upper lip.—Bitter herbs, not aromatic. Flowers generally blue. *May–August.*

§ Flowers large (7–13″ long), racemed above, with bracts....a
§ Flowers large or small, opposite, solitary in the axils of the leaves....8–10
§ Flowers small (3″ long), in slender, axillary, 1-sided racemes....11
    = Bracts ovate, abrupt at base. Lips of the corolla short....1, 2
    = Bracts lance-oblong, acute at base. Leaves notched, petiolate....b
    = Bracts leaf-like, longer than the calyx. Leaves entire, subsessile....7
        b Helmet (upper lip) of the corolla longer than the lower....3, 4
        b Helmet of the corolla not longer than the lip....5, 6

1 **S. versic'olor.** *Variegated S.* Floral leaves sessile, broad-ovate, not cordate. Corolla lower lip scarcely longer than the upper, blue above. M. W.

2 **S. saxati'lis.** *Rock S.* Weak, branched, ascending. Upper leaves oval, obtuse. Corolla lower lip twice longer than the upper, blue above, tube pale. Rocks. W. S.

3 **S. canes'cens.** *Hoary S.* Tall, downy. Leaves petiolate, oblong or ovate. Flowers canescent, tube gradually enlarged. M. W. *c.*

4 **S. villo'sa.** *Woolly S.* Stem woolly. Corolla tube slender, enlarged only at the throat. Helmet much larger than the lip. S.

5 **S. serra'ta.** *Saw-lf. S.* Nearly smooth. Leaves acuminate, both ends. W. S.

6 **S. pilo'sa.** *Hairy S.* Plant hairy. Leaves rhomb.-ovate, obtuse. M. S.

7 **S. integrifo'lia.** *Entire-leaved S.* Erect. Leaves ovate-lance., entire, subsessile. M.

8 **S. nervo'sa.** *Nerve-lf. S.* Lvs. broad-ovate, 3–5-veined. Stem 8–15′. M. W

9 **S. par'vula.** *Pigmy S.* Lvs. oblong, ovate, obtuse, entire, sessile. Stem 3–6′. M. W.

10 **S. galericula'ta.** *Common S.* Leaves lance-cordate, crenate-serrate. Flowers 1′ long. *c.*

11 **S. laterifio'ra.** *Mad-dog S.* Branching, smoothish. Lvs. ovate-lanceolate, acuminate, serrate, petiolate. Racemes lateral, leafy. *a.*

10. SYNAN'DRA. Synandra.

597

Calyx 4-cleft. Upper lips of corolla entire, vaulted, the lower in 3 unequal, obtuse lobes. Throat widened. Stamens ascending beneath the upper lip, the two upper anthers cohering. (Figs. 597, 90.)

1 S. grandiflo'ra. *Great-flowered S.* Grows in woods, West. 6–8' high. Leaves opposite, ovate, cordate, toothed. Fls. few, 1' long, upper lip very large. *June.*

11. PHYSOSTE'GIA. Lion's-heart.

P. Virginia'nii. *Virginian L.* Stem square, erect 2–3f., with very smooth, sessile leaves in four rows, and a terminal, 4-rowed spike of large, showy, purplish-white flowers. *Aug., Sept.* (Figs. 589–591.)

ORDER XC. BORRAGINACEÆ. Borrageworts.

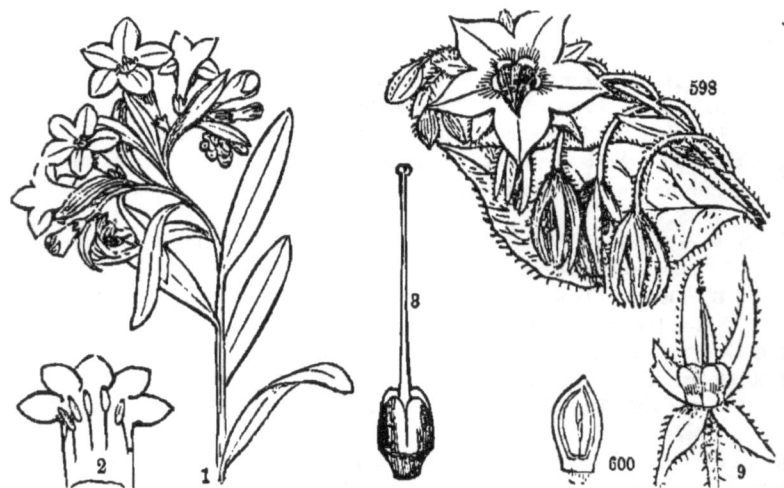

Fig. 598. Borrage (*Borrago officinalis*). 9. The four nuts with the style and calyx. *Fig.* 600. One of the nuts cut open, showing the seed, embryo, &c. 1. Puccoon (*Lithospermum canescens*). 2. Corolla laid open, showing the stamens. 8. Pistil of Comfrey, consisting of the deeply 4-lobed ovary with the slender style arising from between the lobes and near their base.

Herbs, shrubs, or *trees,* with round stems and branches; *leaves* alternate, generally rough with stiff hairs; *stipules* none;

flowers seldom yellow, generally in a coiled (circinate) inflorescence·
sepals 5; *petals* 5, united below, almost always regular;
stamens 5; *ovary* deeply 4-lobed, forming in fruit 4 separate, 1-seeded
nuts or nutlets, generally without albumen.

Analysis of the Genera.

§ Ovary not 4-lobed, but separating when ripe into several achenia....a
§ Ovary 4-lobed or parted, becoming 4 achenia around the style....2
  2 Corolla irregular, with unequal lobes or a bent tube....b
  2 Corolla perfectly regular....3
    3 Achenia or ovary prickly. Corolla throat closed with 5 scales....f
    3 Achenia and ovary not prickly....4
      4 Corolla throat closed by 5 scales....c
      4 Corolla throat open, no scales, sometimes 5 ridges....d
a Corolla tube with open throat. Achenia 4. *Heliotrope. Turnsol.* HELIOTRO'PIUM. 1
a Corolla tube with constricted throat. Achenia 2. *False Helio.* HELIOTROPH'YTOM.
  b Corolla irregularly 5-lobed. Throat open. Blue. *Viper's Bugloss.* ECH'IUM.
  b Corolla with the slender tube bent, closed. Blue. *Bugloss.* LYCOP'SIS.
  c Corolla wheel-form, anthers exserted. Blue. *Borrage.* BORRA'GO.
  c Corolla tubular bell-form. Style exserted. White. *Comfrey.* SYM'PHYTUM.
    d Cor. tubular, with erect, acute lobes. White. *False Gromwell.* ONOSMO'DIUM.
    d Corolla lobes rounded, spreading,....e
      e and imbricated in the bud. White or yellow. *Gromwell.* LITHOSPER'MUM. 2
      e and imbricated in the bud. Purple or blue, large. MERTEN'SIA. 3
      e and convolute in the bud. Blue or white, small. MYOSO'TIS. 4
f Corolla salver-form. Ach. prickly on the edge. *Burr-seed.* ECHINOSPER'MUM.
f Corolla funnel-form. Achenia prickly all over. *Hound's-tongue.* CYNOGLOS'SUM. 5

1. HELIOTRO'PIUM. Turnsol. Heliotrope.

Corolla salver-form, lobes shorter than the tube. Anthers sessile. Style
short, terminal. Ovary entire, splitting at length into 4 achenia.—Herbs
or shrubs. Flowers in one-sided, coiled spikes.

1 H. Europæ'um. *Wild H.* Herb downy. Leaves oval, obtuse. Spikes single or
  forked. White. S. [o'btuse. Blue. W.
2 H. curassav'icum. *Glaucous H.* Herb smooth, glaucous. Lvs. linear-lanceolate,
3 H. Peruvia'num. *Common H.* Shrubby, whitish-downy. Spikes many, clustered. *w.-p.* †

2. LITHOSPER'MUM. Gromwell or Grammell.

Calyx 5-parted. Corolla funnel-form or salver-form, the limb 5-lobed,
throat open, naked or with 5 projections. Stamens included. Achenia

bony, ovate, smooth or wrinkled, fixed by a flat base.—Herbs generally with thick, reddish roots. Flowers spiked or racemed, with leafy bracts. *May–July.*

§ Flowers white, small....a
§ Flowers yellow, showy. Achenia smooth, polished....5–7
  a Achenia roughened with wrinkles....1
  a Achenia smooth and polished....2–4

1 L. arven'se. *Wheat-thief.* Root ☉, red. Leaves lance-linear. Plant 12–18′ high, hairy. *c.*   [tube. N. M.
  2 L. officina'le. *Gromwell.* Root ♃. Lvs. lanceolate. Calyx equal to corolla
  8 L. latifo'lium. *Broad-leaved G.* Root ♃. Leaves lance-ovate, sharply acuminate. Sepals longer than the corolla, spreading in fruit. Stem 1–2f.
  4 L. angustifo'lium. *Narrow-lv. G.* ♃ Lvs. linear, stiff, edges some revolute. M. W.
5 L. canes'cens. *Puccoon.* Soft-velvety, canescent. Lvs. oblong-linear. Tube of the corolla thrice as long as the very short calyx. Plant 8–12′ erect. W. &c.
6 L. hirtum. *Hairy P.* Rough-hairy. Lvs. lance-linear. Cor. long as calyx. W. S.
7 L. longiflo'rum. *Long-flowered P.* Rough-ashy. Lvs. lance-linear. Corolla tube four times as long as the calyx, lobes crenulate, wavy. W. S.

3. MERTEN'SIA. Lungwort.

A short, 5-cleft calyx; a tubular corolla, slender below, suddenly enlarged above, limb 5-cleft; the 5 stamens inserted at top of the tube, and with smooth achenia.—♃ Plants usually smooth, with terminal racemes.

1 M. Virgin'ica. *Virginian L.* Very smooth, 12–18′ high. Root lvs. large, stalked; stem lvs. sessile. Fls. somewhat trumpet-shaped, blue-lilac, very fine. *May.* W.
2 M. marit'ima. *Sea L.* Smooth, diffuse. Leaves ovate, fleshy. Corolla limb longer than the tube, which shows 5 folds in throat, blue-purple. E.
8 M. panicula'ta. *Panicled L.* Rough. Leaves cordate, acuminate, veiny. Calyx hispid, thrice shorter than the tube, bell-form, blue-white corolla. N.-W.

4. MYOSO'TIS. Forget-me-not. Scorpion-grass.

Calyx 5-cleft. Corolla salver-form, the 5 lobes slightly notched at the end, throat closed with 5 short, concave scales. Nuts smooth, ovate, with a small cavity at base.—Little herbs slightly woolly. Racemes finally becoming long. *May–Aug.*

1 M. palus'tris. *True F.* Flowers in one-sided racemes. Plant smoothish, 8–12′ high. Leaves linear-oblong, obtuse. Flowers blue with a yellow centre.

Fig. 604. Forget-me-not,— a pair of scorpoid cymes.

2 **M. arven'sis.** *Field F.* Fls. in 2-sided, leafless racemes. Plant hairy. Pedicels twice as long as the open, equal calyx. Lvs. oblong-lance., acute. Rare. *w.*

8 **M. stric'ta.** *Rough F.* Flowers in 2-sided racemes, which are leafy at their base. Pedicels as long as the closed, 2-lipped calyx. Leaves oblong. *w.*

5. CYNOGLOS'SUM. Hound's-tongue.

Calyx 5-parted. Corolla short, funnel-form, the throat closed with 5 obtuse scales, lobes rounded. Nuts depressed, covered with short, hooked prickles, fixed laterally to the base of the style.—Coarse herbs, strong-scented, with the flowers in leafless, panicled racemes. *June, July.*

C. officina'le. *Common H.* Velvety. Stem leafy (1-2f.). Flowers reddish purple.

C. virgin'icum. *Stalked H.* Hairy. Stem leafless above (2f.). Flowers pale blue.

C. Morriso'ni. *Morrison's H.* Hairy, leafy (2-3f.), wide-spread. Flowers whitish.

ORDER XCI. HYDROPHYLLACEÆ. The Hydrophylls.

Mostly *herbs* with alternate lobed *leaves*, and regular bluish *flowers;*
calyx 5-cleft, usually with appendages at the clefts, persistent;
corolla 5-lobed, often with 10 honey scales or furrows near the base;
stamens 5, inserted into the corolla, with a single bifid *style;*
ovary simple, free, 1-celled, with 2 usually projecting several-seeded placentæ.

Analysis of the Genera.

1 Corolla with 10 honey scales inside, extending lengthwise....2

1 Corolla destitute of honey scales. Stamens equalling corolla. COSMAN'THUS. 2

2 Fls. in coiled cymes, without bracts. Placentæ large, fleshy. HYDROPHYL'LUM. 1

2 Flowers in one-sided racemes, bractless. Placentæ linear. PHACE'LIA.

2 Flowers (mostly) solitary. Calyx very large. Leaves pinnatifid. ELLIS'IA.

1. HYDROPHYL'LUM. Water-leaf.

Sepals slightly united at base. Corolla campanulate, with 10 linear honey scales running lengthwise, folded inward so as to form 5 grooves. Stamens exserted. Pod globular, 2-celled, 1-4-seeded, with large, fleshy placentæ.—Handsome herbs, with the root leaves on long petioles, and the flowers in clustered cymes, bluish or white.

§ Calyx not appendaged. Stamens much exserted....1-3

§ **H. appendicula'tum.** *Appendaged W.* Calyx appendaged at the clefts. Stamens not exserted. W. S.

1 **H. macrophyl'lum.** *Great-leaved W.* Lvs. pinnately-veined and lobed, rough-hairy. Peduncles long. W. S.

2 **H. Virgin'icum.** *Virginia W.* Leaves pinnately-veined and lobed, smooth. Peduncles long. *c.*

8 **H. Oanaden'se.** *Canada W.* Leaves palmately-veined and lobed, smooth. Peduncles shorter than petals. *r.*

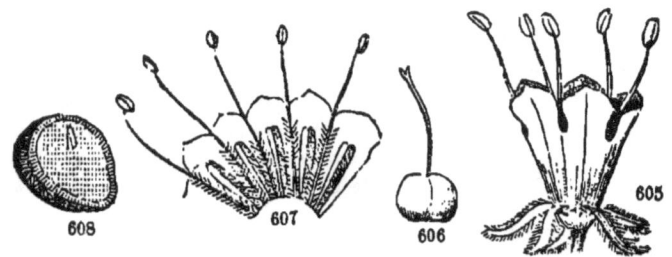

Fig. 605. A flower of Virginian Water-leaf. *Fig.* 606. The Ovary and Style. *Fig.* 607. Corolla cut open, showing the honey grooves. *Fig.* 608. A seed, cut, showing the embryo.

2. COSMAN'THUS. Miami Mist.

Corolla broad-campanulate, soon falling off, throat not appendaged, limb of 5-fringed lobes. Ovary 1-celled, the two projecting placentæ each 2-seeded.—① Delicate herbs with alternate leaves, long, bractless racemes, and small, white or pale-blue flowers.

1 **O. Pui'shii.** *Pursh's M.* Nearly smooth. Leaves pectinately pinnatifid, lobes oblong-acute. Sepals lance-linear. Height 8–12'. Penn., S. and W. Pale blue.

2 **O. fimbria'tus.** *Fringed C.* Downy. Leaves pinnate, segments rounded or oblong, obtuse. Sepals obtuse, oblong-spatulate. Mts. Tenn. S.

ORDER XCII. POLEMONIACEÆ. Phloxworts.

Herbs with alternate or opposite *leaves*, and regular, showy, 5-parted *flowers;* calyx free from the ovary;

corolla of 5 united *petals*, twisted and imbricate in the bud;

stamens 5, inserted into the midst of the corolla tube and alternate with its lobes;

ovary 3-celled; *styles* united into 1; *stigma* 3–cleft;

capsule 3-celled, 8-valved, with few or many albuminous seeds.

Analysis of the Genera.

Corolla salver-form.	Filaments unequal. Leaves simple.	*Phlox.* PULOX. 1
Corolla funnel-form.	Filaments equal. Leaves dissected.	*Gilia.* GI'LIA.
Corolla bell-form.	Filaments equal. Leaves pinnate.	*Polemony.* POLEMO'NIUM. 2

1. PHLOX. Lychnidea

Calyx angular, deeply 5-cleft, corolla salver-form, the tube more or less curved. Stamens quite unequal, inserted in the tube of the corolla above the middle. Capsule 3-celled, cells 1-seeded. — Very beautiful North American herbs. Leaves generally opposite, sessile, simple, entire. Flowers varying from purple to white. *April–July.*

Fig. 609. Flower of a Phlox.

* Lobes of the corolla rounded and entire at the end....10–12
* Lobes of the corolla notched or bifid at the end....a
    a Panicle of cymes oblong or pyramidal, many-flowered....1, 2
    a Panicle of cymes corymbed, level-topped, flowers fewer....b
b Plants glabrous. Calyx teeth shorter than its tube....3, 4
b Plants hairy. Calyx teeth very slender, larger than its tube....c
    c Leaves narrow, linear or nearly so....5, 6
    c Leaves broad, ovate, lanceolate, &c....7–9

1 P. panicula'ta. *Panicled L. Garden P.* Tall. Leaves lance-ovate, acuminate at each end. Calyx-teeth bristle-pointed, nearly as long as the tube. W. S. †
2 P. macula'ta. *Spotted L.* Stem purple-spotted. Leaves lance-ovate, upper cordate. Calyx-teeth lanceolate, acute, half as long as its tube. Fields. W. S. †
3 P. Caroli'na. *Carolina L.* Stem ascending. Leaves lance-ovate. W. S.
4 P. glaber'rima. *Polished L.* Stem erect, simple. Leaves lance-linear. W. S.
5 P. pilo'sa. *Hairy L.* Leaves lance-linear, acute. Calyx segments bristle-subulate, much longer than its tube. Stem slender, 1-2f. W. S. *p.–w.*
6 P. involucra'ta. *Cup-fl. L.* Hoary-downy. Lvs. linear oblong, obtusish at each end, the floral crowded beneath the dense cymes. *p.–r.* S.
7 P. rep'tans. *Creeping L.* Stolons creeping. Stem oblique. Lvs. obovate, obtuse. W. S.
8 P. Laphamii. *Lapham's P.* Slender, erect. Lvs. ovate, acute, thin. W.
9 P. Drummondii. *Drummond's P.* Annual, branched, hairy. Leaves mostly alternate. Calyx segments revolute. Corolla purple, with a star. S. †
10 P. divarica'ta. *Wild L.* Low, diffuse, downy. Lvs. lance-ovate, acute. Panicle corymbed, loose. Corolla grayish-blue. *c.* (No. 8, may be the same.)
11 P. bi'fida. *Beck's L.* Low, diffusely branched. Lvs. lanceolate, stem-clasping.
12 P. subula'ta. *Moss Pink.* Prostrate, much branched. Lvs. linear-subulate. It grows in dense tufts, covered over with rose-colored flowers in May. †

2. POLEMO'NIUM. Polemony.

1 P. oœru'leum. *Blue P. Greek Valerian.* Stem erect, 1–2f. high. Lvs. pinnate, with 11–17 leaflets. Capsule 12–20-seeded. Sometimes wild. †

2 P. rep'tans. *Creeping P.* Stem weak, diffuse. Leaves pinnate, with 7–11 leaflets. Capsule 4–6-seeded. Woods, common. Light blue.

Order XCIII. CONVOLVULACEÆ. Bindweeas.

Herbs twining or trailing. with alternate *leaves ; flowers* snowy :
calyx with 5 much imbricated *sepais.* persistent :
corolla regular, 5-lobed or entire, plaited and twisted in the bud :
stamens 5, and *style* single ; *ovary* free, becoming in
fruit a pod which is 2–4-celled and 2–6-seeded ;
embryo large and leafy, with thin mucilaginous albumen.

The suborder, Cuscutineæ, consists of small orange-colored, leafless plants, living on other plants (parasites), with small flowers, and no cotyledons (Cuscuta, *the Dodder*).

Analysis of the Genera.

§ Ovaries 2, distinct, with 2 distinct styles....f
§ Ovary 1, open. when ripe by 2–4 valves....2
 2 Ovary 2-celled, 2-valved, 4-seeded....3
 2 Ovary 3-celled, 3-valved, 6-seeded....b
 2 Ovary 4-celled, 4-valved, 4-seeded....a
  3 Styles 2, distinct....e
  3 Styles united into 1....4
   4 Calyx enveloped in 2 large bracts....d
   4 Calyx naked....c

Fig. 610 Entire-leaved Cypress-vine (*Quamoclit coccinea.*)

a Stamens exserted. Cor. small (scarce 1' broad). † *Cypress Vine.* Quam'oclit. 1
a Stamens included. Corolla large (2' broad). † *Sweet Potato.* Bata'tas. 2
 b Beautiful twining vines. Cor. bell-funnel. † *Morning-glory.* Phar'bitis. 3
c Stamens included. Stigma capitate. *False Bindweed.* Ipomæ'a. 4
c Stamens included. Stigmas 2, linear. *Bindweed.* Convol'vulus. 5
e Stamens exserted. Corolla tube slender. † *Good-night.* Calynyc'tion.
 d Stigmas 2, obtuse. Corolla bell-form. † *Rutland Beauty.* Calyste'gia. 6
e Peduncles longer than the leaves. Soft-downy. *Stylisma.* Stylis'ma.
 f Capsules 2, each 1-seeded. Plant very small, prostrate. S. Dichon'dra.

1. QUAM'OCLIT. Cypress Vine.

1 Q. vulga'ris. *True C.* Leaves pectinate-pinnatifid. Fls. scarlet, crimson, &c. S.
2 Q. coccinea. *Entire-lv. C.* Lvs. undivided, cordate, acuminate. Crimson. W.S.

2. BATA'TAS. Sweet Potato.

1 B. littora'lis. *Sea-side B.* Peduncle 1-flowered, as long as the sinuate, cordate leaf. S.
2 B. macrorhi'za. *Wild Potato.* Peduncle 1-5-flowered, shorter than the lobed or entire leaf, which is downy beneath. Flowers purplish-white. S.
8 B. ed'ulis. *Sweet Potato.* Peduncle 3-5-flowered, shorter than the palmate or pedate-lobed leaf. Flowers showy, rose-purple. †

3. PHAR'BITIS. Morning-glory. (Fig. 22.)

1 P. purpu'rea. *Common M.* Leaves entire, cordate. Peduncle 2-5-flowered. †
2 P. Nil. *Indigo M.* Lvs. 3-lobed, cordate. Ped. 1-3-flwd. Sepals long. M. S. †

4. IPOMÆ'A. False Bindweed.

A large genus. Some of its tropical species are shrubs and trees; and are all trailing or climbing herbs, chiefly at the South. We mention but one species.

I. pan'durata. *Wild Potato.* Leaves broadly cordate, often fiddle-shaped (panduriform). Corolla large (near 3' long), 4 times longer than the calyx, white, with a purple centre. Root very large. Sandy fields, West and South.

5. CONVOL'VULUS. Bindweed.

1 C. arven'sis. *Field B.* Leaves sagittate. Fls. white, with a tinge of red, small.
2 C. tri'color. *Tri-colored B.* Leaves lance-obovate. Fls. yellow, white, blue. †

6. CALYSTE'GIA. Bracted Bindweed.

1 C. spithamæ'us. *Erect B.* Stem ascending, 8-10' (a span). Leaves lance-oblong. Peduncle as long as the leaves, bearing 1 large, white flower. Fields.
2 C. Sepium. *Rutland Beauty.* Stem twining, long. Leaves cordate-sagittate. Flowers numerous, large, white, sometimes double in cultivation.
8 C. Catesbeia'nus. *Catesby's B.* Plant downy, twining. Flowers purple. S.

Order XCIV. SOLANACEÆ. Nightshades.

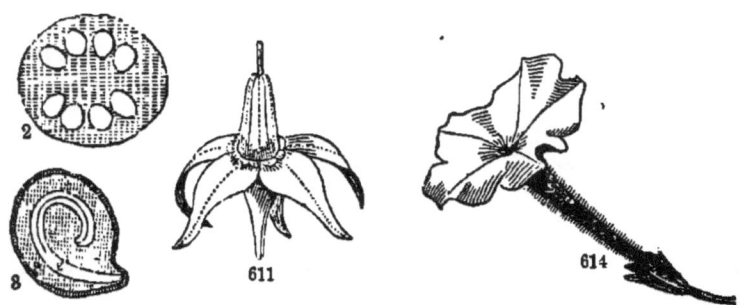

Fig, 611. A flower of Bitter-sweet (*Solanum Dulcamara*). 2. Cross-section of the berry. 8. A seed cut open, showing the long, curved embryo. *Fig.* 614. Flower of Petunia.

Plants herbaceous or shrubby, with alternate *leaves;* and with *flower-stalks* often opposite to the leaves; and the *pedicels* without bracts; *calyx* generally persistent, 5-lobed; *corolla* 5-lobed, mostly regular, valvate and plaited in the bud; *ovary* free, 2-celled (rarely 3 or 4-celled), many-seeded; *style* and *stigma* single; *fruit* a capsule or berry, with many seeds; *embryo* curved, lying in fleshy albumen.

Analysis of the Genera.

§ Corolla wheel-shaped, the tube very short.   Anthers convergent....a
§ Corolla bell-shaped, the broad tube including the erect anthers....b
§ Corolla funnel-form, tube long, and—(2)
   2 The limb somewhat irregular....c
   2 The limb perfectly regular....3
      8 Stamens exserted....d
      8 Stamens included....e

a Anthers connate, opening by slits. Berry lobed.   *Tomato.* Lycosper'sicum.
a Anthers connivent, opening by pores. Berry round.   *Potato.* Sola'num. 1
a Anthers connivent, opening by pores. Pod angular.   *Pepper.* Cap'sicum.
  b Corolla bluish. Berry dry, enveloped in the calyx. *Apple Peru.* Nican'dra.
  b Corolla yellowish. Berry fleshy, inclosed in the calyx.
                        Ground Cherry. Phys'alis.
  b Corolla purplish. Berry black, in the open calyx.   *Belladonna.* At'ropa.
c Stamens exserted, declining. Capsule opening by a lid. *Henbane.* Hyoscy'amus.
c Stamens included, unequal. Capsule opening by valves.   *Petunia.* Petu'nia.

d Stamens growing to the summit of the tube. *Neiremberg.* NEIREMBER'GIA.
d Stamens growing to the bottom of the tube. *Matrimony.* LYC'IUM.
e Calyx 5-angled. Capsule spiny or smooth. *Thorn Apple.* DATU'RA.
e Calyx terete. Stigma capitate. Herbs coarse. *Tobacco.* NICOTIA'NA.
e Calyx terete. Stigma 2-lobed. Delicate shrubs. † *False Tamarisk.* FABIA'NA.

SOLA'NUM. Nightshade.

Calyx 5-parted. Corolla rotate, limb spreading, tube very short, limb plaited in the bud, 5-lobed. Anthers erect, slightly united or converging, each opening at top by 2 pores. Berry globular or depressed, 2-celled. —Herbs or shrubs unarmed or prickly. Leaves often 2 together, a large and a small one. Flowers generally lateral. *May–July.*

§ Plants not prickly. Anthers short, blunt....a
§ Plants prickly. Anthers long, linear, and pointed....b
    a Herbs with pinnatifid leaves, shorter than the racemes....1
    a Herbs with undivided leaves, longer than the racemes....2–4
    a Shrubby plants, erect or climbing. Berries red....5–7
        b Peduncles exceeding the leaves, many-flowered....8, 9
        b Peduncles shorter than the leaves, few-flowered....10, 11

1 S. tubero'sum. *Potato.* Segments of the leaves unequal, some very small. Corolla limb 5-angled. Tubers on the underground branches.
2 S. nig'rum. *Black Nightshade.* Smoothish. Leaves ovate, toothed, and wavy. Flowers small, white, in lateral umbels. Berries black.
3 S. nodiflo'rum. *Knot-flowered N.* Quite smooth. Leaves ovate, entire. Flowers minute, white, the stalk arising from a knot in the stem. S.
4 S. pycnan'thum. Stem hispid. Leaves ovate-acuminate, wavy. Peduncle 2 or 3-flowered. S.
5 S. Dulcama'ra. *Bitter-sweet.* Stem flexuous, climbing. Leaves ovate, cordate, upper ones lobed or gashed. Flowers purple, in lateral cymes, drooping. (Fig. 611.)
6 S. Pseudo-cap'sicum. *Jerusalem Cherry.* A small, handsome, erect shrub, 2–4f. †
7 S. sempervi'rens. *Evergreen N.* Climbing. Leaves thick, cordate, elliptic, obtuse, with a blunt cusp, very smooth and shining. Panicles terminal. †
8 S. Carolinen'se. *Horse Nettle.* Leaves angular-lobed. Racemes leafless. w.
9 S. Virginia'num. *Virginia N.* Leaves pinnatifid. Racemes leafy. Pale. S.
10 S. mammo'sum. *Apple of Sodom.* Woolly and prickly. Leaves roundish-ovate, lobed. Fruit inversely pear-shaped. Flowers violet-colored.
11 S esculen'tum. *Egg-plant.* Leaves ovate, somewhat sinnate, downy. Flowers 6–9-parted. Fruit egg-shaped, from the size of an egg to a water-melon. †

ORDER XCV.—GENTIANACEÆ. The Gentianworts.

Herbs with opposite, entire, smooth *leaves*, and showy regular *flowers;*
corolla usually twisted in the bud, with as many lobes as *stamens*, and alternate with them, mostly persistent and withering;
stigmas 1 or 2;
ovary 1-celled, superior, becoming a 2-valved *pod* with many seeds.

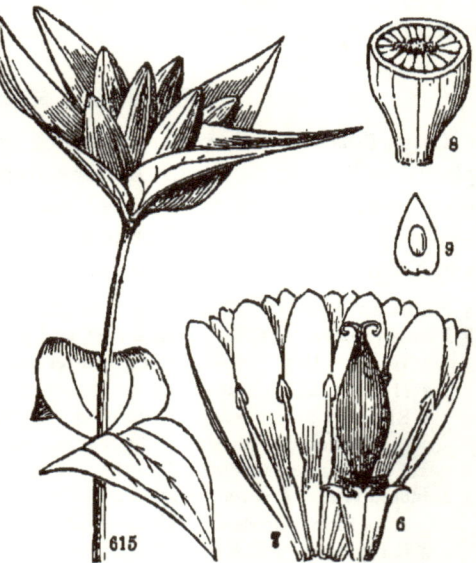

Analysis of the Genera.

§ Leaves opposite or whorled, sometimes minute. Corolla mostly twisted in bud....2

§ Leaves alternate or radical. Corolla valvate in the bud ....d

2 Corolla with a glandular spot on each lobe, sometimes with spurs....c

2 Corolla without glandular spots or spurs....3

3 Corolla tubular, the tube longer than the limb....a

3 Corolla deeply cleft, mostly wheel-shaped, tube very short....b

a Sepals 4 or 5. Corolla fringed, or with folds between lobes. Anthers straight.
GENTIA'NA. 1

a Sepals 4 or 5. Anthers spirally twisted. *European Centaury.* ERYTHRÆ'A.

a Sepals 2, leaf-like. Cor. 4-cleft, white or purplish. *Pennywort.* OBOLA'RIA.

b Leaves very small or mere bracts. Fls. 4-parted. *Screw-stem.* BARTO'NIA.

b Leafy. Fls. 5–12-parted. Anthers curved. *American Centaury.* SABBA'TIA. 2

c Corolla 4-parted, with 4 spurs beneath at base. *Spurred Gentian.* HALE'NIA.

c Cor. 4-parted, without spurs. Tall, with whorled leaves. *Columbo.* FRA'SERA.

d Corolla bearded inside. Leaves 3-foliate, on long stalks.
Buck-bean. MENYAN'THES.

d Corolla smooth inside. Leaves simple, floating.
Floating Heart. LIMNAN'THEMUM.

Fig. 615. Gentiana Andrewsii. 6. The calyx and capsule. 7. The corolla laid open, showing the folds (2-lobed) between the proper petals, and the stamens attached at base. 8. Capsule cut across. 9. Seed magnified, with its large, loose testa.

1. GENTIA'NA. Gentian.

Calyx 4–5-cleft. Corolla 4–5-lobed, regular, usually with plaited folds between the lobes. Stigmas 2, style short or none. Pod oblong, 2-valved, many-seeded.—Leaves opposite. Flowers solitary or in cymes. *Aug.–Oct.*

§ Corolla with folded appendages between the 5 lobes. Anthers cohering....b
Coroila with no appendages between the lobes. Anthers separate....a
   * Segments of the corolla entire, pale-blue, 5 in number....1
   a Segments of the corolla fringed, mostly but 4, bright blue....2, 3
      o Flowers solitary, terminal, blue or white....4
        Flowers clustered, yellowish or cream-white....5, 6
        Flowers clustered, blue....7-9

* G. quinqueflo·ra. *Five-leaved G.* Clusters about 5-flowered. Corolla lobes bristle-pointed.
2 G. crini'ta. *Fringed G.* Leaves lanceolate. Corolla conspicuously fringed. Height 1f.
3 G. det'onsa. *Shorn G.* Lvs. linear. Corolla lobes crenato-ciliate. Height 1f.
4 G. angustifo'lia. *Sand G.* Slender, 1f. erect. Lve. linear. Flower large. *b.* or *p.*
  5 G. ochroleu'ca. *Pale G.* Lvs. lance-oval, narrowed to the sessile base. Corolla greenish-white, a third longer than the sepals. S. M.
  6 G. alba. *Whitish G.* Lvs. lance-ovate, clasping with the broad base. Corolla cream-white, 4 times longer than sepals. W. M.
7 G. Andrew'sii. *Closed Blue G.* Leaves ovate-lanceolate, 3-veined. Corolla never opening, the lobes equalling the 5 fringed folds. (Figs. 615-619.)
8 G. Sapona'ria. *Soapwort G.* Plant smooth. Leaves rough-edged, linear-lanceolate. Corolla open, the lobes twice longer than the cleft folds.
9 G. puber'ula. *Rough G.* Plant scabrous. Lvs. lance-ovate, very rough at edge. Corolla somewhat bell-shaped, folds very short. W. S.

2. SABBA'TIA. American Centaury.

Calyx 5–12-parted. Corolla rotate, limb 5–12-parted. Stamens 5–12. Style 2-parted. Capsule 1-celled.—Beautiful biennials, with mostly roseate flowers.

§ Corolla mostly 9 (rarely 7–12)-parted....1, 2
§ Corolla 5 (rarely 6)-parted....a
   * Branches alternate or forked....b
   * Branches opposite. Flowers with a central star....c
      b Flowers white or nearly white....3, 4
      b Flowers rose-red, with a central star....5, 6
        Flowers white, corymbed....7, 8
        Flowers rose-red, paniculate....9, 10

1 **S. gentianoi'des.** *Gentian C.* Leaves linear, rigid, longer than the internodes. Flowers 8–10-parted, bright flesh-color, clustered. S.

2 **S. chloroi'des.** *Chlora C.* Leaves lanceolate. Branches few, alternate, each bearing at top a solitary, 7–12-parted, bright purple flower. E.

 3 **S. calyco'sa.** *Cup C.* Calyx leafy, as large as the 5–6-parted corolla. S.

 4 **S. panicula'ta.** *Panicled C.* Sepals linear, half as long as 5-parted cor. S.

5 **S. gra'cilis.** *Slender C.* Leaves ovate to linear. Sepals bristle-form, as long as the corolla. M. S.

6 **S. stella'ris.** *Starry C.* Leaves lance-obovate. Sepals linear, much shorter than corolla. c.

 7 **S. corymbo'sa.** *Corymbed C.* Leaves lanceolate, 3-veined. Calyx segments linear, thrice longer than its tube, half as long as the corolla. N.-J. S.

 8 **S. macrophyl'la.** Leaves 5-veined, cusp-pointed. Sepals shorter than calyx tube. S.

 9 **S. angularis.** *Angled C.* Stem square, with winged angles. Leaves ovate, clasping, 5-veined. Flowers many, rose-red, the star greenish. Wet. c.

10 **S. brachia'ta.** *Prairie C.* Stem square, slender, joints 2–4 times longer than the sessile, lance-linear leaves. Panicle oblong. Corolla 6-parted, the star yellow. W. S.

Order XCVI. APOCYNACEÆ. Dogbanes.

Plants with a milky *juice*, entire and mostly opposite *leaves;*
flowers 5-parted and regular, with the *corolla* twisted in the bud;
stamens 5, with distinct filaments, anthers sometimes slightly united;
ovaries 2, distinct, but with their stigmas united at top of the styles;
fruit 2 follicles containing several or many albuminous seeds.

Analysis of the Genera.

* Herbs erect, 2–4f. high, the flowers in cymes....a
* Shrubs twining or trailing, with opposite leaves....b
* Shrubs erect, 3–6f. high, with the leaves in whorls of 3....c
 a Cor. bell-shaped, whitish. Style none. Sds. silky. *Dog's-bane.* Apoc'ynum. 1
 a Corolla funnel-form, bluish. Style 1. Lvs. scattered. *Amson.* Amso'nia.
 b Fls. solitary, blue. Throat 5-angled. Lvs. evergreen. † *Periwinkle.* Vin'ca.
 b Flowers in cymes, yellow, small. Lvs. petiolate. Wet. South. Forstero'nia.
 c Leaves thick, evergreen. Flowers large, rose-colored. *Oleander.* Ne'rium.

APOC'YNUM. Dog's-bane.

Stamens shorter than the corolla, arising from its base, and alternate with 5 glandular teeth. Anthers arrow-shaped, cohering to the stigmas by the middle. Follicles long, slender, separate. Seeds with a tuft of long, silky down. *June–August.*

1 A. androsæmifo'lium. *Tutsan-leaved D.* Corolla rose-white, much longer than the calyx. Leaves ovate. Plant smooth, elegant, about 3f. high.

2 A. cannabi'num. *Hemp D.* Cor. greenish-white, scarce longer than the calyx. Leaves oblong. Bark tough as hemp.

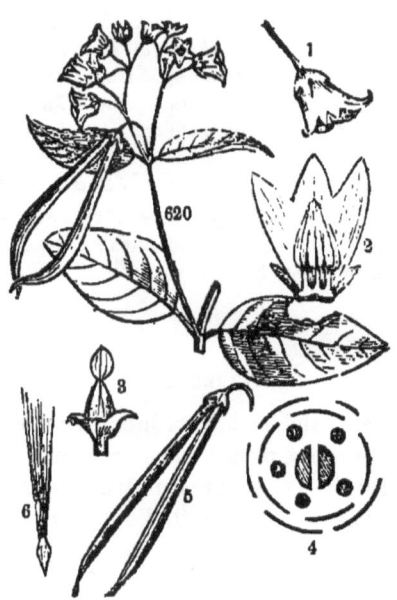

Fig. 620. Common Dog's-bane. 1. A flower of the natural size. 2. The flower cut open, showing the peculiar stamens. 3. The 2 styles and stigmas. 4. The plan of the flower. 5. The 2 follicles. 6. A seed with its tuft of silk.

ORDER XCVII. ASCLEPIADACEÆ. Asclepiads.

Plants (chiefly *herbs* in the United States) with a milky *juice;*
leaves opposite (rarely whorled or scattered), entire, without stipules;
flowers generally umbeled, 5-parted, regular; *sepals* and also the
petals united at base, both commonly valvate in the bud;
stamens united into a fleshy mass with the two stigmas;
pollen cohering in masses; *ovaries* 2, forming *follicles* in fruit

Analysis of the Genera.

§ Plants erect. Stamen-mass crowned with 5 little hoods....2
§ Plants twining or prostrate....3
    2 Hoods each with a little projecting horn....a
    2 Hoods destitute of horns....b
        3 Flowers dark purple....c
        3 Flowers whitish or flesh-colored....

a Petals reflexed. Hoods erect, horns incurved. *Silkgrass. Milkweed.* ASCLE'PIAS. 1
  b Petals reflexed. Hoods erect, adnate to the anthers.   ACERA'TES.
  b Petals spreading, green. Hoods free from the anthers. S. ANAN'THERIX.
  b Petals erect, yellowish. Mass of anthers stalked. S.   PODOSTIG'MA.

12*

c The 5 filaments distinct. Pollen masses 5. N.-Y.   PERIPLO'CA.
c Filaments united as well as the stigmas. Pollinia 10.   GONOL'OBUS.
   d Petals spreading. Hoods erect. Leaves linear. Coast, S.   SENTE'RA.
   d Petals spreading. Hoods flat, spreading. †   *Wax-plant.* HOY'A.
   d Petals erect, white. Hoods erect, 2-awned. S.-W. Common. ENSLE'NIA.

ASCLE'PIAS. Silk-grass. Milkweed.

(We have many species of this genus, blooming in the Summer months. Beginners will find them difficult to distinguish. We omit them here, referring the reader to the Class Book.)

ORDER XCIX. OLEACEÆ. Oliveworts.

Trees and *shrubs* with opposite *leaves,* with
flowers 4-parted, regular, sometimes without petals;
corolla (when present) valvate in the bud; *stamens* 2, rarely 4;
ovary 2-celled, with 2 ovules in each cell; *fruit* fleshy or dry.

Analysis of the Genera.

§ Leaves pinnate. Fruit a dry, winged samara....a
§ Leaves simple. Fruit a dry, 2-celled pod (capsule)....b
§ Leaves simple. Fruit a fleshy drupe or berry....2
   2 Corolla present. Stamens included. White....c
   2 Corolla present. Stamens exserted....d
   2 Corolla none. Fruit an oblong drupe....e
a Trees with imperfect flowers and odd-pinnate leaves.   *Ash.* FRAX'INUS.
  b Corolla salver-form, with short, white or purple lobes. † *Lilac.* SYRIN'GA. 1
  b Corolla bell-form, with long, yellow lobes. †   *Forsythia.* FORSY'THIA.
c Corolla with long, linear, pendulous lobes. *Virginia Fringe-tree.* CHIONAN'THUS.
c Cor. with short lobes. Panicle dense. Berries black. *Privet. Prim.* LIGUS'TRUM.
  d Style 2-parted. Leaves serrate. Shrubs. †   *Osmanth.* OSMAN'THUS.
  d Style simple. Panicles axillary. S.   *American Olive.* OLEA.
  d Style simple. Panicles terminal. Trees. †   *Visian.* VISIA'NA.
e Flowers very imperfect, diœcious. Shrubs. Wet. W. S. *Adelia.* FORESTI'ERA.

SYRIN'GA. Lilac.

1 S. vulga'ris. *Common L.* Leaves cordate-ovate, entire. Flowers lilac-purple.
2 S. al'ba. *White Lilac.* Flowers pure white. Shrub taller. (Variety of No. 1.)
3 S. Per'sica. *Persian L.* Leaves lanceolate, entire or cleft. Flowers in looser panicles, lilac-blue. *Apr. May.*

COHORT III.

THE APETALOUS EXOGENS.

Essential Character.—Flowering plants (PHÆNOGAMIA), with their stems growing by additions to the outside and the wood in circular layers (EXOGENS), with the seeds inclosed in seed-vessels (ANGIOSPERMS), and the flowers destitute of petals (APETALÆ).

ORDER C.—ARISTOLOCHIACEÆ. **Birthworts**

Low *herbs* or climbing *shrubs* with alternate *leaves*, large *flowers ;*
calyx adhering to the ovary, valvate in bud, brown or dull colored;
stamens 6–12, at top of the 6-celled, many-seeded *ovary.*

Analysis of the Genera.

Calyx bell-form, regular, 3-cleft. Stamens 12. Herbs with creeping,
  underground stems.   *Wild Ginger.* ASA'RUM, 1
Calyx tubular, bent, irregular. Anthers 6. Shrubby, erect or climbing,
  with very odd flowers.   *Birthwort.* ARISTOLO'CHIA.

ASARUM. **Wild Ginger.**

1 A. Canadense. *Canada W.* Leaves in pairs, broad-reniform, with the single flower
  between the petioles scarcely above-ground. *May–July.* c.
2 A. Virginicum. *Virginia W.* Leaf solitary, round-ovate, cordate, the single flower
  much shorter than the petiole. Sepals obtuse. Mts. Va. S. *April.*
8 A. arifol'ium. *Arum-lv. W.* Leaf solitary, broadly hastate, with long, angular
  lobes at base. Calyx throat contracted, lobes very short. Va. S. *April.*

ORDER CI.—NYCTAGINACEÆ. **Marvelworts.**

Herbs (shrubs or trees) with swelling joints; entire, opposite *leaves ;*
flowers surrounded by an involucre (which is, of course, calyx-like when
  the flower is solitary):
calyx often colored like a corolla, tubular or funnel-form, breaking off
  above the 1-celled, 1-seeded *ovary.*

Analysis of the Genera.

Involucre just like a calyx, involving a single, large flower. Calyx funnel-form,
corolla-like, the limb entire.   *Four-o-clock.* MIRAB'ILIS. 1
Involucre involving 2–5 small, rose-red flowers. W. S.   OXYB'APHUS.
Involucre none, or minute bractlets. Flowers minute. S.   BOERHAA'VIA.

MIRAB'ILIS. Marvel-of-Peru. Four-o-clock.

1 M. Jala'pa. *Peruvian F.* Leaves ovate, subcordate. Fls. stalked, with a large
border, infinite in variety of color, opening about 4 o'clock, P. M. †
2 M. dichotoma. *Mexican F.* Erect, smooth. Calyx with a small border. †
8 M. longiflo'ra. *Long-fl. F.* Diffuse, viscid. Calyx tube downy, very long. ‡ Mex.

ORDER CII.—POLYGONACEÆ. Knotweeds.

Herbs with alternate *leaves,* swollen joints, and with
stipules sheathing the stem above the joints; *flowers* racemed, perfect;
calyx persistent; *sepals* 4–6, imbricated, distinct or united at base;
stamens 4–12; *ovary* 2 or 3-styled, 1-celled, 1-seeded in fruit.

Analysis of the Genera.

* Calyx 4-parted, regular. Stamens 6. Styles 2.   *Mountain Sorrel.* OXYR'IA.
* Calyx 6-parted. Stamens 9. Sepals all similar.   *Rhubarb.* RHEUM.
* Calyx 6-parted. Stamens 6. Inner sepals large.   *Dock. Sorrel.* RU'MEX.
* Calyx 5-parted (irregularly 4-parted in one species)....a
    a Sep., the 3 inner fringed. Fls. solitary. S. *Fringe Knotweed.* THYSANEL'LA.
    a Sepals not fringed, entire or nearly so....b
b Pedicels solitary. Sep. all open or 3 closed on the fruit. M. S. POLYGONEL'LA.
b Pedicels usually clustered. Sepals all closed on the fruit.   POLYG'ONUM. 1
b Pedicels clustered in the bract. Sepals all open. *Buckwheat.* FAGOPY'RUM.

POLYG'ONUM. Knotweed.

Calyx 5- (rarely 4-) parted, colored or corolla-like, the sepals all erect
and inclosing the fruit. Stamens 4–9. Styles 2 or 3. Nut lens-shaped
or 3-cornered.—Herbs with swollen, sheathed joints. Flowers small,
white, red, or greenish. *May–Aug.*

§ Stems climbing, with reversed prickles. Leaves cordate-sagittate....19, 20
§ Stems unarmed, twining. Leaves cordate-hastate....17, 18
§ Stems erect or decumbent, unarmed. Leaves hardly ever cordate....a

a Calyx unequally 4-cleft. Styles 2, long, deflexed....16
a Calyx equally 5-parted. Styles erect....b
b Sheaths with a spreading border. Stamens 7. Plant tall....15
b Sheaths not bordered. Stamens 5, 6, or 8....c
  c Flowers in leafless, terminal, spike-like racemes....d
  c Flowers axillary, or rarely forming a leafy raceme....f
    d Raceme one, dense. Stems decumbent at base....13, 14
    d Racemes several. Sheaths naked, not fringed....11, 12
    d Racemes several. Sheaths bristly fringe-ciliate. ..e
      e Style 2 (or 3)-cleft. Achenia flat or lens-shaped....8-10
      e Style 3-cleft. Achenia sharply 3-cornered. Wet....5-7
        f Achenium protruding beyond the calyx, 3-angled....3, 4
        f Achenium included in the calyx, 3-angled....1, 2

1 P. avicula're. *Bird K.* Prostrate or erect. Leaves elliptic-lanceolate, acutish at each end. Achenia striate, dull. Very common.

2 P. ten'ue. *Slender K.* Slender, rigidly erect. Leaves lance-linear, erect, acute.

3 P. marit'imum. *Sea K.* Prostrate, diffuse, glaucous, close-jointed. Leaves linear-oblong, fleshy. Achenia smooth, shining. E.

4 P. ramosis'simum. *Lake K.* Erect, much branched, 2-3f. high. Leaves lance-oblong or linear. Achenia smooth, dull. W.

5 P. hirsu'tum. *Hairy K.* Hairy-tawny. Leaves lanceolate from a blunt base. S.

6 P. hydropiperoi'des. *Mild Water-pepper.* Stem smooth. Leaves linear-lanceolate, not acrid, tapering at both ends. Spikes slender. Calyx dotless.

7 P. acre. *Sharp W.* Stem smooth. Leaves biting, dotted as well as the calyx, lanceolate, pointed. Spikes very slender, thread-form.

8 P. hydropi'per. *Water-pepper.* Smooth. Leaves very biting, dotted. Spikes short, nodding. Calyx dotted. Achenia roughened.

9 P. Car'eyi. *Carey's K.* Plant hairy. Spikes nodding, on very long stalks.

10 P. Persica'ria. *Lady's-thumb.* Leaves marked with a brown spot. Spikes short, dense, erect. Achenia shining, flattened.

11 P. Pennsylvan'icum. *Pennsylvania K.* Spikes oblong, dense, with glandular-hispid stalks and pedicels. Achenia with flat sides. *c.*

12 P. incarna'tum. *Flesh-red K.* Spikes linear, nodding, the stalks and branches glandular-dotted. W. S.

13 P. amphib'ium. *Water K.* Stem ascending from a prostrate, rooting base. Leaves lance-oblong. Stamens 5. Spikes large, dense, rose-red.

14 P. vivip'arum. *Alpine Bistort.* Creeping at base. Lvs. lance-linear. Mts. N.

15 P. orienta'le. *Prince's Feather.* Stem stout, tall, with large, drooping spikes. †

16 P. Virginia'num. *Lip-fl. K.* Leaves large. Racemes slender, flowers remote.

17 P. convol'vulus. *Knot-Bindweed.* Roughish. Racemes axillary. Fruit dull.

18 P. cilino'de. *Bearded B.* Sheaths with a hairy ring. Panicles axil. and terminal.

19 P. dumito'rum. *Hedge B.* Calyx with the 3 outer sepals acutely wing-keeled.

20 P. sagitta'tum. *Scratch-grass.* Lvs. lance-sagittate. Stamens 8. Styles 3.

21 P. arifo'lium. *Arum-lv. S.* Lvs. pointed, with pointed lobes. Stam. 6. Sty. 2.

Order CIII. PHYTOLACCACEÆ. Pokeweeds.

Herbs or *shrubs* with alternate *leaves*, no *stipules*, and *flowers* racemed;
sepals colored, 4 or 5; *petals* none; *stamens* few or many;
ovary of one or several carpels, which are united into a ring, forming a
    berry in fruit; *cells* as many as the carpels, each 1-seeded;
embryo curved around the fleshy albumen.

Analysis of the Genera.

Sepals 5, roundish. Sta. 5–20. Ovary 5–12-carpeled and seeded. Phytolac'oa. 1
Sepals 4, persistent. Stamens 4–12. Ovary 1-carpeled and 1-seeded. S. Rivi'na.

PHYTOLAC'OA. Poke.

Character expressed in the
Analysis.—Tall and stout perennials, with greenish flowers
and purple berries.

P. decan'dra. Plant 5–8f. high,
    very smooth, bushy. Leaves
    large, ovate, acute at each end,
    petioled. Racemes at first
    terminal, finally opposite to
    the leaves. Berries oblate, of
    a rich dark purple. *July–Sept.*

Fig. 627. Phytolacca decandria, leaves, flowers, and fruit. 8. A flower, natural size. 9. Its stamens and ovary. *Fig.* 630. Cross-section of the ovary. 1. A seed cut open, showing the embryo coiled around the albumen.

Order CVII. LAURACEÆ. Laurels.

Trees and *shrubs* aromatic, with alternate, simple, dotted *leaves;*
sepals colored, 4–6, slightly united, strongly imbricated;
anthers 2 or 4-celled, opening upwards by as many valves;
ovary 1-ovuled, becoming a drupe in fruit; no albumen.

Analysis of the Genera.

§ Flowers perfect, the calyx persistent. Leaves evergreen....a
§ Flowers imperfect. Calyx deciduous. Leaves deciduous....b
    a Trees. Lvs. thick, lance-oblong. Fls. umbeled. S. *Bay Galls.* Per'sea.

b Involucre none. Anthers 4-valved. Leaves lobed. *Sassafras.* SAS'SAFRAS. 1
b Involucre 4 leaved. Anthers 2-valved. Shrubs. *Spice-bush.* BEN'ZOIN.
b Involucre 4-leaved. Anthers 4-valved. Shrubs. S. *Pond-spice.* TETRAN'THERA.

SASSAFRAS. **Sassafras.**

Flowers diœcious, 6-parted, regular. Stamens 9. Trees with deciduous leaves, expanding after the clusters of yellow flowers.

S. officina'le. *Common S.* Leaves of two forms, ovate and entire, or 3-lobed and
acute at base. Tree aromatic, 10–30f. high.

ORDER CXIX. CUPULIFERÆ. **The Mastworts.**

Trees or *shrubs*, with alternate, simple *leaves*, and deciduous *stipules;*
flowers monœcious, the sterile in aments, which are racemed or head-like;
stamens in the sterile flowers, 6 to 20, on the base of the calyx;
ovary in the fertile flowers with several cells and ovules, but becoming in
fruit a 1-seeded nut surrounded by an involucre (cup, burr, or sac).

Analysis of the Genera.

§ Sterile flowers in aments, fertile flowers solitary or 2 or 3 together....2
§ Sterile flowers and fertile also in aments, the latter loose and large....v
   2 Involucre 1-flowered, cup-like, composed of many little scales....a
   2 Involucre 2 or 3-flowered, composed of few large valves....b
a Sterile aments slender, calyx 5-cleft, stamens 5 or 10. Fertile flowers, consisting of an ovary sitting in a scaly cup, becoming, in fruit, an *acorn*, 1-
celled, 1-seeded. A noble genus of trees (rarely shrubs), always known by
their peculiar fruit, called *acorns.* The timber is of great value, especially
in ship-building. In the Class Book of Botany, 23 species are described.
(See Figs. 32–34, 267.) *Oak.* QUERCUS.
b Involucre of the fruit and fertile flowers a burr with 4 valves. Sterile aments
slender, each flower with 5–15 stamens; 3 fertile flowers in each involucre,
which is beset with slender prickles. We have two species. one a tree, the
other a shrub. Timber excellent. The fruit is sweet and nutritious. (See
Fig. 277.) *Chestnut.* CASTA'NEA.
b Involucre of the fruit a burr with 4 valves. Sterile aments head-like, suspended by a slender stalk. Calyx 6-cleft. Two flowers in each burr, which
is covered by weak spines. Nuts sharply 3-angled. They are tall, valuable forest-trees. *Beech.* FAGUS.
b Involucre a sac, longer than the nuts, torn at the top. Sterile flowers in a
slender ament. Shrubs. Usually but one flower or nut in each involucre.
Hazel. COR'YLUS.

c Involucre a closed, inflated sac, one-flowered, many together in the pendulous, hop-like cluster. Small trees, with very compact, strong timber, called
Hop Hornbeam. Iron-wood. Lever-wood. Os′TRYA.

c Involucre an open, 3-lobed leaf, 1-flowered, Small trees, with a strong, heavy timber.
Hornbeam. CARPI′NUS.

ORDER CXX. BETULACEÆ. The Birchworts.

Trees or *shrubs*, with deciduous *stipules*, with the alternate
leaves simple, having the veinlets running straight to the margin;
flowers monœcious, both kinds contained in scaly
catkins, 2 or 3 under each bract; *calyx* and *corolla* hardly any;
ovary 2-celled and 2-ovuled, but becoming in
fruit a 1-celled and 1-seeded nut, by the suppression of the other seed
and cell.

Analysis of the Genera.

♂ bracts with 12 stam. each; ♀ bracts with mostly 3 ovaries. *Birch.* BET′ULA. 1
♂ bracts with 4–8 stam. each; ♀ bracts with 2 ova. or fls. each. *Alder.* ALNUS.

BET′ULA. Birch.

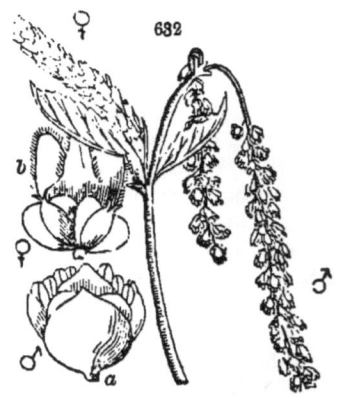

♂ in a cylindrical catkin, bracts each with 3 tetrandrous flowers beneath it. ♀ in an oblong or egg-shaped catkin, bracts 3 lobed, each with 3 2-styled ovaries or flowers, with no calyx. Samara flattened, broadly winged. — Trees and shrubs, mostly with the outer bark in thin layers with horizontal fibres. Catkins appearing in early spring before the leaves.

§ Trees with a yellowish bark, and heart-ovate, serrate leaves....1
§ Trees with reddish-brown bark, and ovate, doubly serrate leaves....2, 3
§ Trees with white bark and long-stalked, long-pointed leaves....4, 5
§ Shrubs with brownish bark, and roundish, crenate leaves....6, 7

Fig. 632. Sweet Black Birch (*Betula lenta*), with staminate and pistillate catkins: *a*, a scale with staminate flowers; *b*, with pistillate flowers. *Fig.* 633. *a*, A winged samara cut lengthwise, showing its fertile and abortive cell; *b*, the same cut across.

1 **B. excel'sior.** *Yellow Birch.* Tree 50–80f. Fertile aments erect, oblong, 1 inch in length, erect, sterile 2–4', pendulous, clustered. *c.* N.

2 **B. lenta.** *Black B. Sweet B.* Tree 40–60f. Fertile aments erect, oval, obtuse, stalked; sterile 2–3', pendulous. Inner bark sweet-spicy. M. N.

8 **B. ni'gra.** *Red Birch.* Tree 30–50f. Leaves rhombic-ovate, acute at both ends, obscurely lobed. Fertile aments sessile, ovoid. M. S. W.

4 **B. populifo'lia.** *White B.* Tree 30–40f. Leaves triangular, long-pointed, smooth, unequally serrate. Sterile aments long, pendulous. N.

5 **B. papyra'cea.** *Canoe Birch.* Tree 50–70f. Leaves ovate, pointed, donbly-serrate. All the aments nodding. Hills and mountains. N.

6 **B. pum'ila.** *Dwarf B.* Shrub erect, 2–6f. Branches warty. Leaves obovate, obtusely serrate above. Fertile aments cylindric. Mountains. N.

7 **B. na'na.** *Tiny B.* Shrub low, trailing, smooth. Leaves round, crenate. Scales of fertile ament deeply 8-parted. 8–12'. Mountains. N.-H.

ORDER CXXII. SALICACEÆ. The Willoworts.

Trees or shrubs, with simple *leaves*, and stipules usually present; *flowers* diœcious, naked, both kinds in aments, each with a bract; *ovary* 1 or 2-celled, with 2 short styles; *capsule* many-seeded; *seeds* with a coma and no albumen.

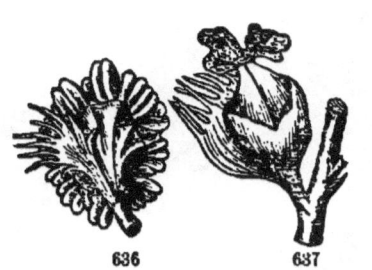  

636 637 634 635

Fig. 634. A fertile flower of a Willow, consisting of a pistil and a bract. *Fig.* 635. Sterile flower, 2 stamens and a bract. *Fig.* 636. A sterile flower of Balm-of-Gilead (*Populus candicans*); many stamens. *Fig.* 637. A fertile flower, consisting of a fringed scale, a calyx holding a double ovary.

Analysis of the Genera.

§ Aments cylindric, bracts entire. Stamens 2 or more. Capsule 1-celled, 2-valved, the seeds very small, clothed with silky hairs. Trees or shrubs. Leaves often long and narrow. (Figs. 12, 17, 88.) We have about 27 species. *Willow. Osier.* SALIX.

§ Aments cylindric, bracts fringed. Stamens 8 or more. Capsule 2-celled, 2-valved. Calyx an entire cup. Buds varnished with a fragrant resin. Leaves broad, large. Trees. *Poplar. Aspen.* POPULUS

COHORT V.

SPADICIFLORÆ, OR THE APETALOUS ENDOGENS.

ORDER CXXXI. ARACEÆ. Aroids.

Chiefly *herbs* with a fleshy rootstock of corm; *leaves* sometimes net-veined; *flowers* mostly without calyx or corolla, arranged on a spadix; *stamens* few or many, hypogynous, very short; *anthers* turned outwards; *ovary* free; *stigmas* sessile; *fruit* a dry or juicy berry, and the *seeds* with or without albumen. Growing in wet places.

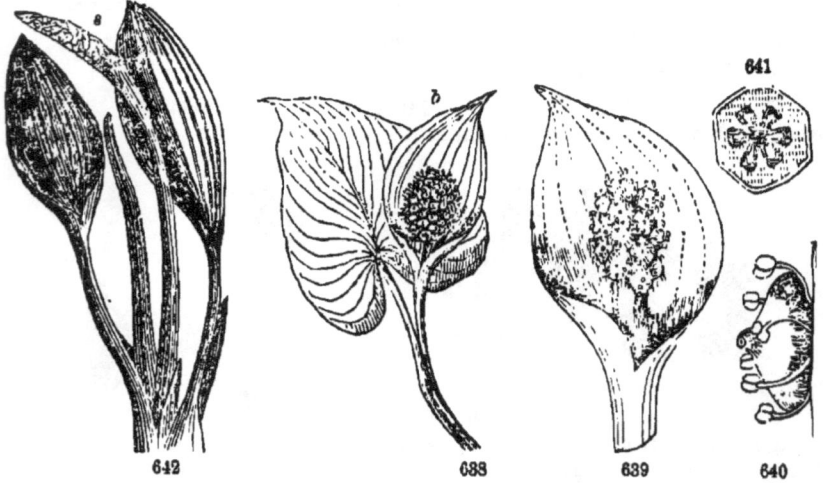

Fig. 638. Wild Calla (*Calla palustris*), a leaf, and a spadix of flowers, with its spathe (*b*). Fig. 639. The same enlarged. Fig. 640. A flower enlarged. Fig. 641. One of the berries cut, showing the 6 cells. Fig. 642. Golden Club (*Orontium aquaticum*); its spadix (*s*) is without a spathe.

Analysis of the Genera.

* Spadix enveloped in a spathe....2
* Spadix destitute of a spathe. Sepals 4-6....d
    2 Flowers covering only the base of the spadix. Perianth 0....a
    2 Flowers covering the whole spadix, monœcious. Perianth 0....b
    2 Flowers covering the whole spadix and perfect....c

a Spathe rolled in at base. Top of spadix club-shaped. *Dragon-root.* ARISÆ'MA. 1
  b Spathe rolled inwards the whole length. *Arrow-lvd. Drogon.* PELTAN'DRA. 2
  'b Spathe rolled backwards above, white.   *Egyptian Calla.* RICHAR'DIA.
c Perianth 0. Spathe open, flattish, white. Lvs. cordate.   *Wild Calla.* CALLA.
e Perianth regular. Spa. shell-form. Lvs. large. *Skunk-cabbage.* SYMPLOCAR'PUS.
  d Flowers terminal, yellow. Scape terete.   *Golden Club.* ORONTIUM.
  d Flowers lateral, green. Scape leaf-like.   *Sweet Flag.* A'CORUS.

1. ARISÆMA. Arum. Indian Turnip.

Spathe rolled inward at base. Spadix covered with flowers below, naked and club-shaped above. Sterile flowers above the fertile, each a clump of 4 stamens. Fertile flowers each a 1-celled ovary, with a flat stigma. Berry red, with 1 or several seeds.—Odd-looking plants, with scape arising from a corm or rootstock, and sheathed with the radical leaves. Taste very acrid.

1 A. triphyl'lum. *Jack-in-the-Pulpit.* Leaves usually 2, trifoliate. Spathe bent and inflected above, covering the obtuse spadix, striped.
2 A. quina'tum. *Five-leaved Jack.* Leaves in pairs, one or both quinate. S.
8 A. Dracon'tium. *Green Dragon.* Leaf mostly solitary, pedate, 7–11-foliate.

2. PELTANDRA. Arrow Dragon.

The sterile flowers consist of 8–12 anthers attached to the border of a shield-shaped (peltate) connectile.—Root fibrous. Leaves sagittate.

1 P. Virgin'ica. *Virginia A.* Spathe green, incurved, long, wavy on the margin. Leaves many, large, hastate-sagittate, very smooth, dark.
2 P. glau'ca. *Glaucous A.* Spathe white, entire, gradually unrolled and widened above. Leaves ovate-sagittate, the base lobes large. S.

COHORT VI.

FLORIDEÆ, OR THE FLOWERING ENDOGENS.

Order CXXXV. ALISMACEÆ. Alismads.

Herbs growing in water, with the *leaves* parallel-veined, and with the
flowers regular and not on a spadix ; the *perianth* consisting of
sepals and *petals*, 3 of each, the former always green ;
ovaries free, 8 or more, separating into as many 1-seeded achenia.

Analysis of the Genera.

§ Both the calyx and corolla greenish. Lvs.
   rush-like (*Arrow-grasses*).....b
§ Corolla colored, white. Leaves mostly
   with a lamina....a
  a Fls. ☿. Sta. 6. Carpels whorled.
        Water Plantain. Alisma. 1
  = Fls. ☿. Stamens 9-24. Carpels in a
    head. *Echinodore.* Echinodorus.
  ▲ Fls. ♂. Sta. many. Carpels in a
    head. *Arrow-head.* Saoittaria. 2
b Lvs. radical. Anthers ovate. Carpels
   1 seeded. *Trigloch.* Triolochin.
b Leaves cauline. Anthers linear. Carpels 1-2-seeded. Soheuchzeria.

Fig. 643. Sagittaria sagittifolia (common form), leaf and flowers. 4. One
of the pistils enlarged. 5. The pistil of
Alisma cut open, showing the seed and
curved embryo.

1. ALISMA. Water Plantain.

Flowers perfect. Stamens 6. Ovaries and styles numerous, collected into
a whorl, becoming in fruit many distinct, flattened achenia.—♃ Stemless
herbs, the leaves all radical. Flowers
in a panicle.

A. planta′go. A common, smooth, handsome
   inhabitant of ponds and ditches. Leaves
   oval or ovate, abruptly acuminate, 7-9-

veined, entire, on long petioles. Scape 1–2f. high, branching in whorls, bearing numerous purplish-white flowers, in July.

2. SAGITTA'RIA. Arrow-head.

Flowers monœcious, rarely diœcious, the ♂ with about 24 stamens, the ♀ with numerous ovaries crowded in a head, and becoming in fruit as many compressed, margined achenia.—♃ Stemless plants, leaves radical, generally arrow-shaped. Flowers in whorls of 3's, the sterile ones above the fertile.

S. variab'ilis. A curious plant, conspicuous with its large white flowers among the rushes and sedges of sluggish waters. The petals are wholly white, and the scape simple, with the stalks 1-flowered. The leaves are generally arrow-shaped (as seen in the figure), but exceedingly variable, sometimes lanceolate, and sometimes even consisting of a petiole only. About 1f. high. *July.*

Order CXXXVIII. ORCHIDACEÆ. The Orchids.

Herbs perennial, with thick, fleshy *roots;* entire, parallel-veined *leaves; flowers* very irregular, but the *perianth* consisting always of 6 parts, viz., of 3 *sepals* and 3 *petals*, all usually colored, the lower petal called the *lip* differing in form from the others, and frequently spurred at base; *stamens* 3, but only 1 or rarely 2 of them perfect, united with the *style* and forming what is called the *column; anthers* 2, 4 or 8-celled; *pollen* powdery, or waxy, or granulated; *ovary* 1-celled, many-seeded.

Analysis of the Genera.

* Stems green, furnished with one or more leaves....2
* Stems green, furnished with sheaths instead of leaves....
* Stems brownish, furnished with sheaths and no leaves, or a late one....c
    2 Corolla lip very large, inflated and sack-like....a
    2 Corolla lip of various forms, but neither very large nor sack-like....3
3 Corolla produced into a spur behind....b
3 Corolla destitute of a spur....4
    4 Flowers small, many, in a loose raceme, beardless....e
    4 Flowers small, many, in a close, twisted spike, beardless....f
    4 Flowers showy, purple or yellow, few or 1 only....g
a Root fibrous. Lip obtuse, spurless. Anthers 2. *Lady's-slipper.* CYPRIPE'DIUM. 1
a Root a corm. Lip 3-lobed, 2-spurred. Anther 1. *Calypso.* CALYP'SO.

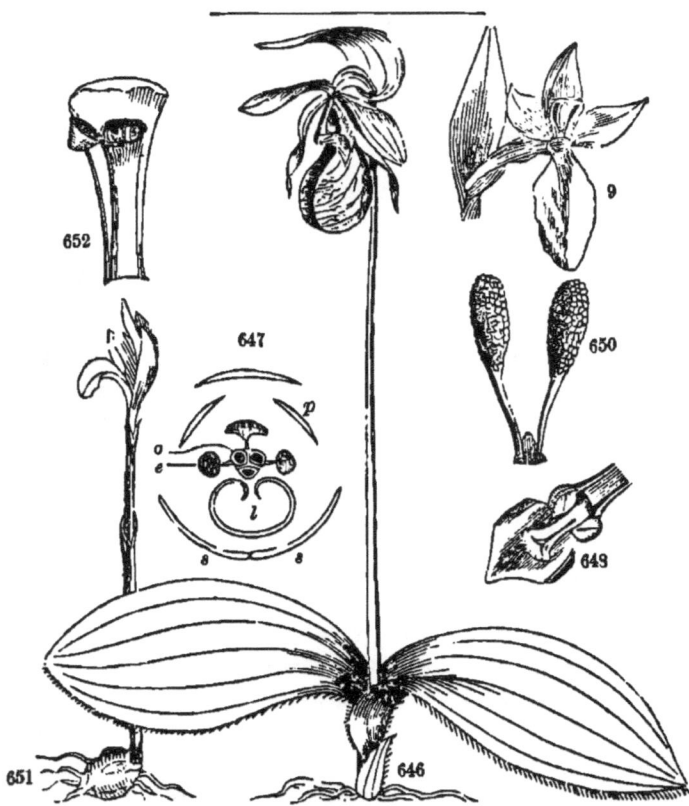

Fig. 646. Lady's-slipper (*Cypripedium acaule*), whole plant with its 2 leaves, scape, and curious flower. 7. Plan of the flower; *s*, sepals (outer circle), the 2 lower united ; *p*, the petals ; *l*, lip (lower petal) ; *a*, the anthers, upper one sterile ; o, the 3-celled ovary. 8. The column seen rom beneath, with the pistil, two stamens. and the leaf-like sterile one. 9. Flower and bract of Orchis spectabilis. *Fig.* 650. Its 2 pollen masses exhibited (enlarged). 1. Arethusa bulbosa ; *f*, he flower. 2. Its column enlarged, with its lid-like anther opening, showing its pollen-masses eneath.

 b Fls. in the axils of bracts. Pollen masses 2. Lvs. 1–∞. *Orchis.* ORCHIS. 2
 h Flowers bractless. Pollen masses 4. Leaf 1 only. *Tipula.* TIPULA′RIA.
 ѕ Root coraline. Spur growing to the ovary. Lvs. none. *Coral-root.* CORALLORHI′ZA. 3
 c Root 2 corms. Spur none. Leaf 1, late, radical. *Putty-root.* APLECTRUM.
  d Flowers 1 only, rose-purple. Lip bearded. *Arethusa.* ARETHU′SA. 4
  d Fls. racemed, dark-purple, beardless. (Lvs. 0 or few.) *Bletia.* BLETIA.
 e Leaf 1. Lip sagittate. Column minute. *Micros′tylis.* MICROS′TYLIS.
 e Leaves 2, radical. Lip flat, ascending. Column winged. *Liparis.* LIP′ARIS.
 ѕ Leaves 2, cauline. Lip pendulous, 2-lobed or 2-cleft. *Tway-blade.* LISTERA.

f Leaves all green. Lip obtuse, erect.   *Ladies' Tresses.* SPIRAN'THES. 5
f Leaves netted with white. Lip pointed, reflexed.
            Rattlesnake Plantain. GOODYE'RA.
f Lvs. all green. Lip 3-lobed, recurved. South. *Cranichis.* CRAN'ICHIS.
g Lip on the upper side of the fl., bearded. Leaf linear. *Grass Pink.* CALOPO'GON. 6
g Lip on the lower side (ovary twisted as in the other genera)....h
h Column free from the lip. Flowers purplish. *Beard Pink.* POGO'NIA. 7
h Column growing to the lip. Yellow. On trees. S. *Tree Orchis.* EPIDEN'DRUM.

1. CYPRIPE'DIUM. Lady's-slipper.

The 2 lower sepals united into one piece or rarely distinct. Lip very large, inflated, sack or slipper form, obtuse. Column terminated by a petal-like lobe (which is the sterile stamen). Fertile stamens 2.—Root fibrous. Leaves large, plaited. Flowers large and showy, one or few. *May–July.*

 * Flowers yellow, one or more. Stems leafy....5, 6
 * Flowers white or rose-purple....1
1 Stem leafy. Flower one or more....2–4
1 C. acau'le. *Stemless L.* (Figs. 642–644.) Scape naked, with 2 leaves at the base, and 1 large flower at top. *c.*
 2 C. can'didum. *White L.* Two lower sepals united. Flowers 1 only, smaller, white. W. S. *r.*
 3 C. spectab'ile. *Showy L.* Two lower sepals united. Flowers few, very large, purplish. *c.*
 4 C. Arieti'num. *Ram's-Head L.* Two lower sepals separate. Flower 1 only, small, purplish. *r.*
5 C. pubes'cens. *Large yellow L. Moccasin Fl.* Sepals narrow-lanceolate. Lip flattened at sides, pale yellow. *c.* (Fig. 89.)
6 C. parviflo'rum. *Smaller yellow L.* Sepals ovate-lanceolate. Lip flat above and below, bright yellow. *c.*

2. ORCHIS. Orchis.

Flower ringent, sepals and petals similar; all, or all but two, ascending and arching over the column. Lip turned downward, entire or lobed, produced at base into a spur beneath, which is distinct from the ovary. Stamen 1, anther 2-celled, pollen-masses 2, consisting of numerous waxy grains.—Flowers generally showy, in spikes or racemes. *June–Aug. mostly.*

NOTE.—Under this genus we include two others, viz.: Gymnadenia and Platanthera. The beginner would find it difficult to separate them. See Class Book of Botany, p. 682, &c. No. 1, is the true *Orchis.* Nos. 8, 9, 10, are *Gymnadenia,* and all the others belong to *Platanthera.*

§ Leaves radical and only 2 (rarely 3). Flowers on a scaps....1-8
§ Leaf radical and only 1. Flowers small, on a scape....4, 5
§ Leaves on the stem, several, upper ones reduced to bracts....a
a Corolla lip entire, neither lobed, fringed, nor toothed....6-9
a Corolla lip 2 or 8-toothed, not fringed nor divided. Flowers greenish....10-12
a Corolla lip cleft into a fringe at the edge, but not divided....13-15
a Corolla lip divided into 8 parts, which are fringed or not....b
     b Flowers white or yellowish, with 5 long bristles, the 2 side petals 2-parted. S....16, 17.
     b Flowers white, the 2 side petals entire or toothed; lip clawed....18, 19
     b Flowers purple, numerous, showy; lip raised on a claw....20-22

1 O. spectab'ilis. *Showy Orchis. True Orchis.* Fls. few, pink-purple, handsome. Leaves oblong-ovate. Height 4-7'. (Figs. 649, 650.)

2 O. orbicula'ta. *Round-lv. O.* Fls. whitish, racemed. Spur very long. Leaves round. Scape bracted.

3 O. Hook'eri. *Hooker's O.* Flowers green, spiked. Spur long as ovary. Leaves round. Scape naked.

4 O. obtusa'ta. *Blunt-lv. O.* Leaf obovate, obtuse. Lip linear, entire. 5-8'. *r.*

5 O. rotundifo'lia. *Small Round-lf. O.* Lf. round. Lip, mid. lobe obcord. 6-9'. *r.*

6 O. hyperbo'rea. *Northern O.* Flowers greenish. Lip, petals, leaves, and bracts, lanceolate.

7 O. dilata'ta. *Broad-lip O.* Flowers whitish. Lip lance-linear, rhombic at base.

8 O. niv'ea. *Snowy O.* Flowers white. Lip oblong-linear. Leaves linear. S.

9 O. nigra. *Black O.* Flowers yellow, close. Lip ovate. Leaves lance-oblong.

10 O. tridenta'ta. *Trident O.* Lip 3-toothed at end. Spur longer than ovary.

11 O. bractea'ta. *Bracted O.* Lip 2-3-toothed at end, spur half as long.

12 O. fla'va. *Small yellow O.* Lip obtuse, with a tooth each side, spur long.

13 O. crista'ta. *Crested yellow O.* Flowers yellow, small, lip as long as the spur.

14 O. cilia'ris. *Large yellow O.* Flowers yellow, large, lip half as long as the spur.

15 O. Blephariglot'tis. *Ox-tongue O.* Fls. white, large; 2 side petals cut-toothed.

16 O. Michauxii. *Michaux's O.* Leaves oval. Spur twice as long as ovary.

17 O. re'pens. *Five-bristled O.* Leaves lance-linear. Spur shorter than ovary.

18 O. leucophæ'a. *White Prairie O.* Bracts shorter than the ovaries. Lip fan-shaped, 3-parted, fringed. Spur club-shaped, twice longer than ovary.

19 O. lac'era. *Rugged O.* Bracts longer than the flowers. Petals notched at end. Lip segments wedge-shaped, fringed. Spur filiform, long as ovary.

20 O. amœna. *Prairie O.* Flowers dark-purple. Lip broad, lobes toothed, not fringed. M. W. S. *c.*

21 O. Psyc'odes. *Fringed O.* Flowers light-purple. Lip wedge-shaped, the 2 petals merely toothed. *c.*

22 O. grandiflo'ra. *Great-fl. O.* Flowers light-purple. Lip semicircular, large, 2 petals fringed. *r.*

3. CORALLORHI'ZA. Coral-root. Dragon's-claw.

Flower ringent. Sepals and petals similar, ascending, the upper arching. Lip produced behind into a short spur, which grows closely to the ovary. Pollen-masses 4, oblique.—Herbs without green herbage, leafless, with coraline roots, and spikes of dull-colored flowers. *May–Sept.*

1 Spur imperceptible. Lip not lobed, often with 2 teeth at base....2, 3
  1 **C. multiflo'ra.** *Many-flowered C.* Spur manifest. Lip 3-lobed (the side lobes small), spotted. Flowers 10–20, purple. Height 10–15′.
2 **C. odontorhi'za.** *Dragon's-claw.* Flowers 9–18, purple. Lip crenulate, spotted. Ovary and pod nearly globular. Scape 9–14′.
3 **C. inna'ta.** *Lesser C.* Flowers 5–10, purplish. Lip obscurely 2-toothed near the base, spotless. Ovary and pod club-shaped. 5–6′.

4. ARETHU'SA.

Sepals and petals cohering at base, similar, ascending, arching. Lip spurless, deflexed at the end, bearded inside, cohering to the petal-like column at base. Anther terminal, closing the 2 pollen cells like a lid. Pollen-masses 2 in each cell.—Small plants, 1-flowered, in wet places. Leaves none, or hidden in the sheaths.

A. bulbo'sa. A beautiful plant 6–12′ high, invested with about 3 long loose sheaths with lanceolate points (hardly leaves). At the top is a large, fragrant purple flower, in June. (See Figs. 651, 652.)

5. SPIRAN'THES. Ladies' tresses.

Flowers in a spiral spike, somewhat ringent. Petals and sepals nearly erect, all tending to the upper side opposite the lip. Lip raised on a short claw, concave, entire, widened at top and furnished with 2 callous processes at base. Column arching, pollen-masses 2.—Stem leafy below or nearly naked, bearing a spike of small, white flowers, which are bent sideways and horizontal. *July–Oct.* (Fig. 240.)

  * Fls. in a single row on one side, and but little twisted. Lvs. radical....1, 2
  * Fls. in several rows all around the short spike. Lvs. on stem below....3, 4
1 **S. grac'ilis.** *Slender L.* Leaves ovate, varying to lance-oblong. Lip obovate, wavy-crisped.
2 **S. tor'tilis.** *Twisted L.* Leaves linear, early withering (like first). Lip 3-lobed, finely crenate.

13

8 S. latifo'lia. *Broad-lv.* L. Leaves oblong-lance. 2-4′ long. Spike dense. Lip oblong, blunt, crisp.

4 S. cer'nua. *Nodding L.* Leaves linear-lanceolate, 3-10′ long. Spike dense. Lip oblong, round, crisp.

6. CALOPO'GON. Grass Pink.

Flower with the sepals and petals similar, spreading, distinct. Lip on the upper side of the flower, stalked at base and bearded above. Column winged at the summit. Pollen-masses 2.—Leaf sheathing the base of the scape, which is bulbous at base. Flowers several. (Fig. 655.)

C. pulchel'lus. A handsome plant, common in moist meadows and in bogs. Scape slender, 1-2f. high. Leaf sword-shaped or broad linear, long. Flowers pink-purple, remarkable for having the lip on the upper side and the column below. *July.*

Fig. 653. Pogonia verticillata. *Fig.* 654. Pogonia ophiglossoides. *Fig.* 655. Calopogon pulchellus.

7. POGO'NIA. Beard-flower.

Flower with its sepals and petals distinct and somewhat spreading. Lip bearded inside, sometimes 3-lobed. Column club-shaped, wingless; anther terminal, pollen-masses 2, mealy.—Leaves 1–5, on the stem. Flowers purple. *June–Aug.*

* Sepals linear, spreading, much longer than the petals....1, 2
* Sepals and petals nearly equal, similar, and nearly erect....3, 4

   1 **P. verticilla'ta.** *Whorl-leaved B.* Leaves 5, in a whorl near the 1 brownish flower. Sepals 2' long. *r.* (Fig. 693.)

   2 **P. divarica'ta.** *Spreading B.* Leaves 2, alternate, distant, lance-oblate. Lip 3-lobed, crenulate. *S.*

  3 **P. ophioglossoi'des.** *Adder-tongue B.* Leaves 2, distant, upper bract-like. Flowers terminal, pink-colored. (Fig. 694.)

  4 **P. pen'dula.** *Nodding B.* Leaves 3–4, alternate, with as many pink-white, drooping flowers.

ORDER CXL. AMARYLLIDACEÆ. The Amaryllids.

Herbs perennial, mostly bulbous, with linear *leaves*, with the *flowers* showy, mostly regular and on scapes, hexandrous; *perianth* of 6 similar pieces united below and adherent to the *ovary*, which is 3-celled, with the *styles* united into 1; *fruit* a capsule or berry, with albuminous *seeds.* (Fig. 137.)

Analysis of the Genera.

§ Perianth bearing a crown on the summit of its tube....a
§ Perianth destitute of a crown....2
   2 Segments united into a tube *above* the ovary....b
   2 Segments distinct down to the ovary. Flowers nodding....3
      3 Perianth irregular....c
      3 Perianth regular....d

a Crown a thin membrane connecting the stamens. S.   PANCRA'TIUM.
a Crown a firm cup containing the stamens.   *Narcissus.* NARCIS'SUS. 1
  b Flr. solitary. Perianth-tube straight, erect. *Atamasco L.* ZEPHYRAN'THUS.
  b Flowers many. Perianth-tube straight.   *American Aloe.* AGAVE. 4
  b Flowers many. Perianth-tube curved.   *Tuberose.* POLYANTHUS.
  c Stamens declined and curved. Scape with 1.   *Jacobea Lily.* SPREKELIA
  d Sepals all white, larger than the petals.   *Snow-drop.* GALANTHUS
  d Sepals green-tipped, as large as the petals.   *Snow-flake.* LEUCOJUM
  d Sepals and petals equal, yellow.   *Star-grass.* HYPOXIS

1. NARCIS'SUS. Jonquil. Daffodil.

Perianth regular, crown of one piece, funnel-form or bell-form, consisting of a whorl of sterile petal-like filaments united by their edges, within which the fertile stamens are inserted.—A beautiful genus of bulbous plants with sword-shaped leaves and yellow or white flowers. None here native. † *April–June.*

* Scape bearing 1–3 large flowers....2–4
* N. Tazet'ta. *Polyanthus.* Scape many-flowered, sep. white, crown yellow, short.
  2 N. Daf'fodil. *Daffodil.* Scape 2-edged. Sepals whitish. Crown yellow, long and large.
  3 N. Jonquil'la. *Jonquil.* Scape terete. Crown yellow, much shorter than the yellow sepals.
  4 N. poet'icus. *Poet's Narcissus.* Scape terete. Crown variegated, rotate, short; sepals mostly white.

2. AGA'VE. American Aloe.

1 A. Virginica. *False Aloe.* Scape simple, 5–6f. high. Flowers in a spike, greenish-yellow. Leaves linear-lanceolate, serrate. Penn. S.
2 A. America'na. *Century Plant.* Scape branched, 15–25f. high, bearing 1 vast panicle of yellow flowers, after many years. Leaves very thick, lanceolate, spinous-dentate, often striped. †

ORDER CXLIII. IRIDA'CEÆ. The Irids.

Perennial *herbs*, arising from bulbs or thickened roots; *leaves* 2-rowed; *flowers* perfect, regular or irregular, spathaceous; *perianth* of 6 petal-like segments; *stamens* 3; *anthers* turned outwards, *ovary* inferior, 3-celled, with 1 *style* and 3 *stigmas*, becoming in *fruit* a 3-celled *capsule* with many albuminous *seeds.*

Analysis of the Genera.

1 Flowers regular, 3 petals unequal to the 3 sepals....2
1 Flowers regular, petals and sepals alike....8
1 Fls. irregular, stamens ascending. Sds. winged. † *Corn Flag.* GLADI'OLUS.
2 Stamens separate. Stigmas petal-like. Petals erect. Sepals reflexed. IRIS. 1
2 Stamens united. Sepals very large. Pets. spreading. † *Tiger-flower.* TIGRID'IA.
  8 Flowers blue, small, rotate. Leaves, &c., grass-like. (Fig. 48.)
        Blue-eyed-grass. SYSIRYN'CHIUM.
8 Fls. purp., white or yellow, tube very long, sessile on the bulb. † CRO'CUS.
8 Fls. yellow, red-spotted, tube short. Height 3–5f. † *Ixia.* PARDAN'THUS.

IRIS. Flower-de-luce.

Perianth 6-parted, the 3 outer divisions (sepals) reflexed, or spreading, the 3 inner (petals) erect. Stamens 3, distinct. Style short. Stigmas 3, petal-like, covering the stamens.—Perennial herbs with thick roots or rootstocks, sword-shaped or grass-like leaves, and large showy flowers. *April–July.*

§ Stems leafy, tall (1–2f. high), mostly bearing several flowers....a
§ Scapes leafless, low (1–6′ high), mostly bearing but 1 flower....c
  a Sepals or perianth bearded. Cultivated exotics in gardens, &c.....11–13
  a Sepals and petals beardless. Wild plants, hardly ever cultivated....b
b Leaves linear, grass-like, less than half an inch wide....1
b Leaves sword-shaped, nearly 1 or 2′ wide....2–4
  c Sepals or perianth bearded, beard crested or not crested....8–10
  c Sepals and petals beardless, but sometimes with a crest....5–7

1 I. Virgin'ica. *Boston I.* Stem slender. Ovary and pod acute, sides 2-grooved. Flowers yellow-blue. E.   [als obtuse, large. *c.*
2 I. versic'olor. *Blue Flag.* Stem 1-angled. Flowers blue-yellow-white. Pet-
3 I. tripet'ala. Stem terete. Flowers blue. Petals very small, 3-toothed. S.
4 I. cu'prea. *Copper-col'd I.* Stem terete. Fls. orange-yellow. Sepals notched. S.
5 I. lacus'tris. *Lake I.* Scape 1-flowered, flower blue and yellow. Lvs. lance. W.
6 I. ver'na. *Early I.* Scape 1-flowered, flower blue. Leaves linear, very long. S.
7 I. ochroleu'ca. *Cream-colored I.* Scape 3-flowered, flowers yellow. Lvs. sword-shaped. Pod 6-angled. †
  8 I. crista'ta. *Crested I.* Scape 1-flowered, 2–4′ high, flower blue and yellow. Leaves lanccolate, 3′ long. S.   [obtuse. Leaves ensiform. †
  9 I. pum'ila. *Dwarf I.* Scape 1-flowered, 6–10′ high, flower deep blue. Petals
10 I. Chinen'sis. *China I.* Scape many-flowered, flattened, flowers pale blue. Stigmas jagged. †
11 I. sambuci'na. *Common Flower-de-luce.* Stem many-flowered, flowers blue or whitish. Petals and sepals notched. *o.* †
12 I. German'ica. *German F.* Stem many-flowered, flowers deep blue, spathes also colored. *r.* †
13 I. Susia'na. *Chalcedonian Iris.* Stem 1-flowered, fl. striped. Petals deflexed. *ƒ*

ORDER CXLVII. TRILLIACEÆ. The Trilliads.

Herbs with tuberous *roots,* simple *stems,* and whorled, net-veined *leaves;* with the *flowers* one or few, terminal, and mostly 3-parted; with the *sepals* green, and the *petals* more or less colored; with the *stamens* 6–10, awl-shaped filaments and linear anthers; with the *ovary* free, 3–5-celled, becoming in *fruit* a juicy, many-seeded *pod.*

Analysis of the Genera.

Plants with 1 whorl of leaves and 1 flower.
  Pod many-seeded. *Wake-Robin.* TRIL'LIUM. 1
Plants with 2 whorls of leaves and several
  greenish flowers. (Fig. 92.)
            Indian Cucumber. MEDE'OLA.

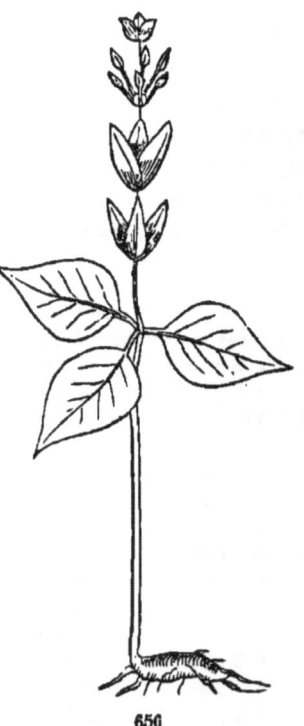

Fig. 656 Trillium erythrocarpum, with the parts of its
flower as if separated : *s*, the 8 sepals ; *p*, the 3 petals ; *st*,
the 6 stamens ; *o*, the 3 carpels.

TRIL'LIUM. Wake-Robin.

Character as expressed in the Order above.
—♃ Low herbs with a simple stem, bearing
at top a whorl of 3 leaves and a single large
flower. *Apr.–June.* (Figs. 108, 110, 656.)

§ Flower sessile, petals dark purple....1, 2
§ Fl. on a peduncle, raised above the leaves....a
§ Fl. on a peduncle, recurved beneath the lvs...7–9
  a Lvs. ses., rhomboidal or rhomb.-ovate...5, 6
  a Leaves petiolate, rounded at the base....3, 4
1 **T. ses'sile.** *Ricket W.* Lvs. sessile. Sepals erect,
    ⅔ as long as the linear-lanceolate petals.
2 **T. recurva'tum.** *Beck s W.* Lvs. petiolate. Sepals
    recurved, long as lance ovate petals.   650
  3 **T. niva'le.** *Snowy W.* Stem 2–4' high. Leaves obtuse. Petals obtuse, wavy,
    snow-white. The smallest species. W.
  4 **T. erythrocar'pum.** *Smiling W.* Stem 8–12' high. Leaves and petals pointed,
    wavy, white, tinged and penciled with purple.
5 **T. grandiflo'rum.** *Great-flowered W.* Petals lance-obovate, recurved, twice larger
    than the sepals, rose-white. Large and showy. M. W. S.
6 **T. erec'tum.** *Bath Flower.* Petals ovate, acute, much broader (not longer) than
    the sepals, dusky purple (or white, W.), ill-scented.
  7 **T. pen'dulum.** *Pendant W.* Style scarcely any. Leaves rhombic. Petals
    lance-obovate, short-pointed, flat, scarce larger than sepals. *w.* M. S. W.
  8 **T. cer'nuum.** *Nodding W.* Style scarcely any. Leaves ovate, petiolate.
    Petals lanceolate, wavy, recurved, much larger than calyx. Rose-white.
  9 **T. stylo'sum.** *Style-bearing T.* Style manifest, as long as the stigmas. S.

ORDER CXLVIII. LILIACEÆ. **Lilyworts**.

Herbs with parallel-veined *leaves*, bulbous or tuberous *stems ;*
flowers perfect, regular, generally large and richly colored ;
perianth 6 (rarely 4)-parted, uniformly colored, free from the ovary ;
stamens 6 (rarely 4) ; *anthers* fixed by a point and turned inwards ;
style single ; *ovary* superior, 2 or 3-celled ; *seed* with fleshy albumen.

Analysis of the Genera.

§ Plants bulbous at the base, or with a thick, woody stem (caudex)....2
§ Plants with a rhizome, creeper, or fibrous roots....4
  2 Perianth segments united, forming a tubular flower....d
  2 Perianth segments separate, not forming a tube....3
    8 Stem (or caudex) leafy at least below, few or many-flowered....b
    8 Stem (scape) sheathed at base, bearing a solitary flower....a
    8 Stem (scape) sheathed at base, leafless, many-flowered....c
  4 Stamens bent to one side, curved-ascending. Flowers showy....e
  4 Stamens straight, and equal in position....5
    5 Perianth segments united to near the summit....f
    5 Perianth segments separate, not forming a tube....6
      6 Flowers in terminal, leafless clusters, small, whitish....g
      6 Flowers axillary, or terminal and solitary....7
        7 Leaves thread-form, &c....h
        7 Leaves ovate, &c....k
a Flowers nodding. Wild plants.   *Erythronium.* ERYTHRONIUM. 1
a Flowers erect. Garden plants. †   *Tulip.* TU′LIPA.
 b Nectary a linear groove at the base of each segment.   *Lily.* LILIUM. 2
 b Nectary a round cavity at base of each seg. † *Crown Imperial.* FRITILLA′RIA.
 b Nectary none. Flowers panicled, large. Seeds many. † *Yucca.* YUCCA.
 b Nectary none. Flowers panicled, small. Seeds 1-3. S. *Nolina.* NOLI′NA.
: Flowers in racemes, blue or purple. †   *Squill.* SCILLA.
c Fls. in racemes or corymbs, yellow or white. *Star-Bethlehem.* ORNITHOG′ALUM.
c Flowers in umbels, white or roseate. Stamens straight.   *Garlic.* AL′LIUM. 3
c Fls. in umbels, blue. Stam. declined and curved. † *Love-flower.* AGAPAN′THUS.
 d Perianth-limb revolute, as long as the tube. †   *Hyacinth.* HYACIN′THUS.
 d Per.-limb spreading, much shorter than tube. † *Grape Hyacinth.* MUSCA′RI.
e Segments distinct. Stamens at base valve-like. †   *Asphodel.* ASPHOD′ELUS.
e Segments half-united. Stamens perigynous (§ 83). † *Day Lily.* HEMEROCAL′LIS.
e Segments half-united. Stamens hypogynous. †   *White Day-Lily.* FUN′KIA. 4
 f Fl. tubular-oblong, greenish, axillary. *Jointed Solomon's Seal.* POLYGONA′TUM.
 f Fl. broad bell-shaped, white, racemed. *Lily-of-the-Valley.* CONVALLA′RIA.

g Stem leafy, bearing a cluster. Flowers 6-parted. *Solomon's Seal.* SMILACI'NA. 5
g Scape leafless, bearing an umbel. Berry 2-seeded.   *Clintonia.* CLINTO'NIA. 6
g Stem leafy, bearing a cluster. Flowers 4-parted. *Tway-leaf.* MAJAN'THEMUM.
    h Stems branching. Flowers small, axillary. Berry red.   ASPAR'AGUS.
k Filaments flat, as long as the sagittate anthers.   *Twist-foot.* STREPTO'PUS.
k Filaments filiform, much longer than the anthers.   *Prosartes.* PROSAR'TES.
k Filaments shorter than the long, linear anthers.   *Bellwort.* UVULA'RIA. 7

1. ERYTHRO'NIUM.   Dog-tooth Violet.

Perianth bell-form, se-
pals recurved, the 3 inner
ones usually with a callous
tooth each side near the
base, and a groove in the
middle.   Pod a little
stalked.   Seeds egg-shaped.—Stem a bulb deep in
the ground.   Scape bear-
ing a single flower, its
base sheathed by the base
of the two smooth leaves.
Apr., May.

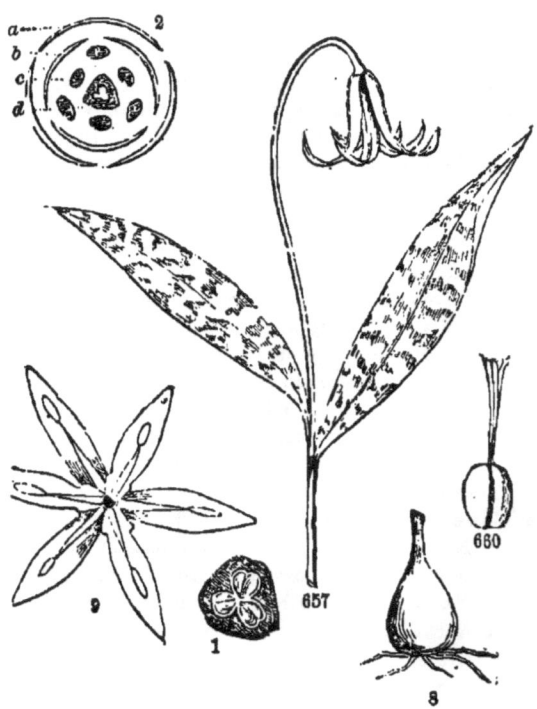

Fig. 657. The Dog-tooth Violet (*E. Americanum*). 8. The
bulb. 9. The flower spread open,
showing the 2 teeth in each petal,
also the position of all the parts.
660. The ovary, style, and stigma.
1. A cross-section of the ovary.
2. The plan of the flower: *a,* the
3 sepals in the outer circle; *b,*
the 3 petals next; *c,* the 6 stamens; and *d,* the 3-celled ovary.

E. America'num.   *Yellow E.*   Flower yellow.   Scape without a bract.   Petals
    toothed.   Leaves spotted, nearly equal.   Common.
E. bractea'tum.   *Bracted E.*   Flower greenish-yellow.   Scape bearing a bract.
    Leaves very unequal.   Mountains.   Vt.
E. albi'dum.   *White E.*   Flower white.   Scape without a bract.   Petals not toothed.
Rare in N. Y. and W.

2. LIL'IUM. Lily.

Perianth bell-shaped, segments mostly recurved, each with a groove running lengthwise within from the middle to the base. Stamens shorter than the style. Valves of the pod connected by latticed hairs.—Herbs with bulbous and leafy stems. Leaves whorled or scattered, sessile. Flowers terminal. *June, July.* (See Figs. 107, 150.)

§ Plants bearing bulblets in the axils. Flowers orange. Gardens....6, 7
§ Plants not bulbiferous in the axils of the leaves....a
    a Flowers erect, orange-red. Sepals raised on claws....4, 5
    a Flowers nodding, white. Sepals sessile. Gardens....6, 7
    a Flowers nodding, yellow or orange. Sepals sessile. Wild plants....1
1 Sepals orange-red, strongly revolute, almost into rings....2, 8
1 **L. Canaden'se.** *Common Meadow Lily.* Sepals yellow, merely recurved, spreading above middle. *c.*
    2 **L. super'bum.** *Superb L. Turk's-cap.* Flowers 3-30, very showy. Leaves lanceolate, lower whorled. *c.* M. W. S.
    3 **L. Carolinia'num.** *Carolina L.* Flower generally but 1. Leaves wedge-lanceolate, partly whorled. S.
4 **L. Philadel'phicum.** *Philadelphia L.* Upper leaves in whorls. Flowers 1-3, purple-spotted. *c.*
5 **L. Catesbæ'i.** *Catesby's L.* Lvs. all scattered. Fl. 1, red and yellow-spotted. S.
    6 **L. bulbif'erum.** *Orange L.* Flowers erect, rough within, bell-shaped. Leaves 3-veined, scattered. †
    7 **L. tigri'num.** *Tiger L.* Flowers nodding. Sepals strongly revolute. Leaves 3-veined, scattered. †
8 **L. can'didum.** *White Lily.* Flowers in a raceme, smooth, large. Lvs. scattered. †
9 **L. Japon'icum.** *Japan Lily.* Flower 1 only, very large. Sepals reflexed at end. †

3. AL'LIUM. Garlic, Onion, &c.

Flowers in a dense umbel with a 2-leaved spathe. Perianth deeply 6-parted, colored, usually spreading, persistent. Stamens 6. Ovary angular. Style thread-like. Pod 3-lobed, containing 1 or 2 black seeds in each cell.—Strong-scented, bulbous, stemless herbs, the leaves radical and the umbel on a scape, sometimes bearing bulblets instead of flowers. *May–July.*

    § Leaves flat, lanceolate, but perishing before flowering....1
    § Leaves flat, lanceolate or linear, present with the flowers....a
    § Leaves terete and hollow, or tubular....c
a Filaments simple. Ovary with a 6-leaved crown. Leaves linear....b
a Filaments 3-forked. Leaves lance-linear. Gardens....

b Stamens longer than the sepals. Umbel nodding....2
b Stamens equalling the sepals. Umbel with bulblets or flowers....3, 4
b Stamens shorter than the sepals. Umbels with flowers only....5, 6
　　c Stem leafy half way up. Filaments 3-forked....7, 8
　　c Stem naked. Filaments simple....

1 A. tricoc'cum. *Lance-leaved Garlic.* Umbel 1f. high, with a thin spathe, 12–20-flowered. Flowers white. Plants strong-scented. Woods. N. W.

2 A. cer'nuum. *Nodding G.* Leaves longer than the 4-angled scape. Rose-colored flowers, 12–20, in the handsome, nodding umbel. M. W. S.

3 A. stella'tum. *Star G.* Umbel erect when in flower (nodding in bud), with many rose-colored flowers. Western.

4 A. Canaden'se. *Canada G.* Umbel a dense head of bulblets and some flowers. Bulblets sessile, bracted, 12–18' high.

5 A. mutab'ile. *Changeable G.* Leaves bristle-form. Scape terete. Flowers many (20–40). S.

4 A. stria'tum. *Striate G.* Leaves striate, linear. Scape 8-angled. Flowers few (3–7).

7 A. sati'vum. *Common G.* Bulb compound. Umbel bearing bulbs. †

8 A. porrum. *Leek.* Bulb simple. Umbel bearing numerous flowers. †

9 A. venea'le. *Crow G.* Sta. exserted. Umbel with bulbs. Slender. M. W.

10 A. Schænopra'sum. *Cives.* Leaves rush-like, as long as the scape. Stamens included.

11 A. fistulo'sum. *Welsh Onion.* Leaves thick, as long as the swollen scape.

12 A. cepa. *Common Onion.* Leaves thick, much shorter than the swollen scape.

4. FUNKIA. Day Lily.

1 F. ovata. *White D.* Flowers white, funnel-form, many in the raceme. Leaves broad-ovate, more or less heart-shaped. † Japan.

2 F. cœru'lea. *Blue D.* Flower blue, rather bell-form. Leaves ovate-pointed, not at all heart-shaped. † Japan.

5. SMILACI'NA. Solomon's-seal.

1 S. racemo'sa. *Clustered S.* Raceme compound. Stamens longer than the perianth. Stem recurved, 1½–2f. Flowers numerous, small, white.

2 S. stella'ta. *Stellate S.* Flowers few, in a simple raceme. Leaves many. N.

3 S. trifolia'ta. *Three-leaved S.* Leaves 3 or 4, lance-elliptic. Flowers few, racemed. N.

6. CLINTO'NIA. Clintonia.

1 O. borea'lis. *Northern C.* Leaves broad, oval-lanceolate. Flowers white, 2-5, nodding in the erect, bractless umbel. Common in woods. N.

2 O. multiflo'ra. *Many-flowered C.* Leaves oblong-lanceolate. Flowers spotted, 12-30 in the corymb, erect or spread. Plant downy. Woods. M. S.

7. UVULARIA. Bellwort.

Perianth 6-parted. Sepals linear-spatulate or lanceolate, with a honey-cavity at the base of each. Filaments very short, anthers half as long as the sepals. Style 3-cleft. Pod (or berry) 3-celled, cells few-seeded.— Root-stock creeping. Stem leafy and usually branched. Flowers mostly solitary, straw-yellow, pendulous. *May.*

1 Leaves perfoliate (§ 220). Pod obovate, 3-lobed at end....3
1 Leaves sessile. Sepals cream-colored, obtusish, ovate, 3-angled....2

Fig. 663. Clintonia borealis.
Fig. 664. A berry cut across to show the 2 cells.

2 U. sessilifo'lia. *Wild Oats.* Leaves glabrous, glaucous beneath. Pod raised on a little stalk. Stem 6-10' high, divided. *c.*
2 U. puber'ula. *Downy B.* Leaves fine-downy, shining green both sides. Pod sessile. Stem 8-12' high. Mountains. S.
3 U. grandiflo'ra. *Great-flowered B.* Sepals smooth within and without, 1¼' long. Anthers obtuse. Stem 1f. high.
3 U. perfolia'ta. *Mealy B.* Sepals granular-roughish within, scarce 1' long. Anthers pointed. Stem 1f. high.

ORDER CXLIX. MELANTHÀCEÆ. The Melanths.

Herbs perennial, often poisonous, with parallel-veined *leaves;* *perianth* double, of six similar pieces, green or colored alike, persistent; *stamens* 6, with their anthers turned outwards (extrorse); *ovary* 3-celled, the styles usually distinct, a capsule in *fruit.*

Analysis of the Genera.

§ Perianth segments united below into a long tube....a
§ Perianth segments distinct, not forming a tube....2
    2 Anthers 1-celled, cordate (shield-form when open)....3
    2 Anthers 2-celled. Flowers in simple racemes....d
        8 Flowers in a panicle, that is, a compound raceme....b
        8 Flowers in a simple raceme or spike....c
a Leaves and flower arising from an underground corm.   *Colchicum.* Col'chicum.
  b Sepals clawed, each claw bearing a stamen.   Melan'thium.
  b Sepals clawed, claw free from stamens.   *Zigadene.* Zi'gadenus.
  b Sepals not clawed, base bearing a stamen.   *Poke.* Vera'trum. 1
c Flowers white, in racemes. Stamens on the sepals. *Fly-poison.* Amian'thium.
c Flowers greenish, in a spike. Stamens free from sepals.   Schænocau'lon.
  d Fls. perfect. Filaments widened at base. Ovary 6-ovuled. Xerophyllum.
  d Flowers perfect. Filaments filiform. Ovary ∞-ovuled.   Helonias.
  d Flowers diœcious, white. Stem leafy.   *Blazing Star.* Chamælir'ium.

VERA'TRUM. Poke. False Hellebore.

Flowers polygamous by abortion in the same plant. Sepals united at base, colored, spreading, sessile, and without glands. Stamens 6, shorter than the sepals, wanting in some of the flowers. Ovaries 3, united at base, often abortive. Pod 3-partible, many-seeded.—Stems leafy more or less. Flowers panicled. *June, July.*

§ Stem stout and very leafy throughout....1
§ Stem slender, nearly naked....2
    2 Sepals rather blunt. Leaves oval and lanceolate....2, 3
    2 Sepals acuminate. Leaves linear....
1 V. viride. *Green-fl. P.* Lvs. large, oval, pointed. Coarse plant with green fls.
2 V. Woodii. *Wood's V.* Scape 3-6f. Leaves lanceolate. Fls. nearly black. W
3 V. parviflo'rum. *Small-fl. V.* Stem 2-5f. Lvs. oval. Fls. dingy green. Mts. S
4 V. angustifolium. *Grass-lv. V.* Very slender, 3f. Lvs. long. Fls. greenish-white. W

Order CLI. JUNCACEÆ. Rushes.

Herbs generally grass-like, often leafless, with small, dry, green *flowers;*
perianth of 6 glume-like pieces, whorled in two circles (sepals and petals)
stamens 6, rarely 2, on the torus; *style* 1;
ovary 3-celled; *seeds* few or many.

Analysis of the Genera.

§ Perianth greenish outside, yellow inside. Stamens 6.
Stigma 1. Seeds many. Leaves sword-shaped.
Scape nearly naked. *Narthecium.* NARTHECIUM.

§ Perianth green or brownish. Stamens 6. Stigmas 3.
Capsule 3-celled, 3-seeded. Stems leafy, jointed.
Leaves linear. *Wood Rush.* LU'ZULA.

§ Perianth green or brownish. Stamens 6, rarely 3.
Stigmas 3. Capsule many-seeded. Leaves terete,
or linear, or none. *Rush. Bullrush.* JUN'OUS.

Fig. 665. Flower of Luzula, much magnified: *p*, the green perianth; *s*, the 6 stamens; *œ*, the 3 stigmas.

ORDER CLII. COMMELYNACEÆ. Spiderworts.

Herbs with flat, narrow *leaves* which are usually sheathing at base;
perianth of 2 circles, outer of 3 green *sepals*, inner of 3 colored *petals;*
stamens 6, on the torus; *ovary* 2 or 3-celled; *style* and *stigma* 1;
capsule 2 or 3-celled, with few seeds.

Analysis of the Genera.

§ Flowers irregular, clustered in a heart-shaped floral leaf. COMMELY'NA.

§ Flowers regular, clustered, floral leaf like the rest. *Spiderwort.* TRADESCAN'TIA. 1

§ Flowers regular, solitary, axillary. Stamens 3. Moss-like herbs. S. MAYA'CA.

TRADESCAN'TIA. Spiderwort.

Flowers regular, in terminal, close umbels, subtended by 2 or 3 leaf-like
bracts. Petals broad, larger than the sepals. Filaments clothed with
jointed hairs. Juice viscid, spinning into cobwebs.

§ Leaves linear, sessile, not narrowed at the base, smooth....1, 2

§ Leaves ovate or lanceolate, narrowed at base, hairy....3, 4

  1 **T. Virgin'ica.** *Common S.* Leaves broad-linear. Umbel many-flowered, sessile, terminal, with 2 leaf-like bracts. Petals large, blue or white.

  2 **T. ro'sea.** *Roseate S.* Leaves linear, long. Umbel few-flowered, with 2 subulate bracts. Petals twice longer than sepals, rose-colored. Penn. S.

3 **T. pilo'sa.** *Hairy S.* Leaves lanceolate, long-pointed. Umbels both terminal and axillary, many-flowers. Petals small, bluish-purple. W.

4 **T. crassifo'lia.** *Thick-lv. S.* Leaves ovate, some petiolate, acute, woolly beneath. Flowers small, rose-purple, terminal. Stem weak. Leaves striped. †

INDEX AND GLOSSARY:

Containing Definitions of Botanical Terms, together with references to those paragraphs in which they are defined in the foregoing Lessons.

Branches, 203.
Branching root, 219.
Branchlets, small branches.
Bristles, stiff hairs.
Bud, 52.
Bulb, 230,
Bulblets, little bulbs borne above ground.
Bulbous, having bulbs.
Bushes, 215.

Caducous, falling off early.
Calyculate, having bracts resembling an outer, additional calyx.
Calyx, 65.
Campanulate, bell-shaped, 91.
Canescent, whitish with fine hairs.
Capillary, very slender, hair-like.
Capitate, head-shaped, globular.
Capsule, a pod, 182.
Carinate, keel-shaped.
Carpels, 124.
Cartilaginous, gristly.
Caryophyllaceous, 88.
Caryopsis, grain or kernel.
Catkin, 149.
Caudate, with a tail.
Caulescent, 235.
Cauline, 146.
Caulis, 235.
Cellular, composed of cells.
Cernuous, nodding.
Chaffy, with chaff.
Character, marks which distinguish a species, genus, &c.
Chartaceous, of the texture of writing-paper.
Ciliæ, hairs, like those of the eye lashes.
Ciliate, furnished with ciliæ.
Circinate, 140.
Circumscissile, opening like a lid.
Cirrhous, furnished with a tendril.
Classification.—Lesson 29.
Clavate, club-shaped.
Claw, 71.
Climbers, 54, 236.
Climbing fern, 12.
Cochleate, resembling the shell of a snail.
Cohering, connected.
Cohesion, 79.
Cohorts, 258.
Colored, not green.
Column, the consolidated stamens and pistils of the Orchis.
Coma, a tuft of hairs, 187, 188.
Complete flower, 110.
Compound flowers, 156.

Compound leaves, 23–35.
Compound petiole, 44.
Compound pistil, 124.
Compressed, flattened lengthwise.
Cone, the same as *strobile*, 183, 185.
Confluent, joining together.
Conjugate, joined in pairs.
Connate, joined together at base, 48.
Connectile, 102.
Connivent, converging together.
Conoids, 258.
Contorted, twisted, 180.
Convex, rising spherically.
Convolute, 132.
Cordate, heart-shaped, 19.
Coriaceous, leathery, thick and tough.
Corm, 230.
Cornute, *Corniculate*, horned.
Corolla.—Lesson 12.
Corona or *Crown*, the expanded, cup-like disk of Narcissus, &c.
Corymb, 151.
Corymbous, arranged like a corymb.
Costate, ribbed.
Cotyledon, 190, 191.
Creeper, 282.
Crenate and Crenulate, 81.
Crisped, *Crispate*, with excess of margin.
Cristate or *Crested*, with raised ridge.
Cruciform, 87.
Cryptogamia, 250.
Cucullate, hood-shaped.
Culm, the stem of grasses.
Cuneate, wedge-shaped, 17.
Cupule, cup of the acorn, &c.
Cuspidate, with a small abrupt point, 88.
Cuticle, the epidermis, scarf-skin.
Cyme, 157.
Cymous, like a cyme.

Decandrous, with 10 stamens.
Deciduous, falling off in autumn.
Decompound, more than once compound-ed, as bi or tri-pinnate.
Decumbent, 224.
Decurrent, extending down the stem as do the leaves of Mullen.
Decussate, crossing at right angles.
Deflexed, bent downwards.
Definite, 106.
Dehiscence, 102.
Dehiscent fruits, 166.
Deltoid, 15.
Dentate, Denticulate, 30.
Depressed, flattened from above.
Descending axis.—Lesson 27.

INDEX TO THE NAMES OF PLANTS,

BOTH LATIN AND ENGLISH:

Also, full references to the Illustrations.

------◄►------

APPENDIX.

THE LANGUAGE OF FLOWERS.

"Then gather a wreath from the garden bowers,
And tell of the wish of thy heart in flowers."

PERCIVAL.

Acacia, Rose (Robinia hispida, 319*).   Friendship.

Adonis, Floss (Adonis autumnalis, 205).   Sad remembrances.

Almond, Flowering (Amygdalus pumila, 329).   Hope.

Aloe (Agave, 694, or Yucca, 709).   Superstition.

Alyssum, Sweet (Alyssum maritinum, 236).   Merit before beauty.

Amaranth, Globe (Gomphrena globosa, 619).   I change not.

Amaryllis (Zephyranthus, 695).   Affectation, Coquetry.

Andromeda (Andromeda, 487).   A cruel fate has fixed me here.

Anemone (Anemone nemorosa, 203).   Anticipation.

Angelica (Archangelica, 381).   These are idle dreams.

Arbor-vitæ (Thuja, 662).   Thy friend till death.

Arethusa (A. bulbosa, 691).   I could weep for thee.

Aspen (Populus tremuloides, 655).   Excessive sensibility.

Asphodel (Asphodelus, 713).   My thoughts will follow thee beyond the grave.

Aster (420).   Cheerfulness in age.

---

* Refers to the page in the Class-Book of Botany, where may be found a more full and complete account of the species or genus than could be consistent with the limits of an elementary treatise. Reference to page and place in this work may be made through the Index.

Auricula (Primula auricula, 502). You are proud.

Bachelor's Button (Centaurea Cyanus, 465). Single blessedness.

Balm (Melissa, 548 ; Monarda didyma, 550). Sympathy.

Balm-of-Gilead (Populus candicans, 656). You have cured my pain.

Balsamine (Impatiens balsamina, 280). Approach not.

Barberry (Berberis, 217). A sour temper is no slight evil.

Basil, Sweet (Ocymum basilicum, 541). Good wishes.

Beech (Fagus, 646). There let us meet.

Bluets (Houstonia cœrulea, 402). Unaspiring beauty.

Box (Buxus, 632). Constancy. I change not.

Broom (Genista, 310). Humility.

Broom Corn (Sorghum saccharatum, 709). Industry.

Bulrush (Scirpus, 738). Indecision.

Burdock (Lappa major, 468). Don't come near me.

Buttercups (Ranunculus, 205). I cannot trust thee.

Cactus (the Cactaceæ, 359). You terrify me.

Canterbury Bells (Campanula Medium, 479). Gratitude.

Carnation (Dianthus caryophyllus, 254). A haughty spirit before a fall.

Catchfly (Silene, 256). I am a willing prisoner.

Cedar (Juniper Virginiana, 664). I live for thee.

Chamomile (Anthemis nobilis, 457). Fortitude.

China Aster (Callistephus Chinensis, 429). I'll think of it.

Chrysanthemum (458). I love.

Clover, Red (Trifolium repens, 312). Industry.

Clover, White ("   "). Truth needs no flowers of speech.

Clover, Yellow ("   "). Slighted love.

Columbim (Aquilegia Canadensis, 210). I cannot give thee up.

Columbine (A. vulgaris, 110). Hopes and fears.

Corn Cockle (Agrostemma Githago, 257). Thou hast more beauty than worth.

Coxcomb (Celosia, 616). You are a fop.

Crocus (700). What an enigma thou art.

Cypress (Cupressus thyoides, 663). Bereavement. Despair.

Daffodil (Narcissus Pseudo-narcissus, 603). Self-esteem.

Dahlia (429).   Forever thine.

Dandelion (Taraxacum Dens-leonis, 473).   You intrude.

Dogbane (Apocynum, 588).   Falsehood.

Dogwood, Flowering (Cornus florida, 390).   False pretensions.

Eglantine Rose (Rosa rubiginosa, 335).   I wound to heal.

Egyptian Calla (Richardia Æthiopica, 669).   Modesty.

Enchanter's Nightshade (Circæa, 856).   I shall beware of your enchantments

Fennel-flower (Nigella damascena, 209).   Love in a mist.   Perplexity.

Fig (Ficus Carica, 635).   It is a secret.

Fir Balsam (Abies Balsamea, 661).   Time will cure.

Flax (Linum usitatissimum, 275).   Domestic industry.

Fleur-de-lis (Iris, 697).   I bring you a message.

Four-o-clock (Mirabilis Jalapa, 603).   Timidity.

Foxglove (Digitalis, 526).   My heart acknowledges your influence.

Geranium, Ivy (P. peltatum, 278).   A bridal decoration.

Geranium maculatam (277).   You burn with envy.

Geranium, Oak-leaf (Pelargonium quercifolium, 279).   There is nothing in a

Geranium Robertianum (277).   Aversion.   [name.

Geranium, Rose (P. graveolens, 278).   Thou art my choice.

Goldenrod (Solidago, 430).   Encouragement.

Hazel-nut (Corylus, 647).   Reconciliation.

Heart's-ease or Pansy (Viola tricolor, 244).   Forget me not.

Hibiscus Syriacus (270).   Thy beauty soon will fade.

Hibiscus Trionum (269).   I would not be unreasonable.

Heliotrope (Heliotropium Peruvianum, 559).   Devotion.

Hellebore (Helleborus, 209).   It is a scandal.

Holly (Ilex opaca, &c., 496).   Am I forgotten ?

Hollyhock (Althæa rosea, 266).   Ambitious only of show.

Honeysuckle (Lonicera, 394).   Seek not a hasty answer

Hop (Humilus lupulus, 638).   You do me injustice.

Hyacinth (Hyacinthus, 712).   Jealousy.

Hydrangea hortensis (373).   Vain boasting.

Ice-plant (Mesembryanth, 265).   Your very looks are freezing.

Indian Tobacco (Lobelia inflata, 477).　Away with your quackery.

Ivy (Hedera Helix, 390).　Nothing can part us.

Japonica, Red (Camellia Jap., 273).　Pity may change to love.

Japonica, White (C. Japonica, 273).　Perfected loveliness.

Jessamine (Jasminum, 596).　Thy gentle grace hath won me.

Jonquil (Narcissus Jonquilla, 693).　Requited love.

Judas-tree (Cercis Canadensis, 301).　Unbelief.　Treachery.

Juniper (Juniperus communis, 663).　I will protect thee.

Lady's-slipper (Cypripedium, 581).　Caprice.

Larkspur (Delphinium, 210).　Fickleness.

Laurel, Sheep (Kalmia angustifolia, &c., 485).　Falsehood.

Lavender (Lavandula, 541).　Owning her love she sent him Lavender.　*Shaks.*

Lemon (Citrus Limonum, 274).　Discretion.

Lilac (Syringa, 598).　My first love.

Lily, White (Lilium candidum, 709).　Purity and sweetness.

Locust, green leaves (Robinia Pseudacacia, 319).　My heart is buried.

Lupine (Lupinus, 311).　Indignation.

Magnolia glauca (214).　He lives in fame who dies in virtue's cause.

Magnolia grandiflora (214).　Thou hast magnanimity.

Marigold (Tagetes, or Calendula, 465).　Cruelty.

Mignonette (Reseda odorata, 241).　Moral worth superior to beauty.

Milkweed (Asclepias, 597).　Conquer your love.

Mistletoe (Phorodendron, 621).　Meanness.　Indolence.

Mock Orange (Philadelphus coronarius, 374).　Deceit.　I cannot trust thee.

Monk's-hood (Aconitum, 211).　Deceit.　Your words are poison.

Morning-glory (Pharbitis purpurea, 571).　You love darkness.

Myrtle (Myrica ceriera, 650).　Thine is the beauty of holiness.

Myrtle (Myrtus communis, 346).　Love's offering.

Narcissus, Poet's (Narcissus poeticus, 693).　Egotists are agreeable only to

Nasturtion (Tropæolum majus, 281).　Honor to the brave.　　[themselves

Nettle (Urtica dioica, 636).　Thou art a slanderer.

Nightshade (Atropa Belladonna, 589).　Death.

Nightshade (Solanum nigrum, 577).　Skepticism.

Oak (Quercus, 642). Thou art honored above all.

Oat (Avena sativa, 790). Thy music charms me.

Oleander (Nerium Oleander, 590). The better part of valor is discretion.

Olive (Olea, 599). Emblem of peace.

Orange Flowers (Citrus Aurantium, 274). Bridal festivity.

Ox-eye Daisy, or Whiteweed (Leucanthemum, 458). Be patient.

Parsley (Apium petroselinum, 388). Thy presence is desired.

Passion-flower (Passiflora, 363). Let love to God precede all other love

Pea (Pisum sativum, 303). Grant me an interview.

Peach blossom (Persica vulgaris, 328). Preference.

Pennyroyal (Hedeoma pulegioides, 544). Flee temptation.

Peony (Pæonia, 212). A frown.

Pepper (Capsicum, 578). Your wit is too keen for friendship

Periwinkle (Vinca, 589). Remember the past.

Phlox (567). Our souls are one.

Pine, Pitch (Pinus rigida, 660). Time and philosophy.

Pine, White (Pinus strobus, 660). High-souled patriotism.

Pink, Single Red (254). A token of pure and ardent love.

Pink, Single White (Dianthus caryophyllus, 254). Artlessness.

Pink, Variegated (254). Frank refusal.

Poppy, Red (Papaver Rheas, 224). Oblivion is the cure.

Poppy, White (Papaver somniferum, 224). 'Twixt life and death.

Primrose (Primula grandiflora, 502). Confidence.

Primrose, Evening (Œnothera, 352). Inconstancy.

Quince (Cydonia, 333). Beware of temptation.

Rocket (Hesperus, 234). Thou vain coquette!

Rose Bud. Thou hast stolen my affections.

Rose, Burnet (Rosa pimpinellifolia, 337). Gentle and innocent.

Rose, Cinnamon (R. cinnamomia, 335). Without pretension. Such as I
am receive me. Would I were more for your sake.

Rose, Damask (R. damascena, 336). Blushes augment thy beauty.

Rosemary (Rosmarinus, 550). Remember me.

Rose, Moss (R. centifolia, B., 336). Thou art one of a thousand.

Rose, White (R. alba, 336). My heart is free.

Rose, White, withered (336). Transient impressions.   [you.

Rose, Wild (R. nitida, 335). Simplicity. Let not this false world deceive

Rue (Ruta, 282). Disdain.   [virtues.

Sage (Salvia, 548). There is nothing lovelier in woman than the domestic

Snap-dragon (Antirrhinum, 519). Thou hast deceived me.

Snow-ball (Viburnum Opulus, 397). Thou livest a useless life.

Snow-drop (Galanthus, 694). I am no summer friend.   [friend.

Sorrel (Polygonum acetosella, 606). Ill-timed wit. A jester is a dangerous

Speedwell (Veronica, 526). My best wishes.

Spiderwort (Tradescantia, 727). You have my esteem ; are you content ?

Star-of-Bethlehem (Ornithogalum, 710). Look heavenward.

Stock (Matthiola, 229). Too lavish of smiles.

Sumac (Rhus, 283). Splendid misery.

Sweet Pea (Lathyrus odoratus, 304). Must you go ?

Sweet-scented Shrub (Calycanthus, 345). Benevolence.   [villain too.

Sweet William (Lychnis chalcedonica, 257). A man may smile and be a

Thistle (Cirsium, 467). Misanthropy.

Thorn Apple (Datura, 581). Thou scarcely hidest thy guilt.

Thyme (Thymus, 547). The prize of virtue.

Tulip, Variegated (Tulipa, 707). Thy spell is broken.

Tulip, Yellow (707). I dare not aspire so high.

Venus' Looking-glass (Specularia, 479). Flattery hath spoiled thee.

Vervain (Verbena hastata, 537). I see thy arts, and despise them.

Violet, Blue (Viola cucullata, 243). Faithfulness. I shall never forget.

Violet, White (V. blanda, 242). Retirement. I must be sought to be found.

Virgin's Bower (Clematis, 200). Filial affection.

Wall-flower (Cheiranthus, 232). A friend in need is a friend indeed.

Water Lily (Nymphæa odorata, 220). Be silent.

Weeping Willow (Salix Babylonica, 655). Mourning for friends departed.

Zinnia (444). To the prude.